Shape Memory Alloys for Civil Engineering

Shape Memory Alloys for Civil Engineering

Editors

Cheng Fang
Canxing Qiu
Yue Zheng

MDPI • Basel • Beijing • Wuhan • Barcelona • Belgrade • Manchester • Tokyo • Cluj • Tianjin

Editors
Cheng Fang
Department of Structural Engineering
Tongji University
Shanghai
China

Canxing Qiu
Department of Civil Engineering
Beijing University of Technology
Beijing
China

Yue Zheng
Department of Bridge Engineering
Tongji University
Shanghai
China

Editorial Office
MDPI
St. Alban-Anlage 66
4052 Basel, Switzerland

This is a reprint of articles from the Special Issue published online in the open access journal *Materials* (ISSN 1996-1944) (available at: www.mdpi.com/journal/materials/special_issues/shape_memory_alloys_for_civil_engineering).

For citation purposes, cite each article independently as indicated on the article page online and as indicated below:

LastName, A.A.; LastName, B.B.; LastName, C.C. Article Title. *Journal Name* **Year**, *Volume Number*, Page Range.

ISBN 978-3-0365-6607-8 (Hbk)
ISBN 978-3-0365-6606-1 (PDF)

Cover image courtesy of Cheng Fang

Contents

About the Editors

Cheng Fang

Dr. Cheng Fang is currently a Research Professor and Assistant Dean in the College of Civil Engineering, Tongji University. He received his BEng in Civil Engineering from Tongji University in 2007 and subsequently obtained his MSc and PhD in Structural Engineering from Imperial College London in 2008 and 2011, respectively. Prior to joining Tongji University in 2015, he worked as a structural engineer in ARUP; a Post-Doctoral Fellow at The Hong Kong Polytechnic University; and an Assistant Professor at Newcastle University, UK. His main research areas include steel and composite structures, smart materials and structures, and seismic resilience. He has authored 4 books, published 90 refereed articles in leading SCI-indexed journals, and secured more than 20 research projects as the Principle Investigator (PI).

Canxing Qiu

Dr. Canxing Qiu is currently a Professor of Structural Engineering at BJUT. He earned his doctoral degree from The Hong Kong Polytechnic University. His research interests include self-centering structures, shape memory alloy-based damping devices, and the performance-based seismic design method. He served as the principle investigator of numerous competitive research fundings. He published over 50 SCI journal papers and was authorized for over 10 invention patents.

Yue Zheng

Dr. Yue Zheng is an Associate Professor of novel materials and structural systems at the Department of Bridge Engineering at Tongji University. His main research focuses on the resilience and sustainability-informed design, assessment, and life-cycle maintenance of bridges under the effects of deterioration and hazards to aid the development of a next generation of bridge engineering systems using novel materials and structural systems.

materials

Editorial

Shape Memory Alloys for Civil Engineering

Cheng Fang [1,*], Canxing Qiu [2] and Yue Zheng [3]

1 Department of Structural Engineering, College of Civil Engineering, Tongji University, Shanghai 200092, China
2 Key Laboratory of Urban Security and Disaster Engineering of Ministry of Education, Beijing University of Technology, Beijing 100124, China
3 Department of Bridge Engineering, Tongji University, Shanghai 200092, China
* Correspondence: chengfang@tongji.edu.cn

Shape memory alloys (SMAs) are a unique class of metals capable of recovering large strains either spontaneously or upon heating, depending on their thermal-mechanical state. Its two most common phases are the martensite and austenite phases, consisting of three different crystal structures: twinned martensite, detwinned martensite, and austenite. The prevalence of these phases is temperature dependent; the austenite phase is stable at higher temperatures and the martensite phase is stable at lower temperatures. Two different martensite crystal structures exist due to the different orientation directions (variants) of the martensitic crystal. The strain recovery properties of SMAs result from reversible phase transformations between the austenite and martensite phases, a mechanism which is attributed to shear lattice distortion rather than dislocation-based plasticity.

Since their early development in the 1960s, SMAs have been successfully applied in the medical, aerospace, robotic, and automobile industries. In the 1990s, SMA emerged as a potential material in civil engineering, and great progress in this area has been made since then. With the increasing need for more hazard resilient structural systems, the knowledge on SMAs has been widely circulated in the civil engineering community over the last two decades. A large number of relevant research projects have been granted by major funders, especially in seismically active countries. These significant investments made over the last two decades have led to considerable increases in vital knowledge that may prepare engineers for the more widespread use of SMAs in civil engineering.

However, to date, the practical application of SMAs in the construction industry has been lacking, partially due to shortcomings in engineering-oriented design approaches and a lack of effective knowledge exchanges between material scientists and civil engineers. There are still knowledge barriers across materials science and structural engineering. This Special Issue plans to present an overview of the most recent advances in the field of SMA research and its applications in civil engineering. It aims to help remove the interdisciplinary knowledge barriers and to shed considerable light on the commercialization of SMA products in the construction industry by detailing their potential applications in buildings, nuclear power plants, television transmission towers, and bridges.

Paper [1] is an element-level study that proposes a machine learning-based approach for efficient identification of thermodynamic parameters, considering the dynamic behavior of NiTi SMA wires. A feedforward artificial neural network (ANN) architecture was developed and strain rate effects were considered in a macroscopic constitutive SMA model. After training, the ANN could identify the searched model parameters from cyclic tensile stress–strain tests. The proposed approach was validated by experiments. Papers [2–5] focus on the applications of SMAs in braces and connections in residential buildings/industrial structures. Zhang et al. [2] present a novel type of hybrid self-centering braces incorporating tension-only superelastic NiTi shape memory alloy (SMA) cables and integrated viscoelastic dampers (VEDs). Their new design demonstrates an enhanced energy dissipation ability and self-centering tendency compared with existing SMA-based

Citation: Fang, C.; Qiu, C.; Zheng, Y. Shape Memory Alloys for Civil Engineering. *Materials* **2023**, *16*, 787. https://doi.org/10.3390/ma16020787

Received: 6 January 2023
Accepted: 9 January 2023
Published: 13 January 2023

self-centering solutions, where its improved behavior mainly stems from the participation of the VEDs. Jia et al. [3] have developed an innovative self-centering SMA brace based on low friction slip. This study comprehensively evaluates the effects of the loading rate and initial strain on the seismic performance of the SMA braces. Both the loading rate and the initial strain are shown to have a great influence on the seismic performance of the self-centering SMA brace. They have also developed improved numerical models combined with the Graesser and Bouc–Wen models in MATLAB. Pei et al. [4] examine a new type of self-centering reinforced concrete (RC) frame beam-column joint, equipped with super-elastic SMA bars. It is shown that by using the SMA reinforcement and a steel plate, the load carrying capacity is improved, the stiffness degradation is delayed, and the ductility and self-centering ability of the joints is improved. Qian et al. [5] explore the potential of a reinforced concrete (RC) beam strengthened by SMAs and engineered cementitious composites (ECC). The test results show that, compared with ordinary reinforced concrete beams, strengthening an existing RC beam with an enlarged section area of a SMA-reinforced ECC can improve the self-recovery capacity, ductility, and deformability of the specimens.

Papers [6,7] discuss the use of SMAs in controlling the effects of seismic activity and wind on television transmission towers. Wu et al. [6] consider using SMA dampers for response reduction of a flexible television tower. A two-dimensional dynamic model was developed for dynamic computations, leading to the introduction of a mathematical model for an SMA damper. The structural dynamic responses were examined with respect to time and frequency, investigating the effects of damper stiffness, service temperature, hysteresis loops, and earthquake intensity on control efficacy. The analysis indicates that SMA dampers with optimal parameters can substantially reduce the vibrations of TV transmission towers under seismic events. Chen et al. [7] propose a wind vibration control method using SMA dampers for tower line coupled systems. Detailed parametric studies were conducted to examine the effects of the physical parameters of SMA dampers on structural responses.

With respect to bridges, Chen et al. [8] discuss a novel self-centering rocking (SCR) bridge system equipped with SMA-based piers, with a particular focus on the benefits of the SCR bridge system in the context of life cycle. Based on a life cycle loss and resilience assessment, the analysis results reveal that the novel SCR pier bridge system slightly increases the bearing displacement but extensively reduces the pier curvature ductility due to its rocking mechanism. The SCR bridge system exhibits a lower life cycle loss and exhibits a more resilient performance than a conventional bridge, especially in the regions with higher seismic intensities. Indirect loss can be significantly larger than the direct loss, specifically for earthquakes, which have a relatively low probability of occurrence. The SCR bridge system outperforms the conventional systems in terms of recovery time. Li et al. [9] examine a resilient bridge system incorporating engineered cementitious composite (ECC)-reinforced piers and SMA energy dissipation components, i.e., SMA washers. The analysis results indicate that this system has superior resilience and damage control compared to conventional bridges.

The above nine papers [1–9] have focussed on NiTi SMAs. Iron-based SMAs (Fe-SMAs), also known as Fe–Mn–Si alloys, are another member of the family of smart metals for civil engineering. In the civil engineering community, Fe-SMAs are well-known for their shape memory effect (SME), for which it has been widely investigated in the area of prestressing. Marinopoulou and Katakalos [10] investigate the basic thermomechanical responses of Fe-SMAs. In particular, their study focuses on the application of prestress and on the alloy's behavior under fatigue. The effect of loading frequency on the recovery stress of the material has been thoroughly investigated. Four dog bone specimens were prepared and tested under low-cycle fatigue. Recovery stress was monitored after pre-straining and heating were applied under strain-control conditions. The measured recovery stress values are satisfactory high, verifying the prestress feasibility. In the context of earthquake engineering, Fe-SMAs are attractive due to their excellent low-cycle fatigue (LCF) resistance, in contrast to NiTi SMAs (especially large-scale SMA elements), which are brittle and exhibit poor LCF performance. This Special Issue concludes with a comprehensive review of the literature compiled by Zhang et al. [11], who provide a detailed summary of recent developments in the research and design of Fe-SMAs. The basic mechanical properties are presented and compared with conventional structural steel, and some necessary explanations are given on the metallographic transformation mechanism. Newly emerged applications such as Fe-SMA-based prestressing/strengthening techniques and seismic-resistant components/devices are discussed. It is believed that Fe-SMAs can offer a wide range of applications in the construction industry, but there are still problems that remain to be addressed and areas to be further explored. Required research on the material-, component-, and system-level is highlighted. With the systematic information provided, this work is not only of benefit to professionals and researchers who have been working in this area for a long time who want to gain an in-depth understanding of the state-of-the-art innovations in this field, but also helps enlighten a wider audience who intend to become more informed on this exciting topic.

Conflicts of Interest: The authors declare no conflict of interest.

References

1. Lenzen, N.; Altay, O. Machine Learning Enhanced Dynamic Response Modelling of Superelastic Shape Memory Alloy Wires. *Materials* **2022**, *15*, 304. [CrossRef] [PubMed]
2. Zhang, Z.-X.; Ping, Y.; He, X. Self-Centering Shape Memory Alloy-Viscoelastic Hybrid Braces for Seismic Resilience. *Materials* **2022**, *15*, 2349. [CrossRef] [PubMed]
3. Jia, Y.; Zhang, B.; Zeng, S.; Tang, F.; Hu, S.; Chen, W. Effect of Loading Rate and Initial Strain on Seismic Performance of an Innovative Self-Centering SMA Brace. *Materials* **2022**, *15*, 1234. [CrossRef] [PubMed]
4. Pei, Q.; Wu, C.; Cheng, Z.; Ding, Y.; Guo, H. The Seismic Performance of New Self-Centering Beam-Column Joints of Conventional Island Main Buildings in Nuclear Power Plants. *Materials* **2022**, *15*, 1704. [CrossRef] [PubMed]
5. Qian, H.; Zhang, Q.; Zhang, X.; Deng, E.; Gao, J. Experimental Investigation on Bending Behavior of Existing RC Beam Retrofitted with SMA-ECC Composites Materials. *Materials* **2022**, *15*, 12. [CrossRef] [PubMed]
6. Wu, J.; Chen, B.; Zhi, L.; Song, X. Seismic Response Mitigation of a Television Transmission Tower by Shape Memory Alloy Dampers. *Materials* **2021**, *14*, 6987. [CrossRef] [PubMed]
7. Chen, B.; Song, X.; Li, W.; Wu, J. Vibration Control of a Wind-Excited Transmission Tower-Line System by Shape Memory Alloy Dampers. *Materials* **2022**, *15*, 1790. [CrossRef] [PubMed]
8. Chen, J.; Liang, D.; You, X.; Liang, H. Performance-Based Assessment of Bridges with Novel SMA-Washer-Based Self-Centering Rocking Piers. *Materials* **2022**, *15*, 6589. [CrossRef] [PubMed]
9. Li, X.; Chen, K.; Chen, J.; Li, Y.; Yang, D. Seismic Response of Resilient Bridges with SMA-Based Rocking ECC-Reinforced Piers. *Materials* **2021**, *14*, 6500. [CrossRef] [PubMed]
10. Marinopoulou, E.; Katakalos, K. Thermomechanical Fatigue Testing on Fe-Mn-Si Shape Memory Alloys in Prestress Conditions. *Materials* **2023**, *16*, 237. [CrossRef] [PubMed]
11. Zhang, Z.-X.; Zhang, J.; Wu, H.; Ji, Y.; Kumar, D.D. Iron-Based Shape Memory Alloys in Construction: Research, Applications and Opportunities. *Materials* **2022**, *15*, 1723. [CrossRef] [PubMed]

materials

MDPI

Article

Machine Learning Enhanced Dynamic Response Modelling of Superelastic Shape Memory Alloy Wires

Niklas Lenzen and **Okyay Altay** *

Lehrstuhl für Baustatik und Baudynamik, Department of Civil Engineering, RWTH Aachen University, 52074 Aachen, Germany; lenzen@lbb.rwth-aachen.de

* Correspondence: altay@lbb.rwth-aachen.de

Abstract: Superelastic shape memory alloy (SMA) wires exhibit superb hysteretic energy dissipation and deformation capabilities. Therefore, they are increasingly used for the vibration control of civil engineering structures. The efficient design of SMA-based control devices requires accurate material models. However, the thermodynamically coupled SMA behavior is highly sensitive to strain rate. For an accurate modelling of the material behavior, a wide range of parameters needs to be determined by experiments, where the identification of thermodynamic parameters is particularly challenging due to required technical instruments and expert knowledge. For an efficient identification of thermodynamic parameters, this study proposes a machine-learning-based approach, which was specifically designed considering the dynamic SMA behavior. For this purpose, a feedforward artificial neural network (ANN) architecture was developed. For the generation of training data, a macroscopic constitutive SMA model was adapted considering strain rate effects. After training, the ANN can identify the searched model parameters from cyclic tensile stress–strain tests. The proposed approach is applied on superelastic SMA wires and validated by experiments.

Keywords: machine learning; artificial neural networks; shape memory alloys; superelastic; parameter identification; constitutive model; thermodynamic parameters

Citation: Lenzen, N.; Altay, O. Machine Learning Enhanced Dynamic Response Modelling of Superelastic Shape Memory Alloy Wires. *Materials* **2022**, *15*, 304. https://doi.org/10.3390/ma15010304

Academic Editors: Cheng Fang, Canxing Qiu and Yue Zheng

Received: 30 November 2021
Accepted: 27 December 2021
Published: 1 January 2022

1. Introduction

Shape memory alloys (SMAs) are superelastic two-phase polycrystal metals. During dynamic loading, repeated forward- and reverse-phase transitions occur allowing the material to dissipate energy. Besides this key property, SMAs exhibit also other unique characteristics, such as large deformation recovery, corrosion resistance, and low fatigue. Therefore, SMA-based vibration control devices have been a particular research field in civil engineering. Qiu and Zhu [1] presented a self-centering steel frame, which utilizes superelastic Ni-Ti wires within a SMA-based damper. Moreover, Liu et al. [2] proposed a base isolation system by incorporating springs made of superelastic SMA wires. Apart from this, Liang et al. [3] used cables composed of SMA wires to enhance the effectiveness of a friction sliding bearing. A review and detailed summary of the related applications can be found in [4–6]. SMAs are also being used in other dynamic systems, such as in aeronautic [7] and automotive [8] engineering.

For the design of SMA-based control devices, accurate constitutive models are required, whereas due to their numerical efficiency, macroscopic models are generally preferred. Furthermore, for an efficient heat transfer, most control devices incorporate SMAs as wires, due to which uniaxial models are particularly necessary. Brinson [9] and Auricchio and Sacco [10], among others, developed fundamental one-dimensional macroscopic models. Since then, several one-dimensional thermomechanically coupled constitutive models were designed, such as in [11,12]. Improved versions of these models are also proposed in [13,14]. For a detailed review on other modelling techniques, we refer the interested readers also to [15] and references therein.

A common characteristic of constitutive SMA models is that their accuracy relies on the utilized material parameters, which need to be identified by experiments. In case of SMA wires, uniaxial tensile tests are sufficient to obtain stress–strain relations. The parameter identification (PI) process of thermodynamic properties requires specific instruments and an expert knowledge-based data processing, such as thermomechanical analysis or more specifically the differential scanning calorimetry (DSC) [16,17]. A more important aspect is the fact that even with perfectly identified parameters, constitutive models require a supplementary tuning step for accurate response computations. Machine learning (ML) methods, particularly artificial neural networks (ANNs), provide efficient and versatile solutions, which can circumvent these challenges.

ANNs are capable of predicting highly nonlinear relations with any desired degree of accuracy according to the universal approximation theorem [18] and have already been used as black-box models for the mapping of constitutive relations, such as in [19,20]. In the field of SMAs, in their pioneering study, Ozbulut and Hurlebaus [21] proposed a neurofuzzy model, which predicts SMA responses from strain, strain rate, and temperature inputs. Although black-box models are efficient, they generally require a large set of representative experimental data.

In the context of PI, Huber and Tsakmakis [22] successfully implemented a feedforward ANN, which determines the parameters of a finite deformation viscoplasticity model. Furthermore, for superelastic SMAs exposed to quasistatic loading, Helm [23] developed a constitutive model and proposed a three-layered feedforward ANN architecture to identify the required model parameters from stress inputs. In this approach, similar to Huber and Tsakmakis, the ANN is trained by data, which are numerically generated by the model using parameters that are randomly sampled within a predefined space. Similarly, Henrickson et al. [24] trained an ANN to identify transformation temperatures of SMAs by using strain–temperature curves as inputs.

This paper proposes particularly for dynamic applications of superelastic SMA wires an ANN-based PI methodology, which considers the strain rate dependency of the material and focuses on the identification of thermodynamic parameters from stress–strain responses. Within the PI methodology, the identified parameters are already being tuned considering the constitutive model to allow accurate response computations.

The remainder of the paper is organized as follows: In Section 2, the approach is presented and implemented within the macroscopic modelling frame of Zhu and Zhang [12]. In Section 3, the approach is applied to SMA wires and validated by experiments. Finally, the conclusions of the study are drawn in Section 4.

2. Methodology

2.1. Superelastic SMA Response

SMAs are characterized by austenite (A) and martensite (M) phase states. Above the austenite transformation finish temperature A_f, the parent phase is austenite, and SMAs exhibit superelastic behavior. Hence, high mechanical stresses induce a forward-phase transformation (AM), and the SMA crystals reorient their atomic grid from body centered (B2) to a monoclinic (B19) lattice, which is more stable for high stress levels. Upon unloading, a reverse-phase transformation (MA) occurs, and the material returns to its original shape without residual deformation. During both forward- and reverse-phase transformations, SMAs exhibit a pseudoplastic deformation, which can be observed by stress plateaus in the stress–strain response. The AM transformation is an exothermic process. The lattice rearrangement causes internal heating, which is released to the environment via heat convection and conduction. By contrast, the MA transformation is a strongly endothermic reaction, in which the austenite formation is accompanied by a reduction in material temperature.

Dynamic loading patterns generally involve high strain rates, which impair the release of the heat generated within the forward transformation. This affects the material, such that the stress plateau slope increases since the austenite phase is energetically more stable for the increased temperature. Consequently, the reverse transformation is also initiated on higher critical stress levels affecting the stress–strain curve and the associated hysteretic energy dissipation.

To illustrate the strain rate effects, experimentally determined stress–strain responses of an SMA wire are depicted in Figure 1. The alloy composition of the wire is Ni-55.8%-Ti-43.95%. The wire length and diameter are $l = 150$ and $d = 0.2$ mm, respectively. A pre-stress of $\sigma_0 = 134.9$ MPa is applied. The ambient temperature is around $T_{init} = 296.2$ K (23.05 °C), whereas the austenite transformation finish temperature of the wire is $A_f = 285.2$ K (12.05 °C). Accordingly, the SMA is expected to response superelastically. Two cyclic loading patterns are applied with the strain rate amplitudes of $\dot{\varepsilon}_a = 1.26\%\ \mathrm{s}^{-1}$ and $\dot{\varepsilon}_b = 50.27\%\ \mathrm{s}^{-1}$. The *AM* transformation plateau changes from horizontal to a steeper slope with the increasing strain rate as the generated heat cannot be released directly. High strain rates also reshape the *MA* transformation, which starts at a higher stress with a steeper slope.

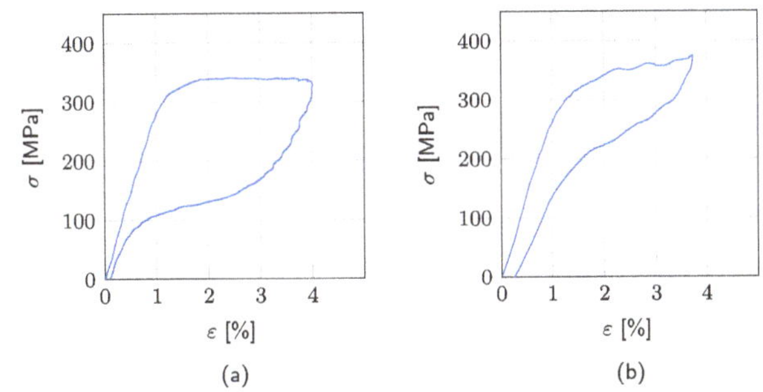

Figure 1. Cyclic tensile tests on an SMA wire. Strain rate amplitudes: (**a**) quasistatic $\dot{\varepsilon}_a = 1.26\%\ \mathrm{s}^{-1}$ and (**b**) dynamic $\dot{\varepsilon}_b = 50.27\%\ \mathrm{s}^{-1}$.

2.2. Machine-Learning-Based Parameter Identification

Macroscopic models of superelastic SMA wires generally compute the stress response σ, temperature T, and martensite volume fraction ξ from strain ε and strain rate $\dot{\varepsilon}$. Accordingly, such models can be represented as $\mathcal{M}((\boldsymbol{\varepsilon}, \dot{\boldsymbol{\varepsilon}}), \mathbf{p})$, where the vectors $\boldsymbol{\varepsilon}$ and $\dot{\boldsymbol{\varepsilon}}$ are time histories of strain and strain rate, respectively. The vector $\mathbf{p}$ contains model parameters, which need to be determined from experiments. Conventionally, as shown in Figure 2a, the stress–strain (σ-ε) experiments are conducted to investigate the cyclic tensile stress and strain response characteristics. Furthermore, thermodynamic experiments are required to investigate the model parameters representing the thermodynamic characteristics. Alternatively, in this study, as shown in Figure 2b, to reduce the experimental effort, an ML-based procedure is proposed for the identification of the thermodynamic parameters from stress–strain experiments by using an ANN. Different types of ML models could realize a better performance as well. However, in this study, ANNs are preferred considering their capability and efficiency in representing the extreme nonlinearities of SMAs.

The procedure consists of forward and reverse steps. As shown in Figure 3a, in the forward step, a constitutive SMA model generates training data for the ANN. In this study, due to its robustness and accuracy, the constitutive model by Zhu and Zhang [12] is chosen and adapted considering dynamic effects, as described in Section 2.3. Other models, such as by Auricchio and Sacco [11], could be implemented as well.

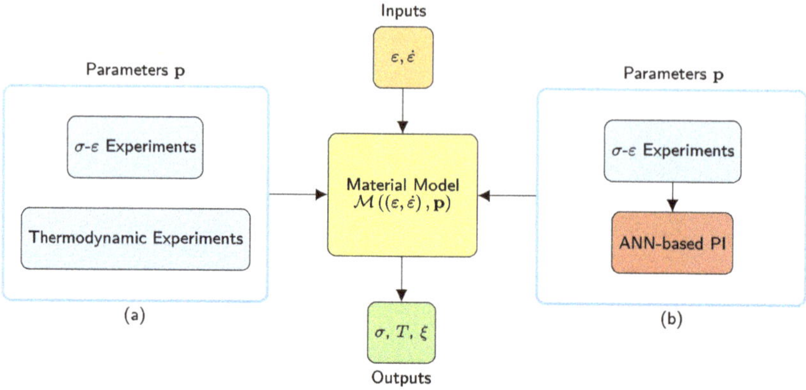

Figure 2. (**a**) Conventional experimental identification and (**b**) proposed machine-learning-based identification of thermodynamic parameters for use in constitutive modelling. The constitutive model is driven by the strain ε and strain rate $\dot{\varepsilon}$ time histories. Stress σ, temperature T, and the martensite volume fraction ξ time histories are computed as model outputs.

The searched thermodynamic parameters $\mathbf{p}_i, \forall i \in \{1 : N\}$ are sampled from a predefined parameter space by the Latin hypercube sampling (LHS) method [25], where N is the number of samples. In Figure 3, as an example, three parameter types ($\mathbf{p}_i^\top = [p_1\, p_2\, p_3]_i$) are searched. Accordingly, the parameter sampling space is three-dimensional here. At this point, it should be noted that, besides material parameters, during PI process, one should particularly consider those parameters as variable, which directly affect the reverse transformation, cf. Figure 1 and $A_{s,d}$ in Section 3.2. Each sampled parameter set $\mathbf{p}_i$ is then fed separately into the constitutive model, such that the corresponding SMA stress response vector $\boldsymbol{\sigma}_i$ is generated as the training data set $\mathcal{D} = \{(\boldsymbol{\sigma}_i, \mathbf{p}_i), \forall i \in \{1 : N\}\}$. Each computation is conducted for one load cycle, which is represented by the strain $\boldsymbol{\varepsilon}_i \in \mathbb{R}^{n\times 1}$ and strain rate $\dot{\boldsymbol{\varepsilon}}_i \in \mathbb{R}^{n\times 1}$ time histories, where n is the number of time instants. Accordingly, each generated stress time history $\boldsymbol{\sigma}_i \in \mathbb{R}^{n\times 1}$ consists of loading $\boldsymbol{\sigma}_i^{AM}$ and unloading $\boldsymbol{\sigma}_i^{MA}$ paths.

As shown in Figure 3b, the reverse step is built as a feedforward ANN, in which the output $\mathbf{p}^{(1)}$ of the first hidden layer and the output $\mathbf{p}^{(l)}$ of the final layer read

$$\mathbf{p}^{(1)} = f^{(1)}(\mathbf{W}^{(1)\top}\boldsymbol{\sigma} + \mathbf{b}^{(1)}), \qquad \mathbf{p}^{(l)} = \hat{\mathbf{p}} = f^{(l)}(\mathbf{W}^{(l)\top}\mathbf{y}^{(l-1)} + \mathbf{b}^{(l)}), \tag{1}$$

where $l-1$ is the number of hidden layers, which depends on the complexity of the material model. Theoretically, one layer is enough according to the universal approximation theorem [18]. For constitutive models using a limited number of parameter types, shallow ANNs with 2–3 layers are suggested, as shown later in Section 3. In Equation (1), f is the activation function, $\mathbf{W}$ is the weight matrix, $\boldsymbol{\sigma}$ is the stress input vector, and $\mathbf{b}$ is the bias vector. The vector $\hat{\mathbf{p}}$ contains thermodynamic parameters, which are predicted by the ANN. For the training, the mean squared error function is applied, which is optimized within the backpropagation algorithm by the Adam optimizer [26]. In total, 10% of the generated data $\mathcal{D}$ is spared for validation. The hyperbolic tangent activation function is used in the hidden layers, whereas the sigmoid function is used in the final layer to scale the outputs between 0 and 1. To avoid vanishing gradients, a batch normalization algorithm [27] is applied. Additionally, the dropout algorithm [28] is used after activation of the hidden layers to prevent the overfitting of the training data.

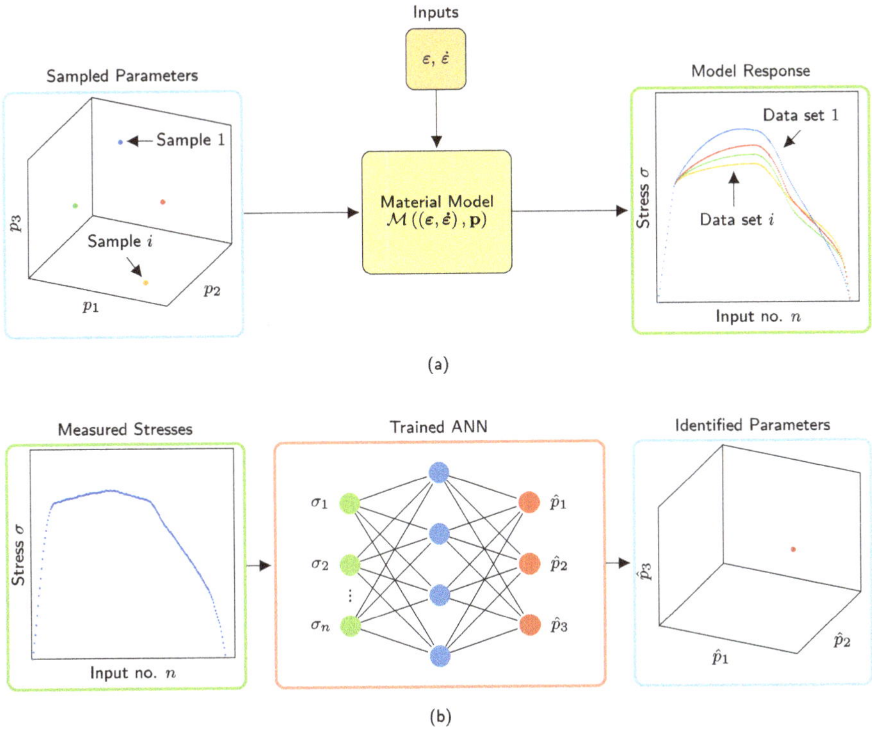

Figure 3. (**a**) Forward step of the proposed parameter identification procedure: Thermodynamic parameter samples $\mathbf{p}_i$ are fed into the material model to produce n stress outputs. (**b**) Reverse step of the proposed parameter identification procedure: Trained ANN estimates searched thermodynamic parameters $\hat{\mathbf{p}}$ from experimental stress–strain response.

2.3. Constitutive Modelling of Superelastic SMA Wire Response

The constitutive material model by Zhu and Zhang [12] is based on the first and second law of thermodynamics. Similar to the work of Tanaka [29], the material model is strain-driven and uses the state variables temperature T and martensitic volume fraction ξ to compute the one-dimensional tensile stress behavior of SMA wires. In analogy to Sadjadpour and Bhattacharya [30], the Helmholtz free energy is computed per unit mass by

$$\psi = \frac{E}{2\rho}\varepsilon_{el}^2 + \frac{L}{T_{cr}}(T - T_{cr})\xi - CT\ln\left(\frac{T}{T_{init}}\right), \tag{2}$$

where ρ is the density, ε_{el} is the elastic strain, T_{cr} and T_{init} are the transformation and initial (ambient) temperatures, E is the Young's modulus, L is the latent heat of the phase transition and C represents the specific heat. Here, the Young's modulus is expressed according to Liang [31] and Sato and Tanaka [32] as a function of the martensite volume fraction to allow for a transition between the two phases as

$$E(\xi) = E_A + \xi(E_M - E_A), \tag{3}$$

where E_A and E_M are constants representing the corresponding elastic moduli of austenite and martensite phases, respectively. The elastic ε_{el} and inelastic ε_{in} strains are expressed according to Brinson [9] and Auricchio and Sacco [10] as

$$\varepsilon_{el} = \varepsilon - \varepsilon_{in}, \qquad \varepsilon_{in} = \varepsilon_l \xi, \tag{4}$$

where ε_l is the maximum strain after a complete AM transformation with the martensite portion $\xi = 1$. Furthermore, the heat equation is derived from the first law of thermodynamics as

$$\begin{aligned} C\dot{T} &= -\frac{\partial \psi}{\partial \xi}\dot{\xi} + T\frac{\partial^2 \psi}{\partial \varepsilon \partial T}\dot{\varepsilon} + T\frac{\partial^2 \psi}{\partial T \partial \xi}\dot{\xi} - \frac{k}{V\rho}(T - T_{init}) \\ &= \frac{\sigma}{\rho}\varepsilon_l \dot{\xi} + L\dot{\xi} - \frac{k}{V\rho}(T - T_{init}), \end{aligned} \tag{5}$$

where k and V are the heat-transfer coefficient and the specimen volume, respectively. The stress response of the SMA wire is then computed by

$$\sigma = \rho\frac{\partial \psi}{\partial \varepsilon} = E\varepsilon_{el} \tag{6}$$

Moreover, four critical stress levels are defined to indicate the start and finish stresses of the phase transformations:

$$\begin{aligned} \sigma_s^{AM} &= c_M(T - M_s), \qquad & \sigma_f^{AM} &= c_M\left(T - M_f\right), \\ \sigma_{s,d}^{MA} &= c_A(T - A_{s,d}), \qquad & \sigma_f^{MA} &= c_A\left(T - A_f\right), \end{aligned} \tag{7}$$

where $M_{s/f}$ and $A_{s/f}$ refer to the start/finish temperatures of martensite and austenite transformation, respectively, as shown in Figure 4. Here, $c_{M/A}$ are material constants indicating the critical stress–temperature slopes.

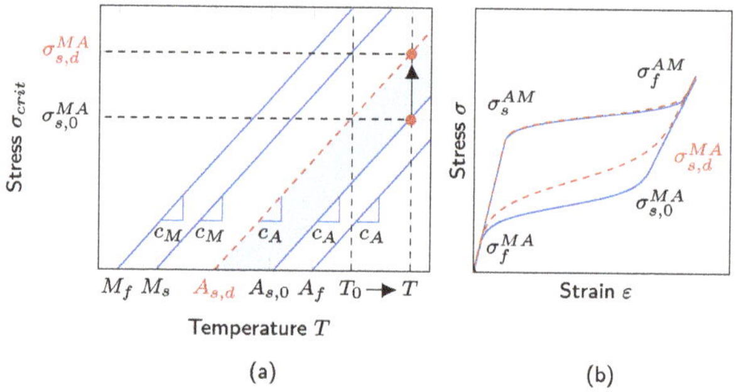

Figure 4. (**a**) The relation between the critical stress levels σ_{crit} and the material temperature T. The austenite transformation start temperature A_s is strain rate dependent. Here, $A_{s,0}$ and $A_{s,d}$ denote the austenite transformation start temperature for quasistatic and dynamic cases, respectively. (**b**) The corresponding stress–strain response of both quasistatic (solid line) and dynamic (dashed line) loading.

To model the dynamic response more accurately, in the formulation above, we use T as the material temperature and not as the environmental temperature, which was originally proposed by Zhu and Zhang, cf. [12]. In this way, the material model is able to consider both the quasistatic and dynamic cases, such that the critical stress levels are increased for high strain rates with increasing material temperatures. However, experiments show that a change in material temperature solely is not enough to initiate the MA transformation accurately. Hence, $A_{s,d}$ and $\sigma_{s,d}^{MA}$ are introduced, where the subscript $_d$ represents the modelling of the austenite transformation start temperature and the corresponding critical stress level with respect to dynamic effects.

This adaptation is also shown in Figure 4. Here, T_0 denotes the material temperature for the quasistatic load case. For dynamic load cases, the material temperature increases, such that a supplementary upward shift in the critical stress level is necessary to replicate the SMA response accurately. This behavior is achieved in the model by decreasing the austenite transformation start temperature from $A_{s,0}$ to $A_{s,d}$.

In the model, the evolution of the martensite fraction ξ is based on the Liang–Rogers model [33]. Its rate form is computed by

$$\text{A} \rightarrow \text{M:} \qquad \dot{\xi} = (1-\xi_0)\left(a_M \dot{T} - \frac{a_M}{c_M}\dot{\sigma}\right) g\left[a_M\left(T - T_M - \frac{\sigma}{c_M}\right)\right], \tag{8}$$

$$\text{M} \rightarrow \text{A:} \qquad \dot{\xi} = \xi_0\left(a_A \dot{T} - \frac{a_A}{c_A}\dot{\sigma}\right) g\left[a_A\left(T - T_A - \frac{\sigma}{c_A}\right)\right], \tag{9}$$

where $g = -e^x(1+e^x)^{-2}$ and

$$T_M = \frac{M_s + M_f}{2}, \quad T_A = \frac{A_s + A_f}{2}, \quad a_M = \frac{\ln(10{,}000)}{M_s - M_f}, \quad a_A = \frac{\ln(10{,}000)}{A_f - A_s} \tag{10}$$

Here, ξ_0 refers to the initial martensite fraction at the beginning of the current transformation. In the model of Zhu and Zhang [12], the outputs σ, T, ξ as well as their time derivatives, are solved simultaneously from the heat equation (Equation (5)), mechanical stress equation (Equation (6)), and the kinetic rules (Equations (8) and (9)) by time integration algorithms, such as the fourth-order Runge–Kutta method. Figure 5 illustrates the input and output parameters of the material model.

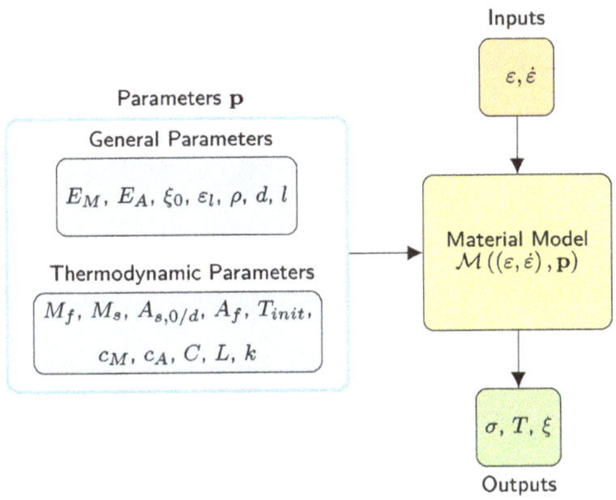

Figure 5. Input and output parameters of the material model. The constitutive model is driven by the strain ε and strain rate $\dot{\varepsilon}$ time histories. Stress σ, temperature T, and the martensite volume fraction ξ time histories are computed as model outputs.

Parameter Influence on Stress–Strain Response

The parameters introduced in Section 2.3 have to be determined accurately as each of them affects the material model response. It is obvious that the Young's moduli, E_M and E_A, have a major influence on the stress–strain behavior. This effect can be easily seen in the elastic parts of the stress–strain response, where an unsuitable parameter would lead to inaccurate slopes in the σ-ε diagram. Furthermore, ε_l denotes the maximum strain at $\xi = 1$, such that it directly affects the strain level, where the phase transformation from austenite to martensite is finished, cf. Equation (4). In this study, purely superelastic material behavior is assumed since the ambient temperature is above the austenite transformation finish temperature ($T_{init} > A_f$). As a result, ξ_0 is assumed to be 0 in the initial austenite state. Nevertheless, this parameter is variable and changes as the martensite transformation proceeds. The ambient temperature T_{init} has only a minor influence on the model response since the material temperature is used to compute the critical stress levels, cf. Equation (7). From Equation (6), it is obvious that the stress response increases for high specimen densities ρ. In addition, from Equation (5), it follows that greater specimen volumes V, or diameters d, lead to higher temperature evolutions. By contrast, high heat-transfer coefficients k result in greater heat transfer and thus in lower temperature evolutions. A major impact on the stress–strain response results from the critical stress levels given in Equation (7). These stress levels are influenced by the parameters c_M and c_A, as well as M_f, M_s, $A_{s,d}$, and A_f. With increasing c_M and c_A, stress levels also increase. Whereas greater transformation temperatures (M_f, M_s, $A_{s,d}$, A_f) lead to lower stress levels.

For the modelling of dynamic SMA response, particularly, the introduced dynamic austenite transformation start temperature $A_{s,d}$ has a high influence as with it an earlier reverse transformation can be started, cf. Figure 4. Furthermore, in case of high strain rates, the material temperature is increased due to self-heating effects. These effects are modelled by the specific and latent heat (C and L), which have the greatest influence on the dynamic model response as they are directly linked to the temperature evolution, cf. Equation (5). For the dynamic load case, finally, k is also important, which controls, as mentioned above, directly the heat transfer of the SMA.

3. Results and Discussion

3.1. Stress–Strain Experiments

For the generation of test data, cyclic tensile tests are conducted by a uniaxial shaking table, as schematically depicted in Figure 6. A pre-stress of $\sigma_0 = 134.9$ MPa was applied. The wire stress response was measured by a load cell. The applied strain was determined by a laser position sensor. Both sensors were sampled at 1000 Hz. The ambient temperature was around $T_{init} = 296.2$ K (23.05 °C). In this study, a Ni-55.8%-Ti-43.95% SMA wire with $l = 150$ mm length and $d = 0.2$ mm diameter was investigated. For further information on the test setup, the reader is referred to [34]. All material parameters are reported in Table 1. Here, E_M, E_A, and ε_l are determined via the stress–strain relation obtained under quasistatic load. Furthermore, ρ, M_f, M_s, $A_{s,0}$, and A_f are provided by the manufacturer, whereas c_M and c_A are determined using Equation (7) from the provided parameters. Ambient temperature and geometric parameters (diameter and wire length) are measured before testing.

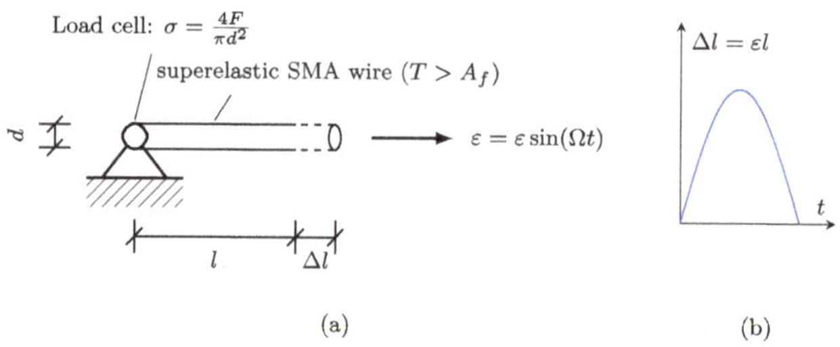

Figure 6. (**a**) Schematic representation of stress–strain experiments. The wire of diameter d and length l is fixed on one side and elongated by Δl. The resulting tensions and the applied strains are measured by a load cell and a laser position sensor, respectively. The ambient temperature is above the austenite transformation start temperature ($T > A_f$). (**b**) Applied displacement with respect to time.

3.2. Identification of Thermodynamic Parameters

As introduced in Section 2.3, the constitutive SMA model of Zhu and Zhang relies on three thermodynamic parameters: the specific heat C, the latent heat L, and the heat-transfer coefficient k. All other thermodynamic parameters (M_f, M_s, A_f, c_M, and c_A) are generally provided by manufacturers (cf. Table 1) and can also be used in constitutive models with a sufficient accuracy for dynamic applications. Furthermore, $A_{s,0}$ is also provided, which is accurate enough for the quasistatic case and is now tuned as $A_{s,d}$ to model the dynamic response more accurately.

Accordingly, the searched parameter set reads:

$$\mathbf{p}_i^{\top} = [C,\ L,\ k,\ A_{s,d}]_i \tag{11}$$

where the parameter space of C, L, and k is determined based on their influence on the response of the constitutive model. For this purpose, the cyclic stress response of the SMA wire is computed by the material model for the strain amplitude $\varepsilon = 4\%$ and the strain rate amplitude $\dot{\varepsilon} = 50.27\%\ \mathrm{s}^{-1}$. The parameters are alternated around the reference setting $C = 800$ J(kgK)$^{-1}$, $L = 8000$ J kg^{-1}, and $k = 0.021$ W K^{-1}, where the works of Kato [35] and Zhu and Zhang [12] are taken as reference. The chosen parameter space is listed in Table 2. The remaining material parameters that are used for this study, including $A_{s,d} = A_{s,0}$, correspond to Table 1.

Table 1. Constitutive model parameters of the SMA wires used in this study.

Parameter		Value	Unit
Young's modulus of martensite	E_M	14,100	MPa
Young's modulus of austenite	E_A	29,000	MPa
Critical stress–temperature slope of martensite	c_M	6.13	MPa K^{-1}
Critical stress–temperature slope of austenite	c_A	6.57	MPa K^{-1}
Maximum strain at $\xi = 1$	ε_l	0.04	-
Martensite transformation finish temperature	M_f	231.4	K
Martensite transformation start temperature	M_s	247.7	K
Austenite transformation start temperature	$A_{s,0}$	263.2	K
Austenite transformation finish temperature	A_f	285.2	K
Ambient temperature	T_{init}	296.2	K
Density	ρ	6500	kg m^{-3}
Diameter	d	0.0002	m
Wire length	l	0.15	m

Table 2. Parameter space of the searched thermodynamic parameters $\mathbf{p}_i$.

Parameter		min	max	Unit
Specific heat	C	800	2000	J(kgK)$^{-1}$
Latent heat	L	400	40,000	J kg^{-1}
Heat-transfer coefficient	k	0.001	0.100	W K^{-1}
Austenite transformation start temperature	$A_{s,d}$	247.0	263.2	K

The results are depicted in Figure 7. The smaller the specific heat C, the less energy is required to raise the material temperature. This results in greater temperature amplitudes and consequently leads to steeper transformation plateaus, as shown in Figure 7a. On the other hand, the latent heat provokes similar effects in the hysteresis of the stress–strain curve, as shown in Figure 7b. In fact, the latent heat is included in Equation (5) and thus directly linked to the temperature evolution. A high latent heat value results in an increase in temperature during the austenite-to-martensite phase transformation. A higher material temperature, in turn, leads to a more stable austenite phase, such that a higher mechanical stress is needed to proceed the phase transformation. Accordingly, C and L affect the model response oppositely and with different parameter combinations the same material response can be replicated. Finally, the higher the heat-transfer coefficient k, the more heat is transferred to the environment. Small k values thus lead to rising temperatures after each cycle as there is less heat transferred from the wire to the environment, as shown in Figure 7c.

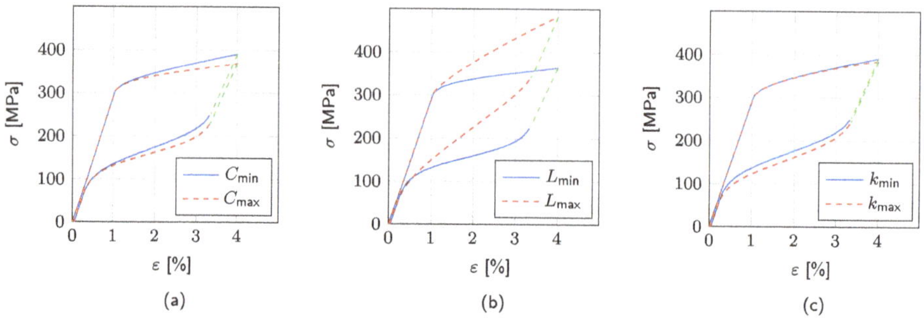

Figure 7. Influence of thermodynamic parameters on the model response: (**a**) Influence of specific heat C, (**b**) influence of latent heat L, and (**c**) influence of heat-transfer coefficient k. The green dashed line illustrates the linear elastic part of the unloading path.

Accordingly, the first three parameters of $\mathbf{p}_i$ influence the slope of both forward- and reverse-phase transformations. However, these parameters do not affect the critical stress σ^{MA} directly, which initiates the reverse transformation. As illustrated in Figure 7 by the green dashed lines, C, L, and k do not affect the length of the linear elastic martensite unloading path. This effect can be efficiently covered by alternating A_s. Its parameter space is also introduced in Table 2 and is determined similar to the prior ones considering the influence on the response of the constitutive model. Accordingly, $A_{s,d}$ is chosen, such that it covers the temperature range below $A_{s,0}$ until M_s, as described previously in Figure 4. It should be noted that the size of the parameter space does not have a major influence on the neural network accuracy as the best-fitting parameter *combination* is identified. On the other hand, with increasing parameter ranges, the required computational effort also increases. In this study, the ranges are chosen based on both numerical and experimental expertise from [12,34,35].

For the PI, a feedforward ANN is utilized with two hidden layers, where the first and second layers contain 128 and 64 neurons, respectively. The learning rate is initialized as 1×10^{-3}. Training is completed after 50 epochs and is conducted with a batch size of 32. A Gaussian noise with zero mean and a variance of 1×10^{-4} is added to the training input data to make the network more stable to measurement errors. For the training of the ANN, $i = 2000$ sets are sampled from the parameter space. The network architecture and its training parameters are determined after a parametric study by trial and error. Deeper network architectures lead to slightly more accurate predictions but at the cost of computational effort. Therefore, in this study, the simplest possible network architecture is chosen, which provides sufficient results. The generated data correspond to $\mathcal{D} = \{(\sigma_i, \mathbf{p}_i), \forall i \in \{1:2000\}\}$, where 10% of them is used for validation. To consider strain rate effects, the data set is generated using the material model repeatedly for five different strain rate amplitudes of $\dot{\varepsilon} \in [1.26, 2.51, 12.57, 25.13, 50.27]\%\ \mathrm{s}^{-1}$ with a constant strain amplitude of $\varepsilon = 4\%$. The generation of signal data and neural network training take less than 30 min using an AMD Ryzen 9 5950× CPU and computing on an NVIDIA GeForce RTX 3090, which is significantly lower than the experimental effort considering the time required for the specimen preparation, tests, and data processing.

After training the ANN, the σ-ε experiments are conducted, as described in Section 3.1 for the same strain rate and strain amplitudes. From the experimental stress–strain response, the parameters $\hat{C}$, $\hat{L}$, $\hat{k}$, and $\hat{A}_{s,d}$ are identified. Table 3 shows the estimated $\hat{\mathbf{p}}$ after training with $\mathcal{D}$. Here, the experimentally measured loading stress responses are divided into 200 equidistantly distributed points. The final values of the identified parameters are determined from the mean over all investigated strain rate cases, except for $A_{s,d}$, where only the dynamic load cases ($\dot{\varepsilon} \in [12.57, 25.13, 50.27]\%\ \mathrm{s}^{-1}$) are considered.

Table 3. Thermodynamic parameter set $\hat{\mathbf{p}}$ identified by the ANN after training with the data set $\mathcal{D}$.

$\dot{\varepsilon}$ [% s^{-1}]	ε [%]	$\hat{C}$ [J(kgK) $^{-1}$]	$\hat{L}$ [J kg^{-1}]	$\hat{k}$ [W K^{-1}]	$\hat{A}_{s,d}$ [K]
1.26	4	1146	5404	0.084	$A_{s,0}$
2.51	4	1824	18,773	0.089	$A_{s,0}$
12.57	4	1827	18,583	0.090	247.3
25.13	4	1888	23,278	0.089	247.4
50.27	4	1896	25,108	0.089	248.0
Final result $\hat{\mathbf{p}}$:		1716	18,229	0.088	247.6

Figure 8 compares the experimental results with the numerical calculations, which use the identified parameter set $\hat{\mathbf{p}}$, where for the quasistatic cases ($\dot{\varepsilon} \in [1.26, 2.51]\%\,\mathrm{s}^{-1}$) the manufacturer provided $A_{s,0}$ is used in numerical calculations. The numerical calculations using $\hat{\mathbf{p}}$ accurately match both the loading and unloading paths of experimental results. It is noteworthy that the accuracy of the numerical calculations is limited by the capability of the constitutive model. Therefore, the effects, such as residual deformations and hardening, are not replicated accurately. We observe from the experimental results that with an increasing strain rate, the residual wire strain becomes significantly visible. This effect is currently not replicated by the model. With these results, we would like to emphasize that the accuracy of the proposed PI is limited by the constitutive model precision. The effects that are not included in the basis material model cannot be considered in the PI process.

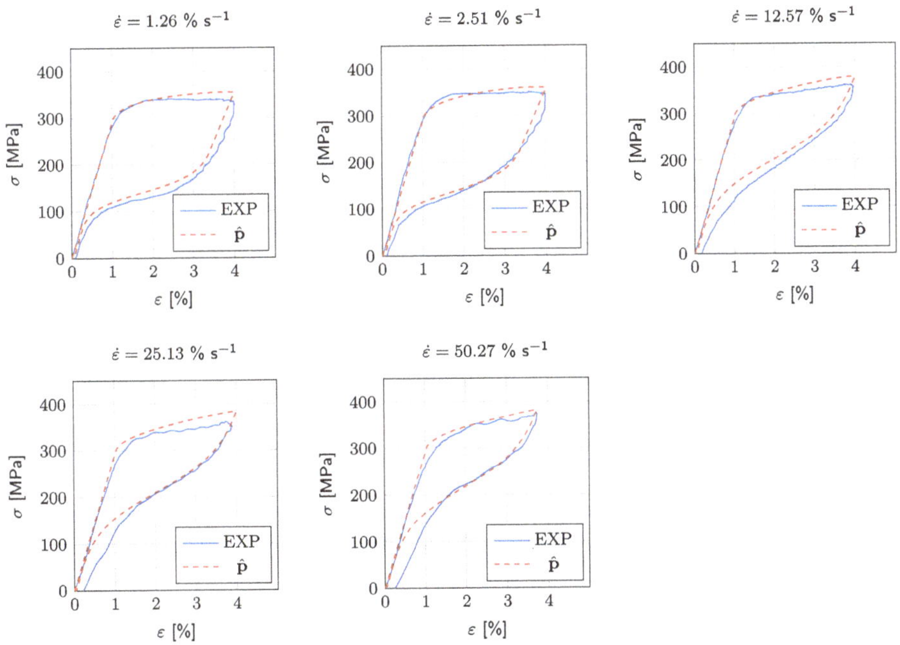

Figure 8. Comparison of the experimentally determined stress–strain responses (EXP) with numerical calculations, which use the identified parameter set $\hat{\mathbf{p}}$.

From the results, the mean value of the relative cumulative stress error ϵ_{n} is computed as

$$\epsilon_{\mathrm{n}} = \sum_{\mathrm{m}}^{5} \left(\frac{|\Delta\sigma_{\mathrm{m,n}}|}{\sigma_{\mathrm{EXP,m,n}}} \right) / 5 \quad (12)$$

where $\Delta\sigma_{\mathtt{m},\mathtt{n}}$ is the difference between the estimated and experimentally determined stress $\sigma_{\mathrm{EXP},\mathtt{m},\mathtt{n}}$ of the corresponding strain intervals $\mathtt{n} \in [1,2,3,4]\%$ for both forward and reverse transformations. Here, the mean is computed over m cases corresponding to the number of investigated strain rate amplitudes. As shown in Figure 9, the error value is located around 5%, neglecting the value for 1% strain on the reverse transformation. In this case, the error value is high, which is due to the material model's inability of modelling residual deformations, cf. Figure 8.

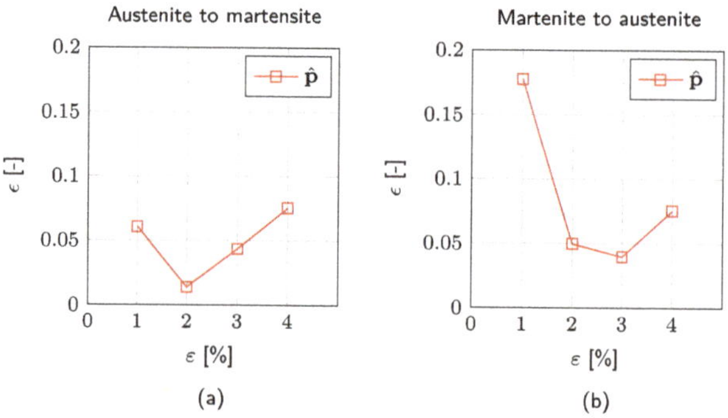

Figure 9. Mean value of the relative cumulative stress error corresponding to the used parameter set $\hat{\mathbf{p}}$ for both forward (**a**) and reverse (**b**) transformations.

4. Conclusions

In the present paper, for the identification of thermodynamic parameters of superelastic SMA wires, an ANN-based PI methodology is presented. The proposed approach was coupled with a macroscopic constitutive model. Here, to consider strain-rate-dependent response effects more accurately, two austenite transformation start temperatures were defined distinguishing between the quasistatic and dynamic cases. Furthermore, in the model, the current material temperature was considered for a strain rate sensitive calculation of critical stress levels. In an initial step, training data were generated by the constitutive model. Here, the searched thermodynamic parameters were sampled from a parameter space, and for each sampling, the stress responses were computed. After training, the ANN estimated the searched parameters from conventional stress–strain experiments. Finally, to validate the accuracy of the proposed method, numerical simulations were conducted, and results were compared with experimental data.

Author Contributions: N.L.: Conceptualization, methodology, software, validation, formal analysis, investigation, resources, data curation, visualization, and writing—original draft preparation. O.A.: writing—review and editing, supervision, project administration, and funding acquisition. All authors have read and agreed to the published version of the manuscript.

Funding: This research was funded by the German Research Foundation (Deutsche Forschungsgemeinschaft, DFG) under the grant number 322268262.

Institutional Review Board Statement: Not applicable.

Informed Consent Statement: Not applicable.

Data Availability Statement: The data presented in this article are available on request from the corresponding author.

Acknowledgments: The authors appreciate the financial support of this study by the German Research Foundation. The authors also wish to thank Andreas Kaup for providing the numerical implementation of the constitutive model used in this study.

Conflicts of Interest: The authors declare no conflict of interest. The funders had no role in the design of the study; in the collection, analyses, or interpretation of data; in the writing of the manuscript, or in the decision to publish the results.

Abbreviations

The following abbreviations are used in this manuscript:

ANN	Artificial Neural Network
LHS	Latin Hypercube Sampling
SMA	Shape Memory Alloy
AM	Austenite to Martensite
MA	Martensite to Austenite
ML	Machine Learning
PI	Parameter Identification

References

1. Qiu, C.; Zhu, S. Shake table test and numerical study of self-centering steel frame with SMA braces. *Earthq. Eng. Struct. Dyn.* **2017**, *46*, 117–137. [CrossRef]
2. Liu, Y.; Wang, H.; Qiu, C.; Zhao, X. Seismic Behavior of Superelastic Shape Memory Alloy Spring in Base Isolation System of Multi-Story Steel Frame. *Materials* **2019**, *12*, 997. [CrossRef]
3. Liang, D.; Zheng, Y.; Fang, C.; Yam, M.; Zhang, C. Shape memory alloy (SMA)-cable-controlled sliding bearings: Development, testing, and system behavior. *Smart Mater. Struct.* **2020**, *29*, 085006. [CrossRef]
4. Ozbulut, O.E.; Hurlebaus, S.; DesRoches, R. Seismic response control using shape memory alloys: A review. *J. Intell. Mater. Syst. Struct.* **2011**, *22*, 1531–1549. [CrossRef]
5. Fang, C.; Wang, W. *Shape Memory Alloys for Seismic Resilience*; Springer: Singapore, 2020. [CrossRef]
6. Tabrizikahou, A.; Kuczma, M.; Nowotarski, P.; Kwiatek, M.; Javanmardi, A. Sustainability of Civil Structures through the Application of Smart Materials: A Review. *Materials* **2021**, *14*, 4824. [CrossRef]
7. Costanza, G.; Tata, M.E. Shape Memory Alloys for Aerospace, Recent Developments, and New Applications: A Short Review. *Materials* **2020**, *13*, 1856. [CrossRef] [PubMed]
8. Jani, J.M.; Leary, M.; Subic, A.; Gibson, M.A. A review of shape memory alloy research, applications and opportunities. *Mater. Des.* **2014**, *56*, 1078–1113. [CrossRef]
9. Brinson, L.C. One-dimensional constitutive behaviour of shape memory alloys: Thermomechanical derivation with non-constant material functions and redefined martensite internal variable. *J. Intell. Mater. Syst. Struc.* **1993**, *4*, 229–242. [CrossRef]
10. Auricchio, F.; Sacco, E. A One-Dimensional Model for Superelastic Shape-Memory Alloys with Different Elastic Properties Between Austenite and Martensite. *Int. J. Non-Linear Mech.* **1997**, *32*, 1101–1114. [CrossRef]
11. Auricchio, F.; Sacco, E. Thermo-Mechanical Modelling of a Superelastic Shape-Memory Wire Under Cyclic Stretching–Bending Loadings. *Int. J. Solids Struct.* **2001**, *38*, 6123–6145. [CrossRef]
12. Zhu, S.; Zhang, Y. A Thermomechanical Constitutive Model for Superelastic SMA Wire with Strain-Rate Dependence. *J. Smart Mater. Struct.* **2007**, *16*, 1696–1707. [CrossRef]
13. Kaup, A.; Altay, O.; Klinkel, S. Macroscopic modeling of strain-rate dependent energy dissipation of superelastic SMA dampers considering destabilization of martensitic lattice. *Smart Mater. Struct.* **2019**, *29*, 025005. [CrossRef]
14. Kaup, A.; Ding, H.; Wang, J.; Altay, O. Strain rate dependent formulation of the latent heat evolution of superelastic shape memory alloy wires incorporated in multistory frame structures. *J. Intell. Mater. Syst. Struct.* **2020**, *32*, 1198–1214. [CrossRef]
15. Cisse, C.; Zaki, W.; Ben Zineb, T. A Review of Constitutive Models and Modeling Techniques for Shape Memory Alloys. *Int. J. Plast.* **2016**, *76*, 244–284. [CrossRef]
16. Depriester, D.; Maynadier, A.; Lavernhe-Taillard, K.; Hubert, O. Thermomechanical modelling of a NiTi SMA sample submitted to displacement-controlled tensile test. *Int. J. Solids Struct.* **2014**, *51*, 1901–1922. [CrossRef]
17. Xiao, Y.; Zeng, P.; Lei, L.; Du, H. Experimental investigation on rate dependence of thermomechanical response in superelastic NiTi shape memory alloy. *J. Mater. Eng. Perform.* **2015**, *24*, 3755–3760. [CrossRef]
18. Hornik, K.; Stinchcombe, M.; White, H. Multilayer feedforward networks are universal approximators. *Neural Netw.* **1989**, *2*, 359–366. [CrossRef]
19. Ghaboussi, J.; Garrett, J.H., Jr.; Wu, X. Knowledge-based modeling of material behavior with neural networks. *J. Eng. Mech.* **1991**, *117*, 132–153. [CrossRef]
20. Furukawa, T.; Yagawa, G. Implicit constitutive modelling for viscoplasticity using neural networks. *Int. J. Numer. Methods Eng.* **1998**, *43*, 195–219. [CrossRef]
21. Ozbulut, O.; Hurlebaus, S. Neuro-fuzzy Modeling of Temperature- and Strain-rate-dependent Behavior of NiTi Shape Memory Alloys for Seismic Applications. *J. Intell. Mater. Syst. Struct.* **2010**, *21*, 837–849. [CrossRef]
22. Huber, N.; Tsakmakis, C. A neural network tool for identifying the material parameters of a finite deformation viscoplasticity model with static recovery. *Comput. Methods Appl. Mech. Eng.* **2001**, *191*, 353–384. [CrossRef]

23. Helm, D. Pseudoelasticity: Experimental observations, thermomechanical modeling, and identification of the material parameters. In Proceedings of the SPIE 5387, Smart Structures and Materials 2004: Active Materials: Behavior and Mechanics, San Diego, CA, USA, 21 July 2004; pp. 198–209. [CrossRef]
24. Henrickson, J.; Kirkpatrick, K.; Valasek, J. Characterization of Shape Memory Alloys Using Artificial Neural Networks. In Proceedings of the 51st AIAA Aerospace Sciences Meeting Including the New Horizons Forum and Aerospace Exposition, Grapevine, TX, USA, 7–10 January 2013; p. 129. [CrossRef]
25. McKay, M.D.; Beckman, R.J.; Conover, W.J. A comparison of three methods for selecting values of input variables in the analysis of output from a computer code. *Technometrics* **2000**, *42*, 55–61. [CrossRef]
26. Kingma, D.P.; Ba, J. Adam: A method for stochastic optimization. *arXiv* **2014**, arXiv:1412.6980.
27. Ioffe, S.; Szegedy, C. Batch normalization: Accelerating deep network training by reducing internal covariate shift. *arXiv* **2015**, arXiv:1502.03167.
28. Srivastava, N.; Hinton, G.; Krizhevsky, A.; Sutskever, I.; Salakhutdinov, R. Dropout: A simple way to prevent neural networks from overfitting. *J. Mach. Learn. Res.* **2014**, *15*, 1929–1958.
29. Tanaka, K. A thermomechanical sketch of shape memory effect: One-dimensional tensile behavior. *Res. Mech. Int. J. Struct. Mech. Mater. Sci.* **1986**, *18*, 251–263.
30. Sadjadpour, A.; Bhattacharya, K. A micromechanics inspired constitutive model for shape-memory alloys: The one-dimensional case. *Smart Mater. Struct.* **2007**, *16*, 51–62. [CrossRef]
31. Liang, C. The Constitutive Modeling of Shape Memory Alloys. PhD Thesis, Virginia Polytechnic Institute and State University: Blacksburg, VA, USA, 1990.
32. Sato, Y.; Tanaka, K. Estimation of energy dissipation in alloys due to stress-induced martensitic transformation. *Res. Mech.* **1988**, *23*, 381–393.
33. Liang, C.; Rogers, C.A. One-dimensional thermomechanical constitutive relations for shape memory materials. *J. Intell. Mater. Syst. Struct.* **1997**, *8*, 285–302. [CrossRef]
34. Kaup, A.; Altay, O.; Klinkel, S. Strain amplitude effects on the seismic performance of dampers utilizing shape memory alloy wires. *Eng. Struct.* **2021**, *244*, 112708. [CrossRef]
35. Kato, H. Latent heat storage capacity of NiTi shape memory alloy. *J. Mater. Sci.* **2021**, *56*, 8243–8250. [CrossRef]

materials

Article

Self-Centering Shape Memory Alloy-Viscoelastic Hybrid Braces for Seismic Resilience

Zhe-Xi Zhang [1], Yiwei Ping [1,*] and Xiuzhang He [1,2]

[1] Department of Structural Engineering, College of Civil Engineering, Tongji University, Shanghai 200092, China; zzx888@tongji.edu.cn (Z.-X.Z.); xiuzhanghe@gmail.com (X.H.)
[2] Department of Building and Real Estate, The Hong Kong Polytechnic University, Hong Kong, China
* Correspondence: zjtxpyw@tongji.edu.cn

Abstract: This paper presents a novel type of hybrid self-centering braces incorporating tension-only superelastic NiTi shape memory alloy (SMA) cables and integrated viscoelastic dampers (VEDs). One of our reasons for proposing this new SMA-viscoelastic hybrid brace (SCVEB) is to provide enhanced energy-dissipation ability whilst promoting increased self-centering tendency compared with the existing SMA-based self-centering solutions, where upgrading behavior is mainly benefited from the participation of the VEDs. The configuration and the working principle, along with theoretical equations describing the mechanical behavior of the SCVEB, are described in detail firstly. Experimental verification of individual elements in this SCVEB system, namely the NiTi SMA cables and VEDs, was performed to obtain a basic understanding of their mechanical properties. A proof-of-concept SCVEB specimen was then manufactured, and its cyclic performance was further investigated. Followed by this, a system-level analysis on a series of steel frames equipped with or without SCVEB was conducted. The results showed that the SCVEB system exhibited a moderate damping ratio and a more efficient controlled behavior in terms of its post-event residual deformation and floor acceleration when compared with those of the non-SCVEB system.

Keywords: self-centering; shape memory alloy (SMA); viscoelastic; brace; hybrid control; seismic resilience

Citation: Zhang, Z.-X.; Ping, Y.; He, X. Self-Centering Shape Memory Alloy-Viscoelastic Hybrid Braces for Seismic Resilience. *Materials* **2022**, *15*, 2349. https://doi.org/10.3390/ma15072349

Academic Editor: Carmine Maletta

Received: 17 February 2022
Accepted: 12 March 2022
Published: 22 March 2022

1. Introduction

The past decades have witnessed several major earthquakes [1–4], and the structures designed by modern ductility-based seismic design philosophy are proven to be effective in providing sufficient life safety assurance during the earthquakes. However, this design philosophy may be no longer sufficient when special attention is given to the issue of the economic seismic loss. For example, the 2011 Christchurch earthquake caused damage to thousands of buildings that did not collapse but were unrecoverable due to unacceptable damage and residual deformation, requiring a heavy cost to repair them [4]. It has been suggested that the magnitude of residual inter-story drift should be controlled within a drift limit of 0.5%, beyond which the structures may no longer be economically feasible to repair from the perspectives of building functionality, construction tolerances, and safety [5].

Driven by this demand, the community of seismic engineers has made continuous efforts to improve the post-earthquake structural performance over the past years. A promising solution, which is known as self-centering framed structural system, has received extensive research interests [6,7]. Additional self-centering members/devices, such as beam-to-column connections [8–17], braces [18–27], and dampers [28–41], with the capability of returning to their initial resting position after earthquake, are proven to be effective in eliminating the residual story drifts when they are installed in the corresponding frame systems. Ricles et al. [42,43] developed a post-tensioned (PT) moment-resisting connection where seat angles were utilized as an energy-dissipation source. Under loading, gap-opening occurred at the beam–column interface and the PT tendons provided a restoring force by acting to close the gap. Maurya and Eatherton [44] developed a new self-centering beam for moment frame that eliminated deformation incompatibility with the gravity framing. In the case of braces, Christopoulos et al. [18] proposed a new bracing system that comprised of tensioning elements and friction pads. Chou et al. [6,27] experimentally examined a dual-core self-centering sandwiched buckling restrained brace to increase deformability capacity. However, there still exists some inevitable issue associated with the PT self-centering technology. For example, due to the limited elastic strain (about 2%) of the PT tendons [18,27], there may be a deformation capacity deficiency, which may put the self-centering framed structures at risk when a severe earthquake happens. Without a redundant load-carrying path, yielding or/even rupture of the PT tendons would lead to serious disaster, and the lateral load resistance may be totally lost, which may trigger the collapse of structure. On the other hand, the cyclic response of the PT self-centering system generally exhibits a typical "flag-shape" hysteretic behavior, whose energy-dissipation capability (can be readily understood as the area enclosed in the hysteretic loops) is significantly reduced compared to that of conventional structural systems. The reduced energy dissipation may lead to the amplification of peak-deformation response and floor acceleration [45,46]. It has been recognized that the magnified peak response would cause extensive damage to both structural and non-structural components, which should be addressed in the practical design.

In recent years, various improvement strategies have been proposed to address the issues associated with PT self-centering system mentioned above. For instance, shape memory alloys (SMAs) [47–50], especially superelastic NiTinol SMA, were introduced in self-centering systems [51–55] because of their good energy-dissipation ability, ductility, and fatigue resistance. Superelasticity refers to the capability to spontaneously recover when the load is removed from SMA elements after experiencing a large strain up to 8~10% [56–58]. Evidently, SMA-based tendon provides a very competitive deformation compared with that of the aforementioned conventional PT tendons. Miller et al. [59] investigated a self-centering buckling-restrained brace (SC-BRB) by using SMA rods as a source of restoring force, which successfully achieved appreciable energy dissipation, large deformation capacity, and self-centering ability. Zhu and Zhang [38] developed a self-centering friction damping brace (SFDB). The study showed that the frame equipped with SFDB was capable of achieving a seismic-response level that is comparable with that of BRBF, but with a significant reduction in residual drifts. Chen et al. [60] employed variable friction devices to improve energy-dissipation efficiency of a brace without compromising the self-centering ability of pre-tensioned SMA cables.

With initial confidence gained from these pioneering investigations, a concept of SMA-viscoelastic hybrid braces (SCVEB) is proposed in this paper. The motivation behind this concept is to move a further step to deal with the issues arising from a self-centering system in previous studies, such as the amplified peak deformation and floor acceleration. The concept has been preliminarily examined by the authors and co-workers through a numerical study [61]. In this paper, a proof-of-concept experiment on a prototype SCVEB device was conducted, and the influence of some key parameters on the seismic behavior was discussed. Followed by this, a system-level analysis that investigated the effectiveness of the proposed SCVEB was conducted.

2. Configuration and Working Principle of SCVEB

The configuration of the investigated SCVEB is illustrated in Figure 1. Conceptually, the proposed SCVEB can be decomposed into two systems, i.e., self-centering system (SC system) and energy-dissipation system (ED system). The former is mainly functioned by the SMA elements, and the latter mainly relies on the energy-dissipation property of viscoelastic materials [62–64]. Note that SMA elements also consume some of the input energy when an earthquake occurs. In this section, some basic concepts, as well as the configuration of the SCVEB, are introduced, and the associated analytical expression for the behavior of the SCVEB is presented.

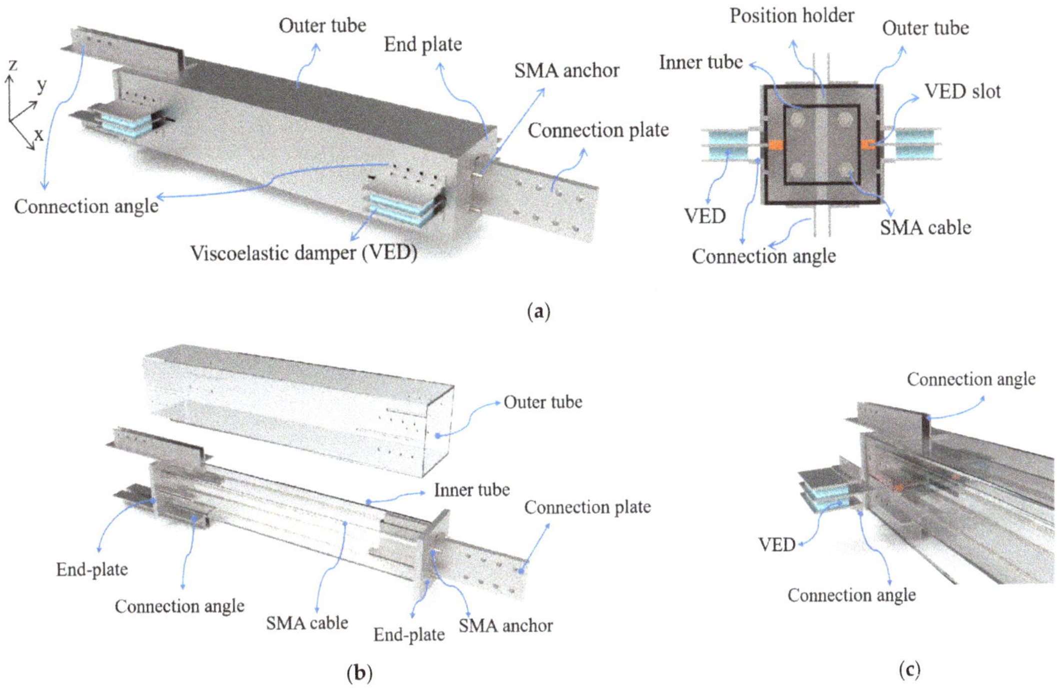

Figure 1. Three-dimensional schematic illustration of (**a**) SMA-viscoelastic hybrid braces (SCVEB), as well as its (**b**) self-centering (SC) system and (**c**) energy-dissipation (ED) system.

2.1. SC System

The studied SCVEB employs SMA cables [65,66] as the core part of the SC system, as shown in Figure 1b. It is worth mentioning that various types of SMA elements (such as the disc spring [67], ring spring [57,68–71], washer spring [72], etc.) have been developed and are welcomed to be incorporated in this SCVEB system with some necessary modifications on the configuration. Other necessary components include the outer tube, inner tube, end-plates, angles, connection plates, position holders, and tightening nuts. The end-plates are not connected to either the outer or inner tubes; therefore, they can slide along the brace length freely when SCVEB works. On one side of the brace, a connection plate passes through the slot cut in the end-plate and is welded to the inner tube, serving as the connecting component to the structural systems. On the other side, two pairs of double angles are welded to the outer tube to accommodate the connection.

During assembly, the SMA cables are firstly housed in the inner tube. Then the inner tube is inserted inside the outer tube concentrically and positioned with two position holders placed between them. SMA cables are connected to the end-plates via a threaded junction. The end-plates are subsequently placed on both side of the outer tube. Finally, the SMA cables are pre-tensioned to obtain sufficient initial stiffness and desired "yield" resistance. It is believed that an appropriate pre-tension level can encourage the SMA cables to reach their full potential of self-centering property [60,73]. The possible method to apply the pre-tension force is explained later.

The working principle of the SC system is further demonstrated in Figure 2. For ease of description, we assume that the outer tube is completely fixed and the inner tube is driven by the connection plate. The terms "tension" and "compression" refer to leftward or rightward movement direction of the inner tube relative to the outer tube, respectively, as the orientation marked in Figure 2. When the brace is in tension, the left end-plate is pushed away from the outer tube by the inner tube, while the right end-plate is blocked by the outer tube. Similarly, when the brace is in compression, the inner tube, together with the connection plate, moves rightward and pushes the right end-plate away from the outer tube, while the left end-plate is blocked by the outer tube. Thus, both compression and tension brace deformations cause the two end-plates to move apart, elongating the SMA cables and increasing their tension force concurrently. Therefore, the SMA cables always remain in tension no matter what condition state the brace is in, giving full play to the performance of the SMA cables. Previous studies have verified that the SMA-based elements/components are expected to exhibit a flag-shaped response [67,74–76]. An idealized/simplified force-displacement relationship is adapted here to capture this unique hysteresis evolutionary path (see Figure 3a), where four stiffness parameters, namely k_1 for loading stage, k_2 for loading plateau stage, k_3 for unloading stage, and k_4 for unloading plateau stage, are considered [77]. This four-segment simplified model was further cooperated in OpenSEES [78] software for the system-level analysis.

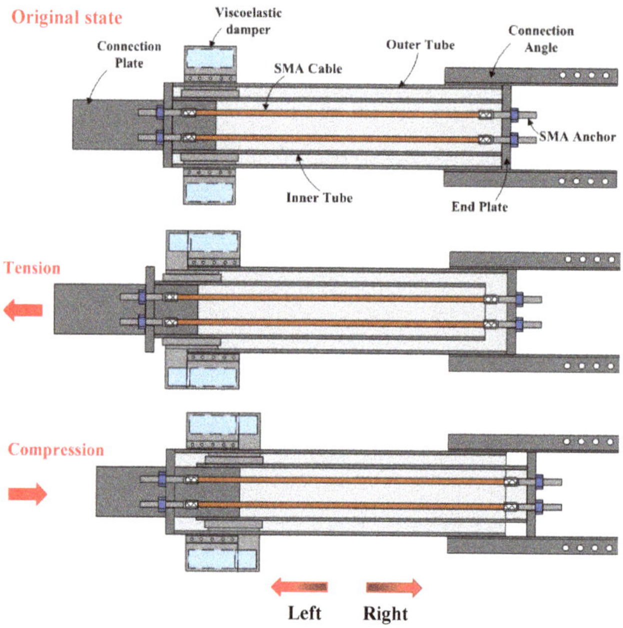

Figure 2. Working principle of SCVEB.

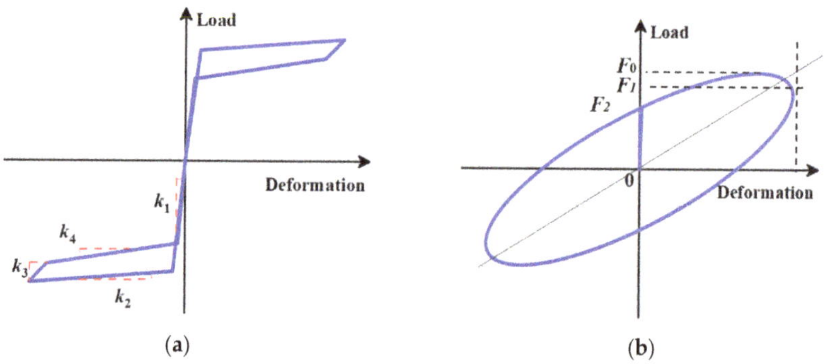

Figure 3. Illustration of hysteretic curves: (**a**) hysteresis model of SMA-based elements/components and (**b**) Kelvin–Voight model for viscoelastic materials.

2.2. ED System

Extra energy dissipation of the SCVEB is provided by the viscoelastic dampers (VEDs). The participation of the viscoelastic material is also expected to provide certain stiffness for the brace especially after the SMA-based elements advance into the post-yield plateau stage. As shown in Figure 1c, the ED system (i.e., the VED) consists of three plates where two rubber layers are sandwiched between them. The two external plates are bolted to the outer tube by steel angles, and the middle plate is inserted into the slot on the inner tube. Slot cuts are made on the outer tube to allow the middle plate to pass through and move freely when the inner and outer tube move apart.

When the ED system functions, viscoelastic materials would act via shear deformation between two steel plates. It is worth noting that the ingredient (e.g., resin content) of viscoelastic materials may strongly affect their linear/nonlinear characteristic; thus, their hysteresis models vary. The so-called linear characteristic refers to the ellipse-shaped hysteresis curve, which is featured by most typical viscoelastic materials (see Figure 3b). Kelvin–Voight model associated with linear viscoelastic materials was adopted herein to derive the theoretical formulas of the proposed SCVEB, since this model is most widely accepted for the viscoelastic materials.

The restoring force, F_{ve}, can be expressed as follows:

$$F_{ve} = K_{eff} \cdot u + C_e \cdot v^{\alpha} \tag{1}$$

where u is the displacement, v is the velocity, α is the velocity exponent, and K_{eff} is the equivalent storage stiffness, which is determined by the following:

$$K_{eff} = \frac{nG'A}{h} \tag{2}$$

where n is the number of viscoelastic material layers, A is viscoelastic material's shear section area, and h is the thickness of a single viscoelastic layer; the storage modulus, G', is determined by the following:

$$G' = \frac{F_1 h}{nAu_0} \tag{3}$$

where u_0 is the maximum displacement during loading (as illustrated in Figure 3b), and F_1 is the damping force corresponding to u_0.

C_e is the equivalent damping coefficient, which is calculated by the loss modulus, G'', and the circular frequency, ω:

$$C_e = \frac{nG''A}{\omega h} \tag{4}$$

The loss modulus, G'', can be determined by the following:

$$G'' = \frac{F_2}{F_1} G' \tag{5}$$

where F_2 is the damping force corresponding to the displacement at zero.

2.3. Theoretical Activation Force of SCVEB

An activation (or sometimes called "decompression") force, F_a, which should be reached firstly before the SMA cables enter the working situation, can be expressed by the following:

$$F_a = \frac{F_a}{k_{br}} \cdot k_2 + F_p \tag{6}$$

where F_p is the total cable preload; k_2 is the "post-yield" stiffness, i.e., that of the forward transformation plateau, as marked in Figure 3a; and k_{br} is the initial stiffness of the brace prior to activation:

$$k_{br} = k_{out} + k_{in} + k_2 \tag{7}$$

where k_{in} and k_{out} are the elastic axial stiffness of the inner and outer tubes, respectively. Since k_{br} >> k_2 for most cases, Equation (6) can be subsequently reduced to the following:

$$F_a \approx F_p \tag{8}$$

Note that there is no relative deformation before activation; therefore, the load resistance provided by VED, i.e., F_{ve}, can be considered as zero. However, when the brace is subjected to dynamic excitations, the velocity-related term in Equation (1) should not be ignored. In this case, the total activation force of SCVEB is as follows:

$$F_a = F_p + C_e \cdot v^{\alpha} \tag{9}$$

Figure 4 shows the theoretical load–deformation relationship of the proposed SCVEB. Ideally, after activation, SCVEB's load at any deformation can be described as the sum of the forces provided by the above two systems.

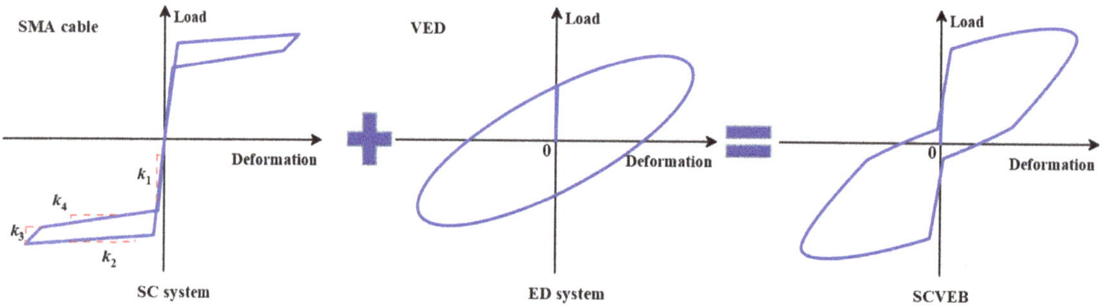

Figure 4. Typical hysteretic behavior of SCVEB.

3. Investigation on Kernel Elements

3.1. Individual SMA Cable Test

Figure 5a shows the geometric configuration of a typical SMA (50.8 at.% nickel–49.2 at.% titanium alloys) cable that was later adopted in the SCVEB. The SMA cable consists of seven helically wrapped strands, each of which contains 19 helically wrapped monofilament SMA wires with a diameter of 1.0 mm (designated as 7 × 19 × 1.0). The effective length of the studied SMA cable is about 300 mm, and special end grips were machined for the SMA cables. The cable segment was first cut from a long cable by local melting. The hot ends of the SMA cable were inserted into the preprocessed hole of the end grips, followed by a machinal squeezing process. As a result, the cable ends were housed in the squeezed end grips tightly. The end grips were further machined to be threaded for connection and pre-tensioning, as shown in Figure 5b.

Pseudo-static tension testing of the SMA cable was conducted on a Universal Test Machine (UTM) to obtain their basic mechanical behavior. The displacement and the applied load were monitored by the grip displacement and the built-in load cell in the UTM. A displacement-controlled incremental loading protocol was applied based on a displacement interval Δ_1 of 2.5 mm, which is equal to an axial deformation of 1% of the test cable sample. The loading began with three cycles each at Δ_1, $2\Delta_1$, $3\Delta_1$, $4\Delta_1$, $5\Delta_1$, and $6\Delta_1$, as shown in Figure 6a.

The force–displacement curve is shown in Figure 6b. It can be seen that the studied SMA cable exhibited a satisfactory self-centering capability, and the flag-shaped hysteresis became stabilized after a few loading cycles. The residual strain was generally small. The aforementioned four segment stiffnesses, k_1, k_2, k_3, and k_4, were measured as 22,500, 1500, 18,000, and 3000 MPa, respectively. A dimensionless index, namely equivalent viscous damping (*EVD*), is employed to evaluate the energy-dissipation capacity and is expressed by the following:

$$EVD = \frac{1}{4\pi}\frac{E_D}{E_S} \quad (10)$$

where E_D is the typical energy dissipation per cycle (i.e., the area within the inelastic force-displacement response curve), and E_S is the recoverable elastic strain energy stored in an equivalent linear elastic system. The calculation results confirm that the studied SMA cables have moderate energy-dissipation capability (EVD = 2.6% at the maximum tested displacement).

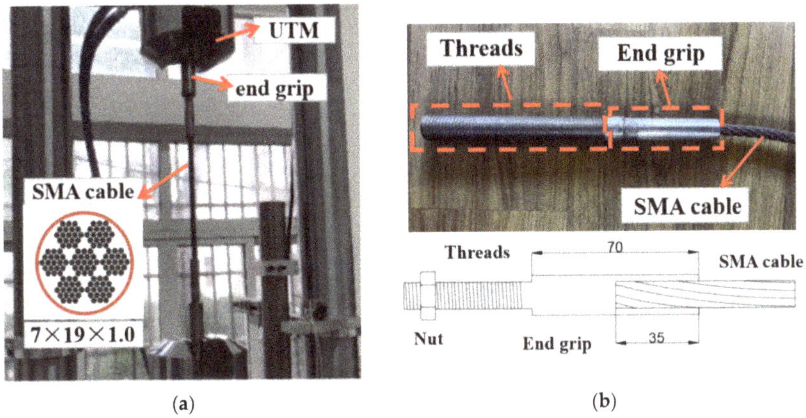

Figure 5. Information of SMA cable: (**a**) test setup and basic construction; (**b**) end grips.

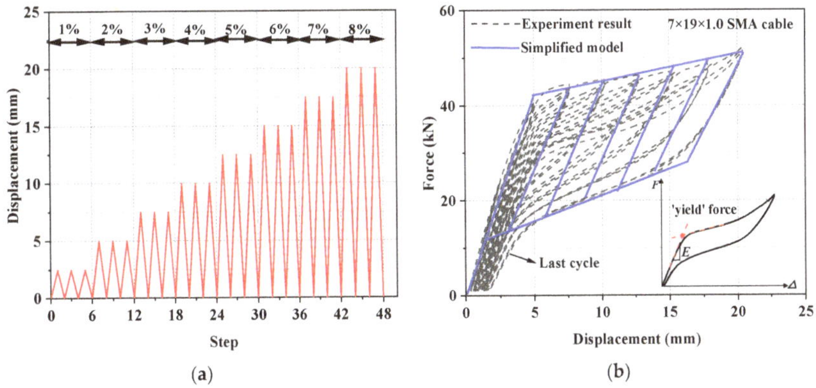

Figure 6. Test of SMA cable specimen: (**a**) loading protocol and (**b**) test results.

3.2. Viscoelastic Material Test

A preliminary test was conducted by using a UTM to investigate the damping behavior of viscoelastic material. The configuration and geometry dimensions of the test setup are shown in Figure 7. The shearing area and thickness of the studied rubber layer are 100×180 mm^2 and 30 mm, respectively. To obtain a comprehensive understanding of the rubber layer under various loading scenarios (e.g., different loading amplitudes and different loading frequencies), a progressive hysteresis loading protocol with six loading steps was conducted, as drawn in Figure 8a.

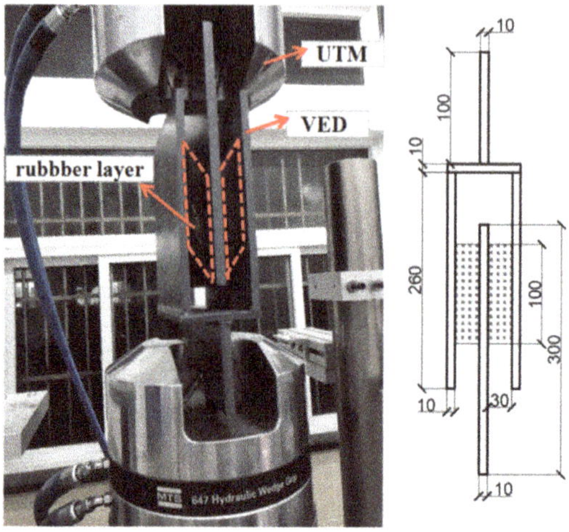

Figure 7. Test setup of viscoelastic material.

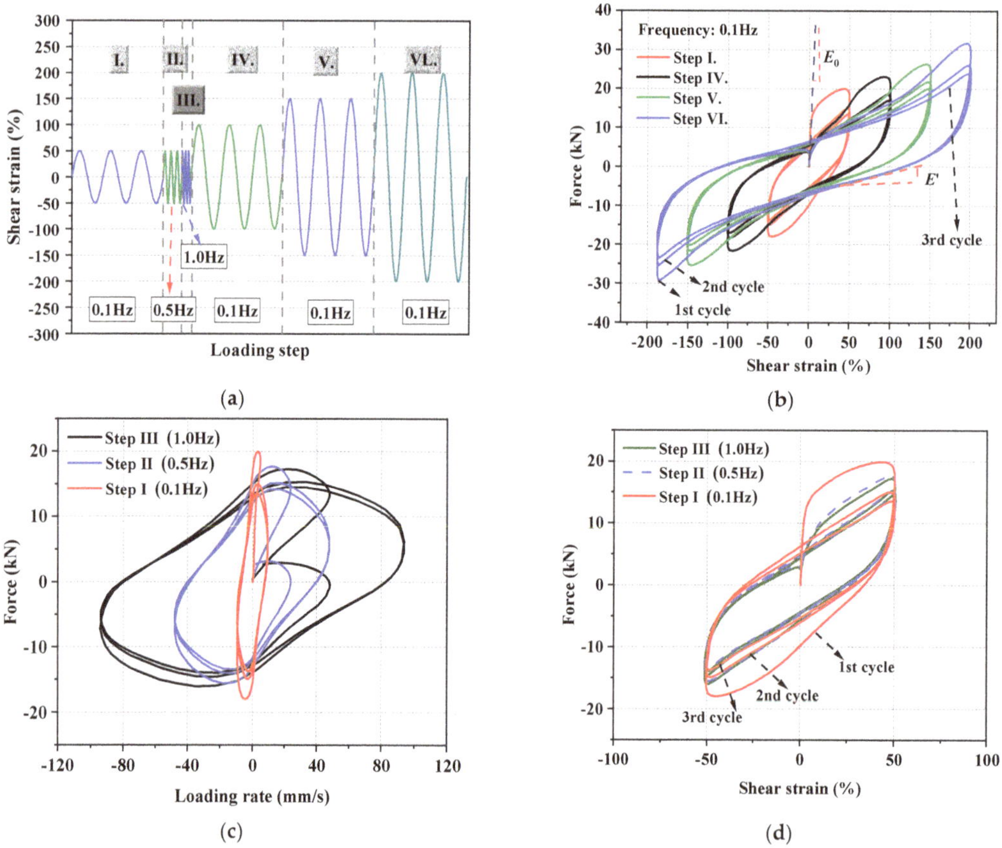

Figure 8. Tests of rubber layer: (**a**) loading protocol and (**b**–**d**) test results.

The test results are plotted in Figure 8b–d. It can be seen that the first cycle of all of these hysteretic loops is plumper than that of the rest cycles, with significant initial stiffness (E_0), regardless of what the loading amplitudes is. The peak strength decreases when the reverse loading is repeated under the same amplitude, whereas it increases as the loading amplitude expands, indicating that the rubber features strong nonlinear characteristics, both cyclic softening and cyclic hardening. Moreover, the rubber's loading/unloading plateau stiffness (E') decreases as incremental loading process progresses. The hysteresis force–shear strain relationship exhibits obvious loading amplitude dependence, where Mullins effect is usually accepted to explain this phenomenon [79]. The hysteretic loop of the studied rubber shows a typical non-linear characteristic [80], a phenomenon which is mainly attributed to the temperature rise, as well as the fatigue performance of this material during cyclic loading [81]. Figure 8c,d shows the relationship between the damping force and loading frequency. It can be seen that the generated peak loads at the maximum loading displacement are not sensitive to the loading frequency, and the load–deformation relationship would remain stable since the second cycle. Therefore, it is reasonable to assume that the velocity-dependent property of the rubber purchased for this study was not significant. Nonetheless, they are still commonly called "viscoelastic material" in the community of civil engineers, as they exhibited many typical behaviors of viscoelastic dampers, such as similar hysteresis, high damping capability, temperature dependence, Mullins effect, etc. A similar phenomenon was also observed in previous works [62,82]. The difference between linear and nonlinear viscoelastic materials probably lies in the energy-dissipation amount and the load under peak deformation (see Figures 3b and 8b for a more clarified understanding). Note that other types of energy-dissipation devices (such as wire rope isolator [83]) also feature similar hysteretic behaviors, so it is feasible to employ those devices in the proposed hybrid braces in future work.

4. Experimental Verification of Proposed SCVEB

4.1. Information of SCVEB Specimen

A proof-of-concept SCVEB specimen was fabricated and tested. The components were fabricated in the shop and assembled in the laboratory. The corresponding configuration and assembling method for the SCVEB specimen have been described in Section 2. The key dimensions of the prototype SCVEB specimen are given in Figure 9.

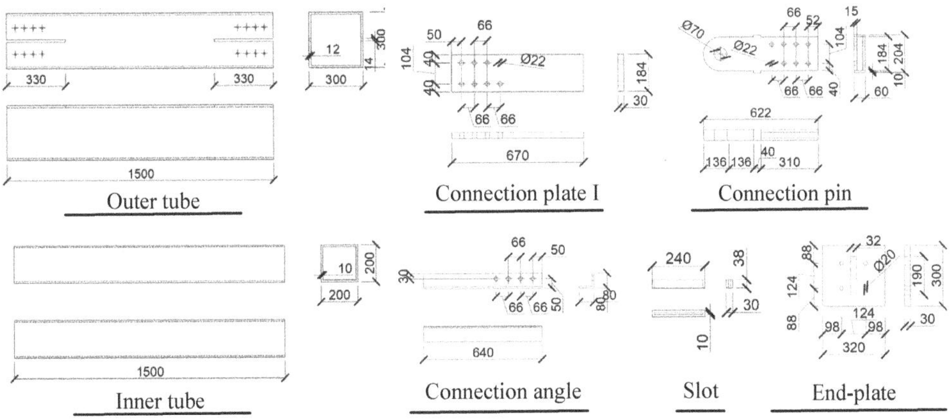

Figure 9. Dimensions of the SCVEB (units in mm).

The main components (i.e., outer tube, inner tube, connecting angles, end-plates, and other necessary accessory components) were manufactured by Grade Q345 steel (nominal yield strength: 345 MPa). The rubbers were custom manufactured to the designed geometric shape and employed in the ED system of the SCVEB. Four SMA cables, whose properties were investigated in Section 3.1, were employed in the SCVEB. The length of the SMA cable was finally designed as 1770 mm. It is worth noting that, in practice, the SMA cables do not need to run the full length between the end-plates, as they have a high capacity of recoverable strain. Alternatively, high-strength steel cables may fill in for the rest of the length. Shortening the length of the SMA cable could promote an increased efficiency of the SMA cable and save the total production cost of the SCVEB at the same time.

The end grips of each cable were threaded, through which design the cables were finally fastened to the end-plates, as shown in Figure 5b. Prior to starting the follow-up experiments, the SMA cables were pre-tensioned through specially designed pre-tensioning equipment. As shown in Figure 10, a reaction base was required due to the limited stroke capacity of the hydraulic jack. When tensioning, the hydraulic jack was operated via a threaded extension rod that was connected to the threaded end grip by a coupler. Once the targeted pre-tensioned level (30 kN) was achieved, the nuts were locked and the pre-tensioning process finished.

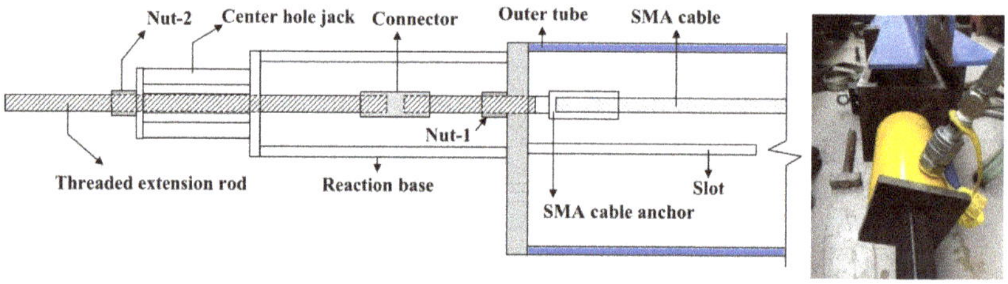

Figure 10. Pre-tension equipment.

4.2. Test Setup, Instrumentation, and Loading Protocol

The test setup for the SCVEB specimen is schematically shown in Figure 11. The whole loading plane was oriented horizontally and the cyclic load was applied axially to the SCVEB specimen. An electro-hydraulic servo actuator with a maximum loading rates of 2000 mm/s was employed to conduct this experiment. Two ends of the SCVEB were pinned to the servo actuator and strong base on the reaction frame, respectively. Through this method, the brace can be viewed as subjected to uniaxial force during cyclic loading. The deformation was measured by two vertical linear variable differential transformers (LVDTs), which were attached to the connection plate of the inner tube and the connection angle of the outer tube, respectively. A series of longitudinal strain gauges were placed on the SCVEB surface to real-time-monitor whether the steel elements stayed elastic during the whole process. The uniaxial force applied to the brace was measured by the built-in load senor of the servo actuator.

The test was conducted in two rounds. Firstly, the complete SCVEB, i.e., with SMA cables + VEDs, was tested. Then the VEDs were disassembled from the SCVEB (denoted as SCB in the following discussions) to examine the hysteretic behavior of the isolated SC system. The loading protocols for the successive component-level tests can be seen in Figure 11 [84]. For the SCVEB test, an incremental amplitude loading protocol with the sequence of $1\Delta_v$, $2\Delta_v$, $3\Delta_v$, and $4\Delta_v$, where Δ_v = 15 mm, was employed. The interval Δ_v corresponds to the deformation amount when the rubber layer experiences a shear strain of 50%, a value commonly adopted for the investigation of viscoelastic material-based damper [80,81]. Such a loading protocol was practiced twice with two loading frequency levels, i.e., 0.1 and 1.0 Hz, successively. For the SCB test, an incremental amplitude-loading protocol with loading frequency of 0.1 Hz was adopted, following the sequence of $1\Delta_b$, $2\Delta_b$, $3\Delta_b$, $4\Delta_b$, and $5\Delta_b$, where Δ_b = 12.7 mm. The interval Δ_b corresponds to the 1% strain level of the SMA cable. All of these amplitudes were input in a sinusoidal wave format, and each amplitude was repeated for three times before moving to the next amplitude.

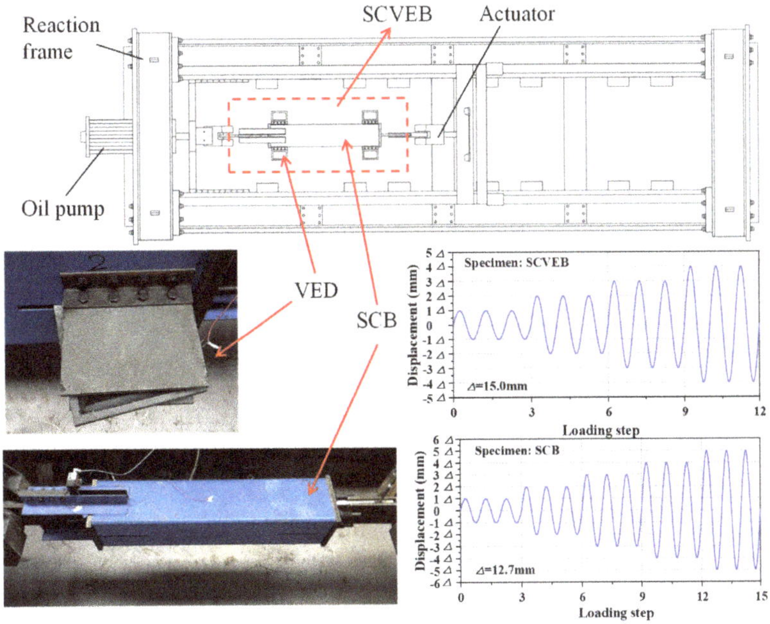

Figure 11. Test arrangement and observation.

4.3. Test Results and Discussions

The load–deformation hysteretic curves of the SCB and SCVEB specimens are plotted in Figure 12. Some important performance indicators are marked in the figure.

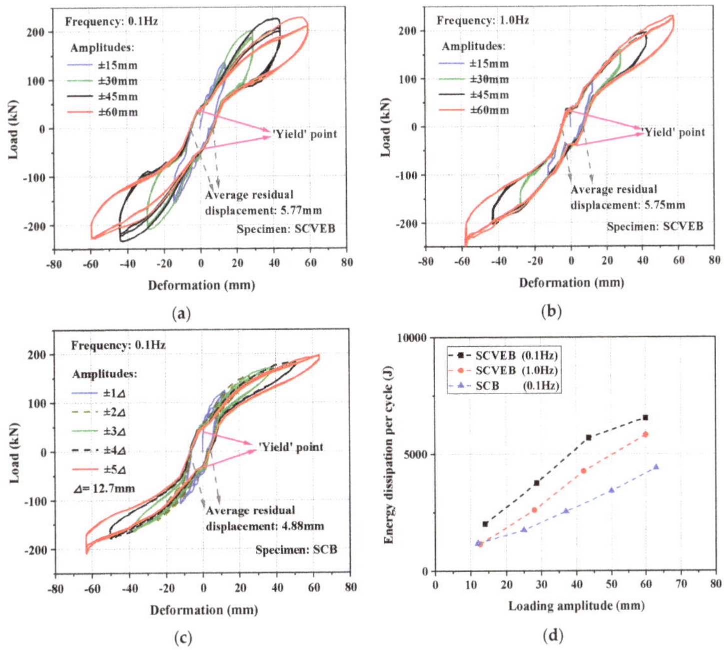

Figure 12. *Cont.*

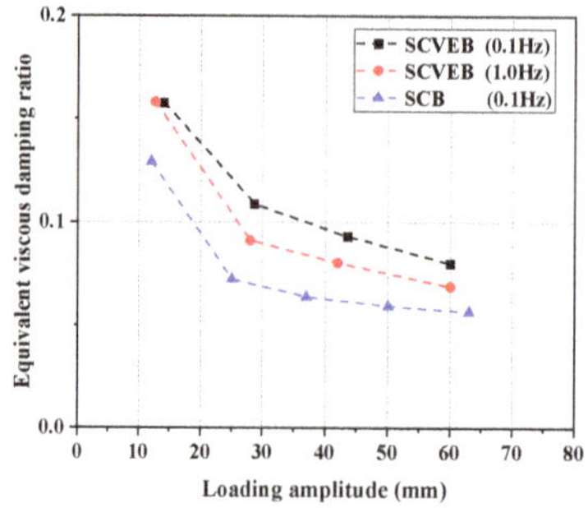

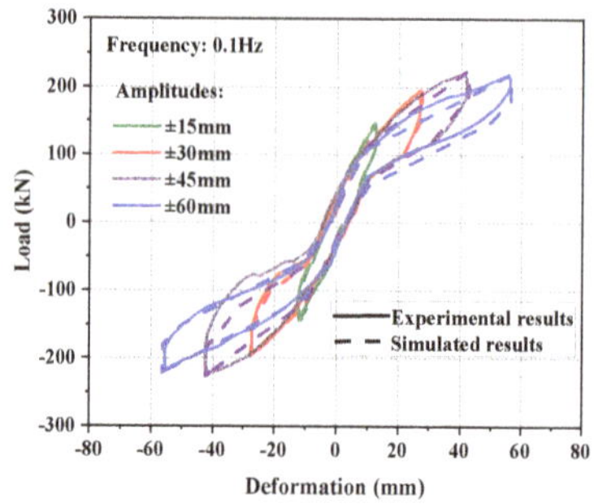

Figure 12. Test results for (**a**) SCVEB-0.1 Hz series, (**b**) SCVEB-1.0 Hz series, (**c**) SCB-0.1 Hz series, (**d**) energy-dissipation per cycle, (**e**) *EVD*, and (**f**) comparison between the experimental hysteresis loops and the simulated results.

It can be seen that a non-linear hardening slope is followed after the elastic range. There is an obvious turning point ("yield" point) in the curve, as depicted in Figure 12a–c. The reason of this phenomenon is mainly attributed to the pre-tensioning status of the SMA cables before decompression. The "yield" loads for SCVEB and SCB are both around 30 kN, a value which is consistent with the amount of the pre-tensioned load. This indicates that the "yield" phenomenon is resulted from the decompression of the pre-tensioned SMA cables. Theoretically, the residual displacement of the SCB would be zero when the load is removed (see Figure 4). Viewing the result of 0.1 Hz series SCB test; however, the residual deformation did not return to zero as expected, but instead remained at a constant value during different loading cycles. This undesirable residual deformation may be due to defects in the brace itself (such as imperfect mechanical factors introduced during machining, manufacturing, and assembly), rather than the performance degradation of the materials (i.e., SMA cables), since SMA cables' self-centering capability has been confirmed by the tests aforementioned in Section 3.1. In fact, the measured length of the outer tube is about 5 mm longer than that of the inner tube, which deviates from the original design that the lengths of the outer and inner tube should be the same. It should be noticed that the residual deformations of SCB and SCVEB were 4.88 mm and 5.77 mm (under 0.1 Hz), respectively; the results were very close to the aforementioned manufacturing error length between outer and inner tube (5 mm). Therefore, it can be concluded that the imperfect self-centering results were mainly resulted from the manufacturing errors. Although the manufacturing errors resulted in a maximum residual displacement of 4.88 mm in the specimen that only contains SMA cables (i.e., specimen SCB, as shown in Figure 12c), it did not much affect the completed brace specimen (i.e., SCVEB) to achieve the predicted hysteretic response, as shown in Figure 4. The "true" residual displacement of SCVEB was 5.77 − 4.88 = 0.89 mm, which was an accepted level.

Figure 12d,e gives the calculated E_D and *EVD* value of the specimens. It can be seen that the energy-dissipation capacity increases with increasing loading amplitudes; however, the *EVD* shows the opposite trend. This is because the *EVD* is only related to the shape of the hysteretic curve, where a greater amount of increase is accumulated for E_S rather than E_D, and, hence, it decreases the *EVD*. The *EVD* indicator also reflects that the larger the deformation, the less plump the hysteresis loop is. Thanks to the extra energy dissipation source provided by the four VEDs, both energy-dissipation indicators (E_D and *EVD*) show that the SCVEB has better energy-dissipation capacity than the SCB, and such superiority is more evident under large loading amplitudes. It seems that the SCVEB exhibited a "better" energy-dissipation performance under low loading frequency. However, it should be noticed that the 1.0 Hz-series comes after the 0.1 Hz series in this experiment. Fatigue damage may be accumulated in the rubber layer as the experiment progresses, and, thus, the energy-dissipation property of ED system may decrease to some extent. In fact, the loading frequency may have negligible influence on the energy-dissipation performance of the SCVEB in this test, since the rubber layer has been proven to be rate-unsensitive in Section 3.2.

A component-level simulation work was conducted to verify the effectiveness of the selected hysteretic models that is further to be adopted in the following-up system-level analysis. A modified SelfCentering model (i.e., the four-segment simplified model mentioned in Section 2.1) was used to capture the hysteresis behavior of the SMA cables and the values calibrated in Section 3.1 was adopted. The hysteresis behavior of VEDs was simulated by BoucWen Material model in OpenSEES. Moreover, the ElasticMultiLinear Material model was used to consider the decompression process of the pre-tensioned SMA cables and the MultiLinear Material model was adopted to consider the extra friction between elements in SCVEB. The simulation results are plotted in Figure 12f. It can be seen that the simulation curves fit well with the tested results, proving the validity of this modeling method.

5. System-Level Analysis

5.1. Prototype Buildings

For an in-depth understanding of the fundamental performance of structural systems employing the proposed SCVEB, a system-level analysis was conducted. Three nine-story steel frames employing different types of braces (i.e., SCVEB, SCB, and conventional buckling restrained brace) were designed and analyzed for comparison. All of these frames employ the concentrically inverted-V-type braces with the same arrangements. These frames were designed according to ASCE 7-16 [85] by the modal-response spectrum analysis method. Figure 13 shows the basic information of the frames.

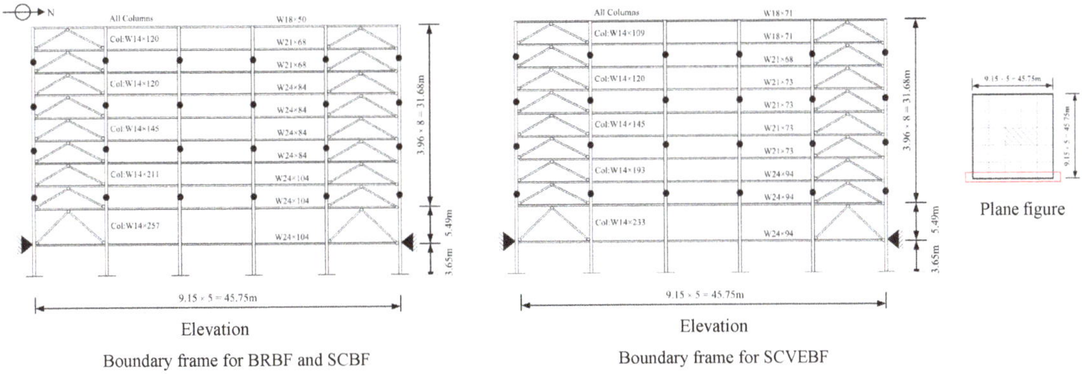

Figure 13. Nine-story prototype building.

These frames were assumed to be located at a stiff soil site (Site Class D) in Los Angeles. Importance factor $I_e = 1.0$, response modification coefficient $R = 8.0$, deflection amplification factor $C_d = 5.0$, design response spectral values $S_{DS} = 2/3S_{MS} = 1.376$ g, and $S_{D1} = 2/3S_{M1} = 0.707$ g, are considered.

5.2. Design and Modeling

For the nonlinear dynamic analysis in OpenSEES, centerline models which represent half of the buildings in the North–South (NS) direction were established (see Figure 13). The basic information of these frames is provided in Table 1. The beam-to-column connections are assumed to be rigid for all of these three buildings to maintain certain redundancy against earthquake [77]. When modeling, the boundary frame members are simulated by the Steel01 material with idealized kinematic hardening. The detailed modeling information of these frames are described in the following sections.

Table 1. Basic information of frames.

	BRBF			SCBF			SCVEBF			
Story	k_{ini} (kN/mm)	F_y (kN)	A (mm^2)	No.	l_{SMA} (mm)	F_p (kN)	No.	l_{SMA} (mm)	F_p (kN)	$L \times W \times H$ (mm × mm)
1	400.4	2448.2	10,418	32	1760	2407	20	1760	1504	500 × 280 × 30
2	329.6	1706.3	7261	22	1500	1655	15	1500	1128	500 × 210 × 30
3	256.2	1326.1	5643	18	1500	1354	13	1500	978	500 × 180 × 30
4	228.9	1185.1	5043	16	1500	1203	12	1500	902	500 × 160 × 30
5	223.6	1157.4	4925	16	1500	1203	10	1500	752	500 × 140 × 30
6	185.1	958.1	4077	12	1500	902	9	1500	677	500 × 125 × 30
7	155.4	804.6	3424	10	1500	752	8	1500	602	250 × 210 × 30
8	102.0	528.3	2248	7	1500	526	4	1500	301	250 × 110 × 30
9	61.9	320.3	1363	4	1500	301	2	1500	150	250 × 70 × 30

Note: "k_{ini}" refers to initial stiffness of BRB, "F_y" refers to yield force of BRB, "A" refers to the area of BRB's steel core, "No." refers to the number of SMA cables, "l_{SMA}" refers to the length of SMA cables, "F_p" refers to the total preload of SMA cables, and "$L \times W \times H$" refers to the geometric dimension of the rubber layer.

5.2.1. Conventional Buckling Restrained Brace Frame (BRBF)

The BRBs were modeled with "truss" elements, and the Steel02 Giuffre–Menegotto–Pinto material model was used. The seismic weight was appropriately distributed to the main frame and the adjacent lean columns. The Rayleigh damping ratio was adopted as 5% for the first and third modes of vibration.

5.2.2. Typical Self-Centering SMA Cable-Based Frame (SCBF)

The information of the employed SMA cables is provided in Table 1. The effective length of the SMA cables was determined by ensuring that the strain not exceeding 10% at an inter-story drift is around 5%. The preload was determined by the "yield" force of the SMA cables (see Figure 6). The SCBs were modeled with "truss" elements, and a modified SelfCentering material model was used to capture the unique flag-shaped hysteresis behavior. An energy dissipation factor (β) of 0.6 was assumed for SMA cables [53].

It is worth noting that the SCBF and BRBF are designed to have the same initial stiffness and boundary frame. This is because experiment results have revealed that the actually measured initial stiffness of SCB specimens is comparable to those of the BRBs with a similar level of load-carrying capacity [18]. Through this design, the main difference between these two frames lies in the energy-dissipation capability, and the energy dissipation provided by the two types of braces can be compared more intuitively.

5.2.3. Frame with SCVEB (SCVEBF)

When modeling the SCVEBF, a modified boundary frame with an approximately 25% reduction in the overall strength of the structure is considered. This is because ASCE 7–16 [85] allows for a 25% reduction in the base shear for damped structures, with the perspective of saving the overall cost of construction. The basic information of the prototype SCVEBF is shown in Figure 13 and summarized in Table 1. The first-modal-added damping ratio, ξ_{add}, [85] of structure SCVEBF provided by the viscoelastic material is assumed to be around 0.1. A total of eight viscoelastic material layers are considered for each brace, and the geometric information of each layer is summarized in Table 1. The VEDs were simulated by using the Kelvin–Voigt model in OpenSEES, which involves paralleled viscous and elastic spring elements [63], and the associated parameters for simulating viscoelastic material properties are chosen from the work performed by Zimmer [86] rather than the test results in this paper. It is noteworthy that the models proposed in References [87,88] can be employed to simulate the nonlinear viscoelastic hysteretic behavior. However, since the linear viscoelastic materials are much more mature than the nonlinear ones and have been adopted by more independent researchers, for ease of comparison with previous existing results, the follow-up system-level analysis is conducted based on the linear viscoelastic hysteretic behavior.

5.3. Structural Performance

A nonlinear time-history analysis was carried out to examine their seismic performance. A total of 20 far-field (FF) and 20 pulse-like near-fault (NF) ground motions at the MCE level were considered herein. The FF records were selected from the FEMA P695 database [89] and were scaled to fit the target spectrum (see Figure 14a). As for NF records, scaling should be cautiously performed for NF ground motions, since some key pulsing characteristics may be violated by uniform scaling [90]. In this paper, the 20 NF ground motions were carefully selected according to the criteria proposed by Baker [91] to match the design spectrum (see Figure 14b). It is confirmed that the mean spectrum of the NF records is not less than the design response spectrum for periods ranging from $0.2T_1$ to $2.0T_1$, which satisfies the ASCE 7–16 requirements.

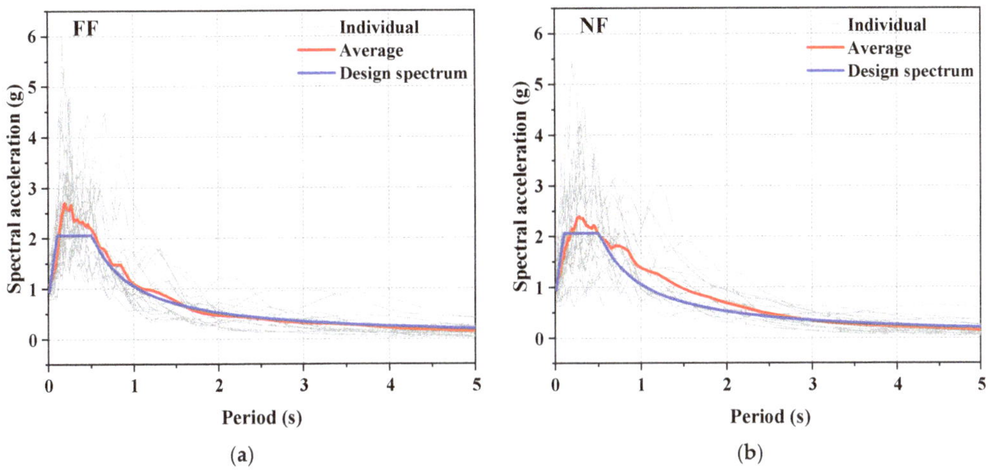

Figure 14. Response spectra of selected ground motions: (**a**) far-field (FF) ground motion records and (**b**) pulse-like near-fault (NF) ground motion records.

5.3.1. Peak Inter-Story Drift (PID)

The PIDs of the frames are shown in Figure 15a. The results show that the PIDs of the SCVEBF are all smaller than that of SCBF, with the maximum PID of the SCVEBF being reduced by 35% and 33% compared to SCBF under FF and NF ground motions, respectively. It demonstrates that the use of the viscoelastic material is highly effective in controlling the PID. Extra energy-dissipation capacity provided by the viscoelastic material is responsible for this improvement. Moreover, the maximum PID of the SCVEBF is even less than BRBF; this, again, confirms the feasibility of the proposed SCVEBF.

5.3.2. Residual Inter-Story Drift (RID)

The RIDs of the frames are shown in Figure 15b. Both the SCBF and SCVEBF exhibit reduced residual deformation compared to the BRBF, thus confirming the effectiveness in RID control for the SCBF/SCVEBF. Due to the pulsing effect, the RIDs of the SCBF increased under the NF earthquakes compared with those under FF earthquakes, where the increased residual deformation of SCBF is mainly attributed to the inelastic deformation of the boundary frame. By contrast, the RIDs of the SCVEBF at each story remain basically the same, and there is almost no difference observed between the results under FF and NF earthquakes. It is believed that the added damping ratio contributed by the viscoelastic material suppresses the RID especially for the upper floors. These results, again, prove that the proposed SCVEBF is a promising solution for controlling residual drifts that is especially effective in near-fault region.

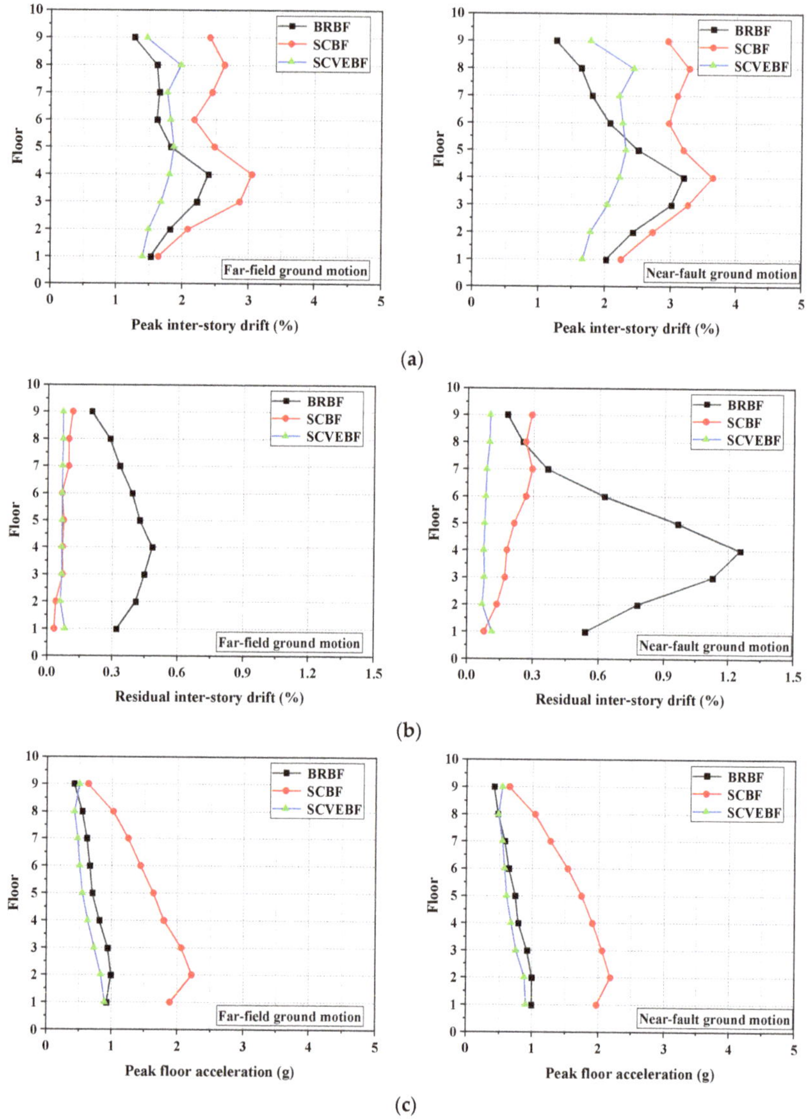

Figure 15. Mean responses of structures: (**a**) peak inter-story drift (PID), (**b**) residual inter-story drift (RID), and (**c**) peak absolute floor acceleration (PFA).

5.3.3. Absolute Peak Floor Acceleration (PFA)

The height-wise peak floor acceleration (PFA) responses of the structures are shown in Figure 15c. It is found that the SCBF exhibits the maximum PFA. This is because the unique flag-shaped hysteresis loop of SCBF that features an abrupt "transition points" under reverse-loading would induce a large difference in the shear force between the adjacent stories. Relevant research works have reported that inconsistent inter-story shear forces of the two adjacent floors may amplify PFA [92]. However, it is of interest to find that the

PFAs of the SCVEBF are significantly smaller than that of the SCBF, and even lower than the BRBF. This is mainly attributed to the participation of viscoelastic material, which can effectively neutralize the sharp "transition points" during unloading [61].

6. Conclusions

A novel type of self-centering brace, namely the self-centering SMA-viscoelastic hybrid brace (SCVEB), was proposed in this study. The energy dissipation was provided by the SMA cables, as well as the viscoelastic dampers (VEDs), whilst the self-centering capacity was provided by the former. The fundamental mechanical behavior of individual SMA cables and viscoelastic dampers was first investigated, followed by a more comprehensive experimental study on a proof-of-concept SCVEB specimen. The main findings and conclusions are summarized as follows:

(1) The SMA cable exhibits typical flag-shaped hysteretic loops with a large recovery strain. Reasonable cyclic pre-training is suggested before anchoring to SCVEB, since this process was shown to help stabilize the hysteretic response.

(2) The VED is capable of providing reliable energy dissipation. The rubber in the test did not show rate-dependence property, such as typical viscoelastic material, and this may be due to the differences in their compositions.

(3) The SCVEB specimen exhibited satisfactory self-centering and energy-dissipation capability, although a certain degree of residual strain was observed due to manufacturing error. The participation of viscoelastic material, indeed, enhanced the energy-dissipation capability.

(4) The system-level analysis shows that the frames employing the proposed SCVEB have satisfied peak inter-story drifts under the MCE and almost negligible residual inter-story drifts. More importantly, the SCVEB can further reduce the peak floor acceleration of the frames. These encouraging findings demonstrate that the proposed SCVEB could be a cost-effective self-centering solution by reducing member size of the boundary frame with less SMA consumption.

Author Contributions: Conceptualization, Z.-X.Z., Y.P. and X.H.; methodology, Y.P.; software, Y.P.; validation, Z.-X.Z., Y.P. and X.H.; formal analysis, Y.P., X.H. and Z.-X.Z.; investigation, Y.P.; resources, Y.P. and Z.-X.Z.; data curation, Y.P.; writing—original draft preparation, Z.-X.Z., X.H. and Y.P.; writing—review and editing, Z.-X.Z. and Y.P.; visualization, Z.-X.Z.; supervision, Y.P., X.H. and Z.-X.Z.; project administration, Y.P.; funding acquisition, Z.-X.Z. and Y.P. All authors have read and agreed to the published version of the manuscript.

Funding: The financial supports from the National Natural Science Foundation of China (NSFC) with Grant Nos. 52078359 and 51778456 are gratefully acknowledged. Support for this study was also provided by the Shanghai Rising-Star Program (20QA1409400).

Institutional Review Board Statement: Not applicable.

Informed Consent Statement: Not applicable.

Data Availability Statement: Data presented in this study are available in this article.

Acknowledgments: Not applicable.

Conflicts of Interest: The authors declare no conflict of interest.

References

1. Matsuyama, M.; Kimura, R.; Hayashi, H. Organizational Structure and Institutions for Disaster Prevention: Research on the 1995 Great Hanshin-Awaji Earthquake in Kobe City. *J. Disaster Res.* **2015**, *10*, 1051–1066. [CrossRef]
2. Nguyen, V.Q.; Nizamani, Z.A.; Park, D.; Kwon, O.S. Numerical simulation of damage evolution of Daikai station during the 1995 Kobe earthquake. *Eng. Struct.* **2020**, *206*, 110180. [CrossRef]
3. Luo, Q.B.; Dai, F.; Liu, Y.; Gao, M.T. Numerical modelling of the near-field velocity pulse-like ground motions of the Northridge earthquake. *Acta Geophys.* **2020**, *68*, 993–1006. [CrossRef]
4. Bruneau, M.; MacRae, G. Building Structural Systems in Christchurch's Post-Earthquake Reconstruction. *Earthq. Spectra* **2019**, *35*, 1953–1978. [CrossRef]

5. McCormick, J.; Aburano, H.; Ikenaga, M.; Nakashima, M. Permissible residual deformation levels for building structures considering both safety and human elements. In Proceedings of the 14th World Conference on Earthquake Engineering, Beijing, China, 12 October 2008; Seismological Press of China: Beijing, China, 2008.
6. Chou, C.C.; Chung, P.T.; Cheng, Y.T. Experimental evaluation of large-scale dual-core self-centering braces and sandwiched buckling-restrained braces. *Eng. Struct.* **2016**, *116*, 12–25. [CrossRef]
7. Fang, C.; Wang, W.; Qiu, C.; Hu, S.; Macrae, G.A.; Eatherton, M.R. Seismic resilient steel structures: A review of research, practice, challenges and opportunities. *J. Constr. Steel Res.* **2022**, *191*, 107172. [CrossRef]
8. Lin, Y.C.; Ricles, J.M.; Sause, R. Seismic performance of a steel Self-Centering Moment Resisting Frame: Hybrid simulations under DBE and MCE. In *Stessa 2012: Proceedings of the 7th International Conference on Behaviour of Steel Structures in Seismic Areas, Santiago, Chile, 9–11 January 2012*; CRC Press: Boca Raton, FL, USA, 2012; pp. 729–735.
9. Lin, Y.C.; Sause, R.; Ricles, J. Seismic Performance of a Large-Scale Steel Self-Centering Moment-Resisting Frame: MCE Hybrid Simulations and Quasi-Static Pushover Tests. *J. Struct. Eng.* **2013**, *139*, 1227–1236. [CrossRef]
10. Li, Z.; He, M.J.; Wang, K.L. Hysteretic Performance of Self-Centering Glulam Beam-to-Column Connections. *J. Struct. Eng.* **2018**, *144*, 04018031. [CrossRef]
11. Yun, C.; Chao, C.; Cong, C. Study on seismic performance of prefabricated self-centering beam to column rotation friction energy dissipation connection. *Eng. Struct.* **2021**, *241*, 112136. [CrossRef]
12. Huang, L.J.; Zhou, Z.; Liu, H.; Si, Y. Experimental Investigation of Hysteretic Performance of Self-Centering Glulam Beam-to-Column Joint with Friction Dampers. *J. Earthq. Tsunami* **2021**, *15*, 2150005. [CrossRef]
13. Cai, X.N.; Pan, Z.F.; Zhu, Y.Z.; Gong, N.N.; Wang, Y.W. Experimental and numerical investigations of self-centering post-tensioned precast beam-to-column connections with steel top and seat angles. *Eng. Struct.* **2021**, *226*, 111397. [CrossRef]
14. Huang, L.J.; Zhou, Z.; Peng, Z.; Ma, J.F. Experimental investigation of a self-centering precast concrete beam-to-column connection with top and bottom friction energy dissipaters. *Struct. Des. Tall Spec. Build.* **2020**, *29*, e1707. [CrossRef]
15. Deng, K.L.; Pan, P.; Lam, A.; Pan, Z.H.; Ye, L.P. Test and simulation of full-scale self-centering beam-to-column connection. *Earthq. Eng. Eng. Vib.* **2013**, *12*, 599–607. [CrossRef]
16. Ergang, X.; Kun, Z.; Jing, W. Seismic performanceanalysis of self-centering concentrically braced steel frame structures. *World Earthq. Eng.* **2020**, *36*, 69–79.
17. Fang, C.; Wang, W.; He, C.; Chen, Y. Self-centring behaviour of steel and steel-concrete composite connections equipped with NiTi SMA bolts. *Eng. Struct.* **2017**, *150*, 390–408. [CrossRef]
18. Christopoulos, C.; Tremblay, R.; Kim, H.-J.; Lacerte, M. Self-centering energy dissipative bracing system for the seismic resistance of structures: Development and validation. *J. Struct. Eng.* **2008**, *134*, 96–107. [CrossRef]
19. Chou, C.C.; Chung, P.T. Development of cross-anchored dual-core self-centering braces for seismic resistance. *J. Constr. Steel Res.* **2014**, *101*, 19–32. [CrossRef]
20. Zhou, Z.; Xie, Q.; Lei, X.C.; He, X.T.; Meng, S.P. Experimental Investigation of the Hysteretic Performance of Dual-Tube Self-Centering Buckling-Restrained Braces with Composite Tendons. *J. Compos. Constr.* **2015**, *19*, 04015011. [CrossRef]
21. Xu, L.H.; Xie, X.S.; Li, Z.X. Seismic Behavior and Design Approach of Variable-Damping Self-Centering Braced Frame. *J. Struct. Eng.* **2021**, *147*, 05021001. [CrossRef]
22. Xiao, Y.; Eberhard, M.O.; Zhou, Y.; Stanton, J.F. Proportioning of self-centering energy dissipative braces. *Earthq. Eng. Struct. Dyn.* **2021**, *50*, 2613–2633. [CrossRef]
23. Sun, G.H.; Zhu, Y. Cyclic testing of an innovative self-centering X-braced ductile shear panel. *Eng. Struct.* **2021**, *244*, 112732. [CrossRef]
24. Rezvan, P.; Zhang, Y.F. Nonlinear seismic performance study of D-type self-centering eccentric braced frames with sliding rocking link beams. *Earthq. Eng. Struct. Dyn.* **2021**, *51*, 875–895. [CrossRef]
25. O'Reilly, G.J.; Goggins, J. Experimental testing of a self-centring concentrically braced steel frame. *Eng. Struct.* **2021**, *238*, 111521. [CrossRef]
26. Chung, P.T.; Chou, C.C.; Ling, Y.T. Mechanics, modeling and seismic behavior of a dual-core self-centering brace in series with a frictional gusset connection. *Eng. Struct.* **2021**, *247*, 113018. [CrossRef]
27. Chou, C.C.; Tsai, W.J.; Chung, P.T. Development and validation tests of a dual-core self-centering sandwiched buckling-restrained brace (SC-SBRB) for seismic resistance. *Eng. Struct.* **2016**, *121*, 30–41. [CrossRef]
28. Kitayama, S.; Constantinou, M.C. Probabilistic collapse resistance and residual drift assessment of buildings with fluidic self-centering systems. *Earthq. Eng. Struct. Dyn.* **2016**, *45*, 1935–1953. [CrossRef]
29. Wang, W.; Fang, C.; Zhao, Y.; Sause, R.; Hu, S.; Ricles, J. Self-centering friction spring dampers for seismic resilience. *Earthq. Eng. Struct. Dyn.* **2019**, *48*, 1045–1065. [CrossRef]
30. Wang, Y.W.; Zhou, Z.; Zhang, L.X.; Zhao, K.S. Hysteretic Behavior of Dual-Self-Centering Variable Friction Damper Braces with Low Pretensioned Force. *J. Earthq. Eng.* **2021**, 1997840. [CrossRef]
31. Wang, Y.W.; Zhou, Z.; Zhang, L.X.; Xie, Q. Quantification of higher mode effects of steel frame and control method using dual self-centering variable friction damper brace. *Eng. Struct.* **2021**, *240*, 112368. [CrossRef]
32. Wang, Y.W.; Zeng, B.; Zhou, Z.; Xie, Q. Hysteretic and seismic performance of dual self-centering variable friction damper braces. *Soil Dyn. Earthq. Eng.* **2021**, *147*, 106774. [CrossRef]

33. Veismoradi, S.; Yousef-beik, S.M.M.; Zarnani, P.; Quenneville, P. Development and parametric study of a new self-centering rotational friction damper. *Eng. Struct.* **2021**, *235*, 112097. [CrossRef]
34. Qiu, C.X.; Liu, J.W.; Du, X.L. Analytical and numerical study on the cyclic behavior of buckling-restrained SMA-based self-centering damper. *Smart Mater. Struct.* **2021**, *30*, 095021. [CrossRef]
35. Falahian, A.; Asadi, P.; Riahi, H.T.; Kadkhodaei, M. An experimental study on a self-centering damper based on shape-memory alloy wires. *Mech. Based Des. Struct. Mach.* **2021**, 1939048. [CrossRef]
36. Ding, Y.; Zhou, Z.; Huang, L.J.; Si, Y. Seismic performance of self-centering glulam frame with friction damper. *Eng. Struct.* **2021**, *245*, 112857. [CrossRef]
37. Zhu, R.Z.; Guo, T.; Mwangilwa, F. Development and test of a self-centering fluidic viscous damper. *Adv. Struct. Eng.* **2020**, *23*, 2835–2849. [CrossRef]
38. Zhu, S.Y.; Zhang, Y.F. Seismic analysis of concentrically braced frame systems with self-centering friction damping braces. *J. Struct. Eng.* **2008**, *134*, 121–131. [CrossRef]
39. Kam, W.Y.; Pampanin, S.; Palermo, A.; Carr, A. Advanced Flag-Shaped Systems For High Seismic Performance. In Proceedings of the First European Conference on Earthquake Engineering and Seismology, Geneva, Switzerland, 3–8 September 2006.
40. Shi, Y.; Xinxin, J.; Changhong, D. Mechanical performances of a new type of resilient SMA damper. *World Earthq. Eng.* **2020**, *36*, 96–102.
41. Zhang, R.B.; Wang, W.; Fang, C.; Zhang, W.J.; Zhuang, L.J. Self-centering Devices with Paralleled Friction Spring Groups: Development, Experiment and System Behavior. *J. Earthq. Eng.* **2021**, 2009059. [CrossRef]
42. Ricles, J.M.; Sause, R.; Peng, S.W.; Lu, L.W. Experimental evaluation of earthquake resistant posttensioned steel connections. *J. Struct. Eng.* **2002**, *128*, 850–859. [CrossRef]
43. Garlock, M.M.; Sause, R.; Ricles, J.M. Behavior and design of posttensioned steel frame systems. *J. Struct. Eng.* **2007**, *133*, 389–399. [CrossRef]
44. Maurya, A.; Eatherton, M.R. Experimental study of the restoring force mechanism in the self-centering beam (SCB). *Front. Struct. Civ. Eng.* **2016**, *10*, 272–282. [CrossRef]
45. Fang, C.; Zhong, Q.M.; Wang, W.; Hu, S.L.; Qiu, C.X. Peak and residual responses of steel moment-resisting and braced frames under pulse-like near-fault earthquakes. *Eng. Struct.* **2018**, *177*, 579–597. [CrossRef]
46. Fan, X.W.; Xu, L.H.; Li, Z.X. Seismic performance evaluation of steel frames with pre-pressed spring self-centering braces. *J. Constr. Steel Res.* **2019**, *162*, 105761. [CrossRef]
47. Zhang, Z.-X.; Zhang, J.; Wu, H.; Ji, Y.; Kumar, D.D. Iron-Based Shape Memory Alloys in Construction: Research, Applications and Opportunities. *Materials* **2022**, *15*, 1723. [CrossRef] [PubMed]
48. Fang, C.; Yam, M.C.H.; Lam, A.C.C.; Xie, L. Cyclic performance of extended end-plate connections equipped with shape memory alloy bolts. *J. Constr. Steel Res.* **2014**, *94*, 122–136. [CrossRef]
49. Yam, M.C.H.; Fang, C.; Lam, A.C.C.; Zhang, Y. Numerical study and practical design of beam-to-column connections with shape memory alloys. *J. Constr. Steel Res.* **2015**, *104*, 177–192. [CrossRef]
50. Fang, C.; Zhou, X.; Osofero, A.I.; Shu, Z.; Corradi, M. Superelastic SMA Belleville washers for seismic resisting applications: Experimental study and modelling strategy. *Smart Mater. Struct.* **2016**, *25*, 105013. [CrossRef]
51. Villa, E. Manufacturing of Shape Memory Alloys. In *Shape Memory Alloy Engineering*; Elsevier: Amsterdam, The Netherlands, 2015; pp. 79–96.
52. Fang, C.; Yam, M.C.H.; Chan, T.-M.; Wang, W.; Yang, X.; Lin, X. A study of hybrid self-centring connections equipped with shape memory alloy washers and bolts. *Eng. Struct.* **2018**, *164*, 155–168. [CrossRef]
53. Fang, C.; Zheng, Y.; Chen, J.; Yam, M.C.H.; Wang, W. Superelastic NiTi SMA cables: Thermal-mechanical behavior, hysteretic modelling and seismic application. *Eng. Struct.* **2019**, *183*, 533–549. [CrossRef]
54. Fang, C.; Yam, M.C.H.; Ma, H.; Chung, K.F. Tests on superelastic Ni–Ti SMA bars under cyclic tension and direct-shear: Towards practical recentring connections. *Mater. Struct.* **2013**, *48*, 1013–1030. [CrossRef]
55. Zheng, Y.; Fang, C.; Liang, D.; Sun, R. An innovative seismic-resilient bridge with shape memory alloy-washer-based footing rocking RC piers. *J. Intell. Mater. Syst. Struct.* **2020**, *32*, 549–567. [CrossRef]
56. Wang, W.; Fang, C.; Liu, J. Large size superelastic SMA bars: Heat treatment strategy, mechanical property and seismic application. *Smart Mater. Struct.* **2016**, *25*, 075001. [CrossRef]
57. Fang, C.; Yam, M.C.H.; Lam, A.C.C.; Zhang, Y.Y. Feasibility study of shape memory alloy ring spring systems for self-centring seismic resisting devices. *Smart Mater. Struct.* **2015**, *24*, 075024. [CrossRef]
58. Qiu, C.X.; Zhu, S.Y. Shake table test and numerical study of self-centering steel frame with SMA braces. *Earthq. Eng. Struct. Dyn.* **2017**, *46*, 117–137. [CrossRef]
59. Miller, D.J.; Fahnestock, L.A.; Eatherton, M.R. Development and experimental validation of a nickel-titanium shape memory alloy self-centering buckling-restrained brace. *Eng. Struct.* **2012**, *40*, 288–298. [CrossRef]
60. Chen, J.; Fang, C.; Wang, W.; Liu, Y. Variable-friction self-centering energy-dissipation braces (VF-SCEDBs) with NiTi SMA cables for seismic resilience. *J. Constr. Steel Res.* **2020**, *175*, 106318. [CrossRef]
61. Fang, C.; Ping, Y.; Chen, Y.; Yam, M.C.H.; Chen, J.; Wang, W. Seismic Performance of Self-centering Steel Frames with SMA-viscoelastic Hybrid Braces. *J. Earthq. Eng.* **2020**, 1856233. [CrossRef]

62. Jha, N.K.; Dobriyal, R.; Kumar, P.; Badhotiya, G.K. On the modelling of nonlinear viscoelastic relation for rubber-like materials. *Mater. Today-Proc.* **2021**, *46*, 10546–10550. [CrossRef]
63. Tsai, C.S.; Lee, H.H. Applications of Viscoelastic Dampers to High-Rise Buildings. *J. Struct. Eng.* **1993**, *119*, 1222–1233. [CrossRef]
64. Shu, Z.; Gan, Z.Z.; Fang, C.; MacRae, G.; Dong, H.L.; Xie, Y.Z. Replaceable Rotational Viscoelastic Dampers for Improving Structural Damping and Resilience of Steel Frames. *J. Earthq. Eng.* **2021**, 2009058. [CrossRef]
65. Zhang, Z.; Liang, D.; Fang, C.; Zheng, Y. Experimental study and dynamic analysis on self-centering rocking piers equipped with shape memory alloy (SMA) cables. *World Earthq. Eng.* **2020**, *36*, 138–146.
66. Fang, C.; Liang, D.; Zheng, Y.; Lu, S.Y. Seismic performance of bridges with novel SMA cable-restrained high damping rubber bearings against near-fault ground motions. *Earthq. Eng. Struct. Dyn.* **2022**, *51*, 44–65. [CrossRef]
67. Wang, W.; Fang, C.; Shen, D.; Zhang, R.; Ding, J.; Wu, H. Performance assessment of disc spring-based self-centering braces for seismic hazard mitigation. *Eng. Struct.* **2021**, *242*, 112527. [CrossRef]
68. Fang, C.; Wang, W.; Ricles, J.; Yang, X.; Zhong, Q.; Sause, R.; Chen, Y. Application of an Innovative SMA Ring Spring System for Self-Centering Steel Frames Subject to Seismic Conditions. *J. Struct. Eng.* **2018**, *144*, 0401811. [CrossRef]
69. Fang, C.; Wang, W.; Zhang, A.; Sause, R.; Ricles, J.; Chen, Y. Behavior and Design of Self-Centering Energy Dissipative Devices Equipped with Superelastic SMA Ring Springs. *J. Struct. Eng.* **2019**, *145*, 04019109. [CrossRef]
70. Feng, W.; Fang, C.; Wang, W. Behavior and design of top flange-rotated self-centering steel connections equipped with SMA ring spring dampers. *J. Constr. Steel Res.* **2019**, *159*, 315–329. [CrossRef]
71. Wang, W.; Fang, C.; Zhang, A.; Liu, X. Manufacturing and performance of a novel self-centring damper with shape memory alloy ring springs for seismic resilience. *Struct. Control. Health Monit.* **2019**, *26*, e2337. [CrossRef]
72. Fang, C.; Liang, D.; Zheng, Y.; Yam, M.C.H.; Sun, R. Rocking bridge piers equipped with shape memory alloy (SMA) washer springs. *Eng. Struct.* **2020**, *214*, 110651. [CrossRef]
73. Liang, D.; Zheng, Y.; Fang, C.; Yam, M.C.H.; Zhang, C. Shape memory alloy (SMA)-cable-controlled sliding bearings: Development, testing, and system behavior. *Smart Mater. Struct.* **2020**, *29*, 085006. [CrossRef]
74. Ping, Y.; Fang, C.; Chen, Y.; Yam, M.C.H. Seismic robustness of self-centering braced frames suffering tendon failure. *Earthq. Eng. Struct. Dyn.* **2021**, *50*, 1671–1691. [CrossRef]
75. Qiu, C.; Fang, C.; Liang, D.; Du, X.; Yam, M.C.H. Behavior and application of self-centering dampers equipped with buckling-restrained SMA bars. *Smart Mater. Struct.* **2020**, *29*, 035009. [CrossRef]
76. Liu, Y.; Wang, H.; Qiu, C.; Zhao, X. Seismic Behavior of Superelastic Shape Memory Alloy Spring in Base Isolation System of Multi-Story Steel Frame. *Materials* **2019**, *12*, 997. [CrossRef]
77. Fang, C.; Ping, Y.; Zheng, Y.; Chen, Y. Probabilistic economic seismic loss estimation of steel braced frames incorporating emerging self-centering technologies. *Eng. Struct.* **2021**, *241*, 112486. [CrossRef]
78. McKenna, F. OpenSees: A Framework for Earthquake Engineering Simulation. *Comput. Sci. Eng.* **2011**, *13*, 58–66. [CrossRef]
79. Diani, J.; Fayolle, B.; Gilormini, P. A review on the Mullins effect. *Eur. Polym. J.* **2009**, *45*, 601–612. [CrossRef]
80. Gong, S.M.; Zhou, Y.; Ge, P.L. Seismic analysis for tall and irregular temple buildings: A case study of strong nonlinear viscoelastic dampers. *Struct. Des. Tall Spec.* **2017**, *26*, e1352. [CrossRef]
81. Zhou, Y.; Aguaguina, M.; Beskos, D.E.; Gong, S.M. A displacement-based seismic design method for building structures with nonlinear viscoelastic dampers. *Bull. Earthq. Eng.* **2021**, *19*, 4535–4585. [CrossRef]
82. Ghosh, K.; Shrimali, B.; Kumar, A.; Lopez-Pamies, O. The nonlinear viscoelastic response of suspensions of rigid inclusions in rubber: I-Gaussian rubber with constant viscosity. *J. Mech. Phys. Solids* **2021**, *154*, 104544. [CrossRef]
83. Vaiana, N.; Spizzuoco, M.; Serino, G. Wire rope isolators for seismically base-isolated lightweight structures: Experimental characterization and mathematical modeling. *Eng. Struct.* **2017**, *140*, 498–514. [CrossRef]
84. Fang, C.; Ping, Y.; Chen, Y. Loading protocols for experimental seismic qualification of members in conventional and emerging steel frames. *Earthq. Eng. Struct. Dyn.* **2019**, *49*, 155–174. [CrossRef]
85. ASCE. *Minimum Design Loads for Buildings and Other Structures*; American Society of Civil Engineers (ASCE): Reston, VA, USA, 2016.
86. Zimmer, M.S. *Characterization of Viscoelastic Materials for Use in Seismic Energy Dissipation Systems*; State University of New York at Buffalo: Buffalo, NY, USA, 1999.
87. Vaiana, N.; Sessa, S.; Marmo, F.; Rosati, L. A class of uniaxial phenomenological models for simulating hysteretic phenomena in rate-independent mechanical systems and materials. *Nonlinear Dyn.* **2018**, *93*, 1647–1669. [CrossRef]
88. Vaiana, N.; Sessa, S.; Rosati, L. A generalized class of uniaxial rate-independent models for simulating asymmetric mechanical hysteresis phenomena. *Mech. Syst. Signal. Process.* **2021**, *146*, 106984. [CrossRef]
89. FEMAP695. *Quantification of Building Seismic Performance Factors*; FEMAP695: Redwood City, CA, USA, 2009.
90. Somerville, P.G. Magnitude scaling of the near fault rupture directivity pulse. *Phys. Earth Planet. Inter.* **2003**, *137*, 201–212. [CrossRef]
91. Baker, J.W. Quantitative classification of near-fault ground motions using wavelet analysis. *Bull. Seismol. Soc. Am.* **2007**, *97*, 1486–1501. [CrossRef]
92. Tremblay, R.; Lacerte, M.; Christopoulos, C. Seismic response of multistory buildings with self-centering energy dissipative steel braces. *J. Struct. Eng.* **2008**, *134*, 108–120. [CrossRef]

materials
MDPI

Article

Effect of Loading Rate and Initial Strain on Seismic Performance of an Innovative Self-Centering SMA Brace

Yigang Jia [1,2], Bo Zhang [1,2], Sizhi Zeng [1,3,*], Fenghua Tang [1], Shujun Hu [1] and Wenping Chen [4]

1 School of Civil Engineering and Architecture, Nanchang University, Nanchang 330031, China; jiayigang999@sina.com (Y.J.); zhangbo_tujian@nerin.com (B.Z.); tfh2450748@163.com (F.T.); hushujun@ncu.edu.cn (S.H.)
2 Design and Research Institute, Nanchang University, Nanchang 330031, China
3 Zhongmei Engineering Group Ltd., Nanchang 330001, China
4 Jiangxi Huaye Special Engineering Technology Co., Ltd., Nanchang 330001, China; huayejiagu2022@163.com
* Correspondence: zengsizhi1986@126.com; Tel.: +86-137-6701-6820

Abstract: In order to improve the energy dissipation capacity and to reduce the residual deformation of civil structures simultaneously, this paper puts forwards an innovative self-centering shape memory alloy (SMA) brace that is based on the design concepts of SMA's superelasticity and low friction slip. Seven self-centering SMA brace specimens were tested under cyclic loading, and the hysteresis curves, bond curves, secant stiffness, energy dissipation coefficient, equivalent damping coefficient, and the self-centering capacity ratio of these specimens were investigated, allowing us to provide an evaluation of the effects of the loading rate and initial strain on the seismic performance. The test results show that the self-centering SMA braces have an excellent energy dissipation capacity, bearing capacity, and self-centering capacity, while the steel plates remain elastic, and the SMA in the specimens that are always under tension are able to return to the initial state. The hysteresis curves of all of the specimens are idealized as a flag shape with low residual deformation, and the self-centering capacity ratio reached 89.38%. In addition, both the loading rate and the initial strain were shown to have a great influence on the seismic performance of the self-centering SMA brace. The improved numerical models combined with the Graesser model and Bouc–Wen model in MATLAB were used to simulate the seismic performance of the proposed braces with different loading rates and initial strains, and the numerical results are consistent with the test results under the same conditions, meaning that they can accurately predict the seismic performance of the self-centering SMA brace proposed here.

Keywords: shape memory alloy (SMA); self-centering SMA brace; loading rate; initial strain; energy dissipation coefficient

Citation: Jia, Y.; Zhang, B.; Zeng, S.; Tang, F.; Hu, S.; Chen, W. Effect of Loading Rate and Initial Strain on Seismic Performance of an Innovative Self-Centering SMA Brace. *Materials* **2022**, *15*, 1234. https://doi.org/10.3390/ma15031234

Academic Editor: F. Pacheco Torgal

Received: 25 November 2021
Accepted: 31 January 2022
Published: 7 February 2022

Publisher's Note: MDPI stays neutral with regard to jurisdictional claims in published maps and institutional affiliations.

In major earthquakes, buckling-restrained brace frames [1] and eccentrically braced frames [2] demonstrate high stiffness, high ductility, and good energy dissipating capacity. However, conventional steel frames may develop severe residual deformations and structural damage when subjected to strong earthquakes [3,4] In recent years, the concept of a re-centering mechanism for civil structures has been proposed; it was considered to be an efficient way to reduce residual deformation and to further improve the energy dissipating capacity of these structures [5,6]. Therefore, many kinds of steel frames with self-centering devices were proposed, and have demonstrated the advantages of high stiffness, low residual deformation, and easy construction [7,8].

Shape memory alloy (SMA) wire is a new type of smart material with a shape memory effect and superelasticity effect that can return to its initial shape after experiencing a strain value of 0.06 with negligible residual deformation upon unloading [9,10]. A great deal of research shows that self-centering dampers equipped with SMAs have emerged as energy-dissipating and re-centering candidates for civil structures [11–16]. For example, Xue et al. [11] proposed a self-centering friction damper with SMA wires and friction devices, and this damper had an excellent energy dissipation capacity. Qiu and Zhu [12] investigated the behavior of novel self-centering SMA braces equipped with SMA as a key component. Xu et al. [13] illustrated an innovative self-centering link beam with steel rods and SMA rods to provide the re-centering force. Fang et al. [12,13] presented a novel type of self-centering steel connection with SMA rings, which showed satisfactory energy dissipation and excellent self-centering capability. Hu et al. [16] indicated that a new self-centering brace had the advantages of good seismic performance, a high self-centering capacity, and zero damage. To examine the influence of self-centering SMA dampers on the seismic performance of civil structures, Li et al. [17] studied the seismic performance of a six-story steel frame with an innovative re-centering damper, which had an outstanding re-centering capacity. Fan et al. [18] further confirmed that the prepressed spring self-centering braces in the steel frame could mitigate post-earthquake residual deformation.

However, some studies revealed that the seismic performance of the self-centering braces mentioned above could be affected by the mechanical properties of the SMA wires themselves. Zhou et al. [19] conducted fatigue testing on SMA wires that were 0.5 mm and 1.0 mm in diameter and indicated that the initial strain and loading frequency had a great effect on the mechanical properties of the SMA wires. Yan et al. [20] revealed the influence of cyclic numbers, the loading rate, and the strain amplitude on SMA wires. Qian et al. [21] carried out the cyclic loading of SMA wires by changing the variable amplitudes and loading rates. Hu et al. [22] investigated the effect of the cyclic number, strain amplitude, initial strain, and loading rate on an SMA wire with a 1 mm diameter, and the results of the mechanical property evaluation indicated that the cyclic number was less clear but that the initial strain and loading rate should be emphasized. Therefore, the loading rate and initial strain are two of the main factors that could be used to determine the seismic performance of a self-centering brace with SMA wires.

This paper presents experimental research on the effect of the loading rate and the initial strain of a self-centering SMA brace under cyclic loading, and the hysteresis curves, bond curve, secant stiffness, energy dissipation coefficient, equivalent damping coefficient, and self-centering capacity ratio (ratio between super-elastic displacement and maximum displacement) of the braces are analyzed in detail. Then, the modified mechanical model of the self-centering SMA brace is developed based on the improved Grasser model program and the Bouc–Wen model, and the MATLAB/SIMULINK toolbox is used to conduct the simulation, allowing the accuracy of the numerical results to be compared to the test results.

1. Basic Properties of Self-Centering SMA Brace

As shown Figure 1, self-centering SMA braces mainly include three parts: a slip component, a fixed component, and SMA wires. The slip component is composed of a moving plate, slip plate I and slip plate II, which are connected by slip bolt I and slip bolt II. The fixed component is constituted by the moving plate, the fixed plate, and slip plate II, which has a fixed bolt connection. The SMA wires are set on both ends of the two slip bolts to provide the energy dissipating capacity and elastic restoring force, which provides the special property of superelasticity. Moreover, rubber shims are placed on both sides of the slip shim to reduce the friction coefficient, and the slip bolt passes through the moving plate, slip shim, slip plateI, slip shim, and slip plate II in sequence. It is suggested that the fixed shims be installed between the moving plate and fixed plate and slip plate II. In addition, two slot holes are located at both ends of the moving plate, and there are two slot holes in the same position on slip plate II and on the fixed plate.

Figure 2 illustrates the work principle of a self-centering SMA brace. As shown Figure 2b, when the fixed plate is fixed and the brace is in tension conditions at the moving plate, slip bolt II is fixed in the moving plate and slipped in the slip plates, and slip bolt I is fixed in the slip plates and slipped into the moving plate simultaneously. As shown Figure 2c, when the brace is in pressure conditions at the moving plate, the slip bolt I is fixed in the moving plate and slipped in the slip plates, slip bolt II is fixed in the slip plates and slipped into the moving plate simultaneously. During the positive and negative movements, the SMA wires are always subjected to elongation, thus increasing the ductility, energy dissipating capacity, and self-centering capacity. In ideal conditions, when the external load is unloaded, the SMA wires can almost force the slip component return back to the initial state, and only slight residual deformation in observed in the self-centering SMA brace, which is mainly caused by the very low friction coefficient of the rubber shims.

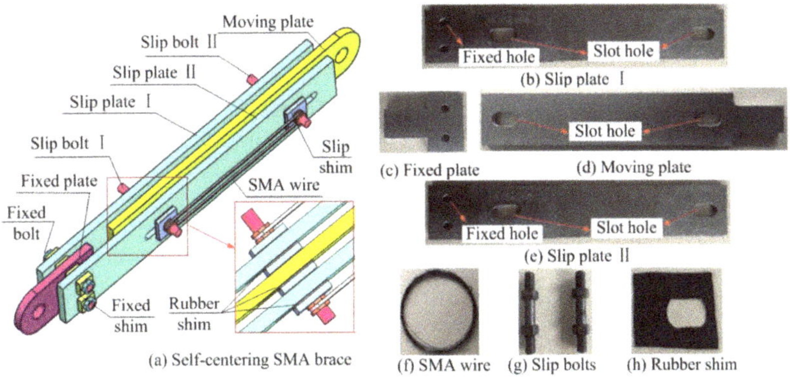

Figure 1. Schematic diagram of self-centering SMA brace.

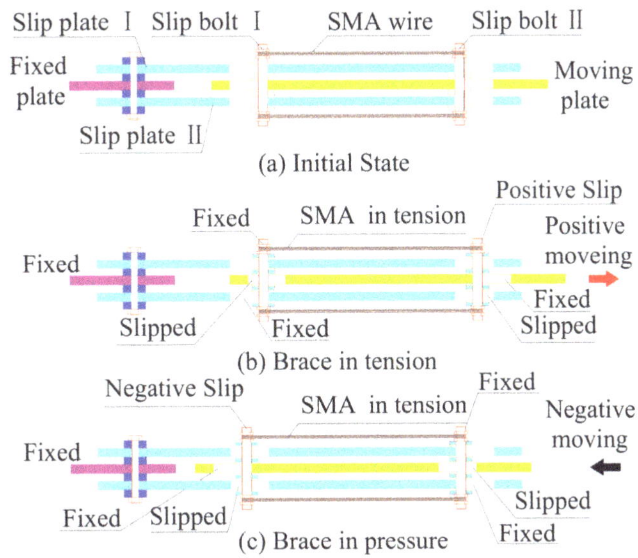

Figure 2. (**a**) initial state, (**b**) brace in tension, (**c**) brace in pressure, Work principle of self-centering SMA brace.

2. Test Investigation

2.1. Test Specimens

In total, the comparison of seven self-centering braces with different loading rates and initial strains are tested in this paper. All of the braces are composed of a slip component, fixed component, and SMA wires, as shown in Figure 3. Slip plateI has a cross-section that is 449 mm × 80 mm in size and a thickness of 15 mm. The moving plate and slip plate II have a cross-section that is 424 mm × 80 mm in size and a thickness of 8 mm. The fixed plate has a cross-section that is 130 mm × 80 mm in size and a thickness of 15 mm. The radii, R, of the fixed hole, slip shim, and fixed shim are 5 mm, 15 mm, and 40 mm, respectively. The thicknesses, t_f, of the slip shim, fixed shim, and rubber shim are 5 mm, 3 mm, and 1 mm, respectively. The radius of the slot hole is 10 mm and has a length of 42 mm.

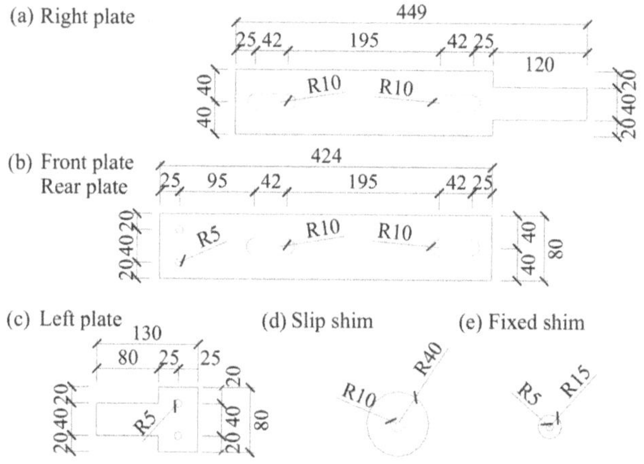

Figure 3. (**a**) right plate (**b**) front plate/rear plate (**c**) left plate (**d**) slip shim (**e**) fixed shim. Details of the self-centering SMA brace.

The parameters of the specimens are listed in Table 1. The specimens consist of four parameters: the torque value of the slip bolt, the SMA area, the loading rate of the SMA wire, and the initial strain of the SMA wire. The torque value and SMA area of the specimens are 10 N·M and 43.96 mm^2. For example, the specimen "SCB-12-25" represents the brace at a loading rate of 0.0012 s^{-1} and an initial strain amplitude of 0.25%. The effect of the loading rate of the self-centering SMA brace can be revealed by specimens SCB-12-0, SCB-18-0, SCB-24-0, and SCB-36-0, while the influence of the initial strain is reflected by specimens SCB-12-0, SCB-12-25, SCB-12-50, and SCB-12-100.

Table 1. Main specimen parameters.

No.	Specimens	Torque Value/N·M	SMA Area/mm^2	Loading Rate/s^{-1}	Initial Strain/%
1	SCB-12-0	10	43.96	0.0012	0
2	SCB-18-0	10	43.96	0.0018	0
3	SCB-24-0	10	43.96	0.0024	0
4	SCB-36-0	10	43.96	0.0036	0
5	SCB-12-25	10	43.96	0.0012	0.25
6	SCB-12-50	10	43.96	0.0012	0.50
7	SCB-12-100	10	43.96	0.0012	1.00

2.2. *Material Properties*

2.2.1. SMA Wire

The tested SMA wire with a diameter of 1.0 mm diameter was obtained from Gao'an SMA Material Co., Ltd. According to data offered by the manufacturer, the chemical composition of the SMA wire in terms of weight was close to Ni-55.96%, Ti-43.9835%, H-0.0005%, Cr-0.0070%, Co-0.003%, C-0.005%, Fe-0.006%, Cu-0.006%, and others-0.029%. Tests on the mechanical properties of the SMA wire at different loading rates and initial strains under cyclic loading were carried out, and the hysteresis curves are shown in Figure 4 [22].

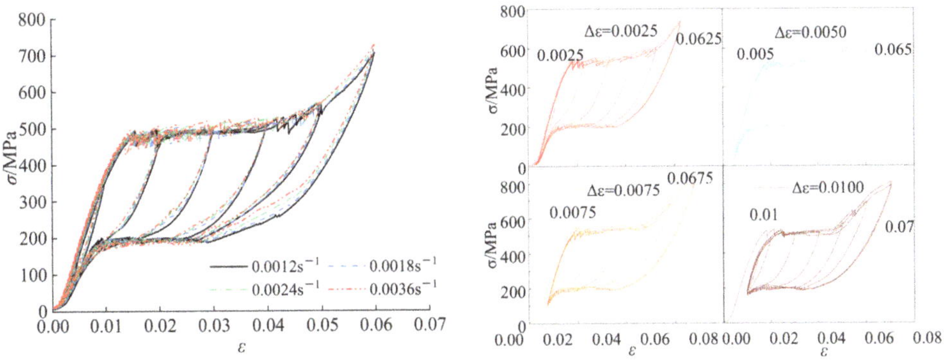

(a) (b)

Figure 4. (**a**) Loading rate; (**b**) initial strain. Hysteresis curves of SMA wires with different influencing factors [22].

The mechanical properties in the constitutive model of the SMA wire comprise six parameters, i.e., σ_s^{AM}, σ_f^{AM}, σ_s^{MA}, σ_f^{MA}, $\overline{\varepsilon_L}$, and E_A [10], as shown in Figure 5. Based on the mechanical properties obtained from Figure 4, the main parameters of SMA wires with different influencing factors are shown in Table 2.

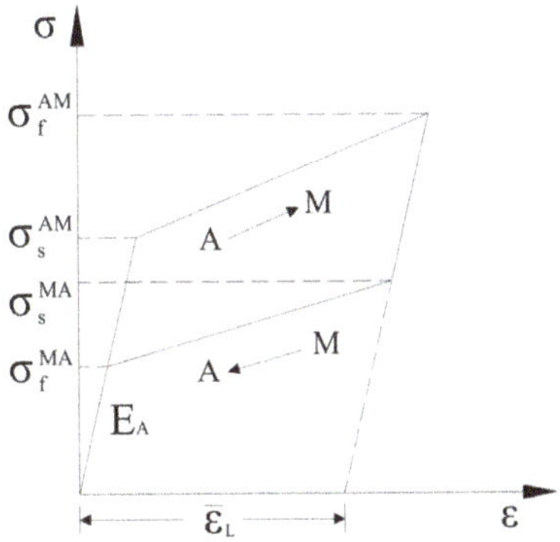

Figure 5. Constitutive model of SMA wire [22].

Table 2. Main parameters of SMA wire with different influencing factors.

Influencing Factor	Value	$\overline{\varepsilon_L}$	E_A/MPa	σ_s^{AM}/MPa	σ_f^{AM}/MPa	σ_s^{MA}/ MPa	σ_f^{MA}/MPa
Loading rate/s^{-1}	0.0012	0.06	47000	492.81	703.70	277.53	117.89
	0.0018	0.06	49000	493.94	712.23	288.85	117.49
	0.0024	0.06	50000	495.30	720.81	295.19	116.55
	0.0036	0.06	51000	496.30	730.62	303.79	116.24
Pre-tensioned/Δε	0.0025	0.06	50000	490.81	742.35	288.29	119.44
	0.0050	0.06	51000	484.68	776.27	285.48	135.37
	0.0075	0.06	52000	475.81	807.56	274.28	146.18
	0.0100	0.06	53000	433.86	813.31	258.74	156.65

2.2.2. Steel Plate

The tested steel plates with thicknesses of 8 mm and 15 mm were all made of Q345B steel and had a nominal design yield strength of 345MPa. The samples that were obtained from the steel plates were subjected to standard metallic tensile tests, and the mechanical properties that were measured are listed in Table 3.

Table 3. Mechanical properties of steel plates.

No.	Thickness/mm	Yield Strength/MPa	Tensile Strength/MPa	Young's Modulus/GPa	Elongation/%
1	8	365	545	206	23.1
2	15	372	556	209	25.2

2.3. Test Setup

The cyclic loading test was carried out using an SDS100 fatigue test machine at the Engineering Mechanics Experiment Center, Nanchang University. The test setup was composed of four parts: the sensor, control station, control terminal, and signal connection, as shown in Figure 6. The maximum hydraulically driven load that the SDS100 can apply is 100 kN, which is applied with a precision of 0.01 kN. The lower fixture and upper fixture were connected to the fixed plate and slip platel of the self-centering SMA brace, respectively. The load and displacement of the test specimens were directly recorded by the sensor. All tests were conducted at 27 °C.

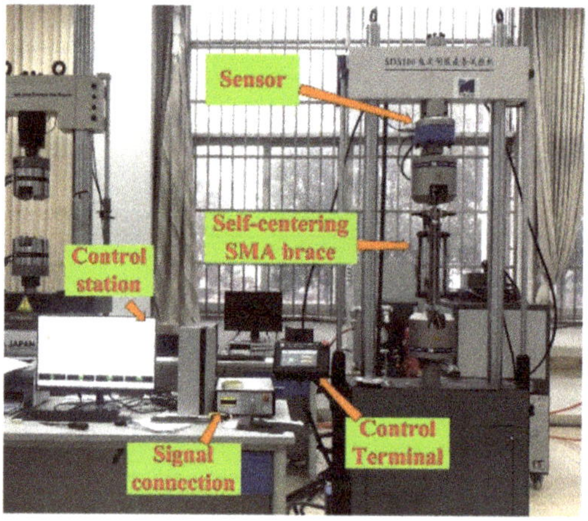

Figure 6. Diagram of the experimental setup.

2.4. Test Cases

The tested cases of self-centering SMA brace specimens are shown in Table 4. The cases one through four studied the effect of the different loading rates, and the loading rates were set as 0.0012 s^{-1}, 0.0018 s^{-1}, 0.0024 s^{-1}, and 0.0036 s^{-1}, respectively, whereas the initial strain was zero, and the loading cycle was one. The first case as well as cases five to seven considered the effect of the initial strains, which were 0.0025, 0.0050, 0.0075, and 0.0100, and all tests were loaded for at a loading rate of 0.0012 s^{-1} for one cycle. In addition, a total of seven loading displacement ranges were set for all of the specimens: 1.20 mm, 2.40 mm, 4.80 mm, 7.20 mm, 9.60 mm, 12.00 mm, and 14.40 mm, which were set successively [22], and the corresponding strain amplitudes were 0.005, 0.01, 0.02, 0.03, 0.04, 0.05, and 0.06.

Table 4. Test cases of self-centering SMA brace specimens.

Test Case	Specimens	Loading Rate/s^{-1}	Initial Strain/%	Loading Displacement/mm	Loading Cycles
1	SCB-12-0	0.0012	0	1.20, 2.40, 4.80, 7.20, 9.60, 12.00, 14.00	1
2	SCB-18-0	0.0018	0	1.20, 2.40, 4.80, 7.20, 9.60, 12.00, 14.00	1
3	SCB-24-0	0.0024	0	1.20, 2.40, 4.80, 7.20, 9.60, 12.00, 14.00	1
4	SCB-36-0	0.0036	0	1.20, 2.40, 4.80, 7.20, 9.60, 12.00, 14.00	1
5	SCB-12-25	0.0012	0.25	1.20, 2.40, 4.80, 7.20, 9.60, 12.00, 14.00	1
6	SCB-12-50	0.0012	0.50	1.20, 2.40, 4.80, 7.20, 9.60, 12.00, 14.00	1
7	SCB-12-100	0.0012	1.00	1.20, 2.40, 4.80, 7.20, 9.60, 12.00, 14.00	1

3. Test Results

3.1. Hysteresis Curves

Figure 7 shows the hysteresis curves of the self-centering SMA brace specimens with different loading rates and initial strains. The hysteresis curves mainly consist of three successive phases: the initial slip phase, the rapidly increasing stress–strain phase, and the rapidly decreasing phase. The figure clearly shows that all of the hysteresis curves exhibit a high self-centering capacity and ideal flag-shape hysteresis with low residual deformation and slip strength. The self-centering capacity and ideal flag-shape hysteresis are primarily caused by the SMA wires, while low residual deformation and slip strength are induced by the slip component. In addition, the rectangular loops around the origin point are caused by low residual deformation and slip strength.

(1) Effect of loading rates: Figure 7a plots the hysteresis curves of the SCBs with different loading rates. It can be seen that the maximum axial force F and residual deformation D_1 are 31.08 kN, 31.49 kN, 31.75 kN, 32.23 kN and 1.97 mm, 1.90 mm, 1.87 mm, 1.85 mm at the loading rates of 0.0012 s^{-1}, 0.0018 s^{-1}, 0.0024 s^{-1}, and 0.0036 s^{-1}, respectively. The ultimate axial force increased gradually, and the residual deformation decreased slightly as the loading rate increased.

(2) Effect of initial strains: The hysteresis curves of the SCBs with different initial strains are shown in Figure 7b. The maximum axial forces F of the SCBs were 31.08 kN, 33.21 kN, 36.80 kN, and 42.24 kN at the loading rates of 0.25%, 0.50%, 0.75%, and 1.00%, respectively, which significantly increased as the initial strain increased. In addition, the results also indicate that the residual deformation D_1 at the loading rates of 0.25%, 0.50%, 0.75%, and 1.00% were 1.97 mm, 1.89 mm, 1.69 mm, and 1.53 mm, respectively, which was mainly caused by the increase in the initial stress and ultimate stress of the SMA wires [22].

3.2. Bond Curves

The bond curves in Figure 8 were obtained from the results in Figure 7. The bond curves in the initial slip phase have a smaller initial stiffness, and the bearing capacity and initial stiffness gradually increased in the increasing phase. Finally, the bearing capacity of the specimens increased slowly and reached the maximum axial force, and the slope of the bond curve began to decrease in the decreasing phase. All of the bond curves are origin-symmetric under axial tension and axial compression. In addition, as shown in Figure 8a,b, the bearing capacity of specimens SCB-12-0, SCB-18-0, SCB-24-0, SCB-36-0 increased as the loading rate increased, and the same conclusion can be obtained for the initial strain rates for specimens SCB-12-0, SCB-12-25, SCB-12-50, and SCB-12-100.

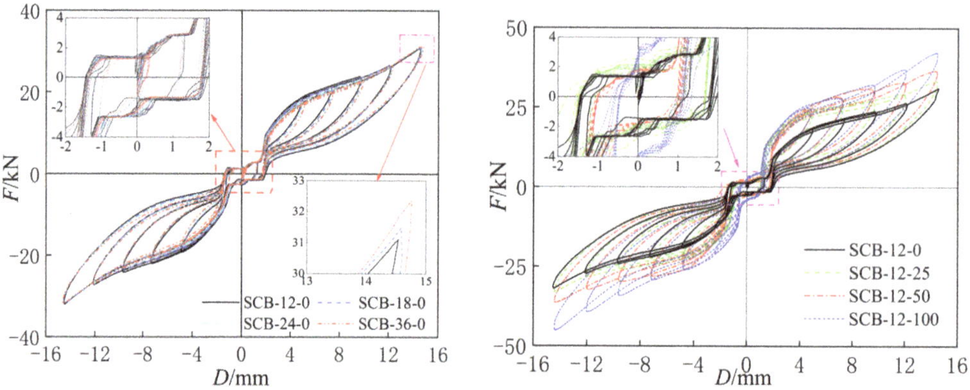

(a) (b)

Figure 7. (**a**) Loading rate; (**b**) initial strain. Hysteresis curves of self-centering SMA brace specimens.

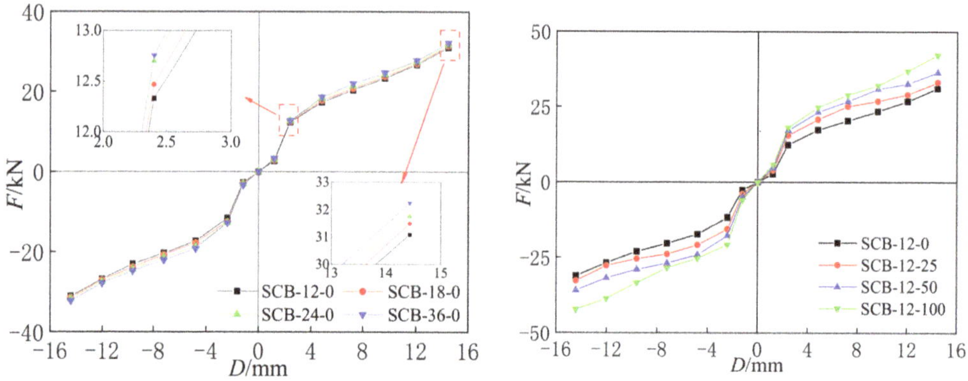

(a) (b)

Figure 8. (**a**) Loading rate; (**b**) initial strain. Bond curves of self-centering SMA brace specimens.

3.3. Secant Stiffness

In this paper, the stiffness degradation of the self-centering SMA brace is presented by the secant stiffness coefficient, K_{si}, which can be calculated by:

$$K_{si} = \frac{|F_{i,\max}| + |-F_{i,\min}|}{|D_{i,\max}| + |-D_{i,\min}|} \quad (1)$$

where F_i,max, F_i,min, D_i,max, and D_i,min represent the maximum axial force, minimum axial compression, maximum displacement, and minimum displacement at the i-th hysteretic cycle under the load displacement of i.

Figure 9 shows the secant stiffness curves of all of the test specimens. The secant stiffness coefficient decreases after reaching the maximum value as the applied displacement increases and the reduction rate gradually decreases. The initial increase in the secant stiffness coefficient is mainly because of the slip friction of the slip components. In addition, as the loading rate and initial strain increase, it can be seen that the secant stiffness coefficients of the self-centering SMA braces increase gradually.

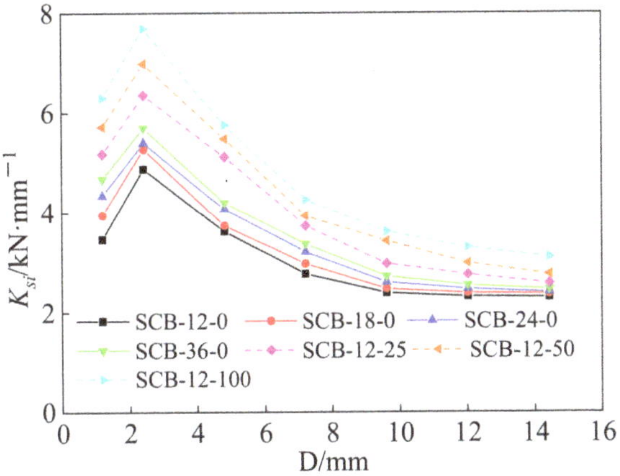

Figure 9. Secant stiffness of self-centering SMA brace specimens.

3.4. Energy Dissipation Coefficient

The different energy dissipation coefficients, E_i, which are calculated by the enclosed area of the hysteresis curve for each specimen, are shown in Figure 10. The maximum E_i of the specimens SCB-12-0, SCB-18-0, SCB-24-0, and SCB-36-0 are 317.37 J, 320.80 J, 310.73 J, and 291.13 J, respectively. An increase in the E_i in each test case can be observed when the loading rates are increased from 0.0012 s^{-1} to 0.0018 s^{-1}, an a gradual decrease occurs as the loading rate increases to 0.0036 s^{-1}, which is influenced by the pinch phenomenon in the hysteresis curves [22]. In addition, an increase in the E_i for specimens SCB-12-0, SCB-12-25, SCB-12-50, and SCB-12-100 can be clearly seen under different test cases where there is an increase of initial strain from 0 to 0.01, which was mainly caused by the increase in the hysteresis areas. Therefore, the energy dissipation capacity can be effectively increased by increasing the initial strain, but the influence law of the loading rate for the self-centering SMA braces is uncertain, meaning that further research should be on the equivalent damping coefficient.

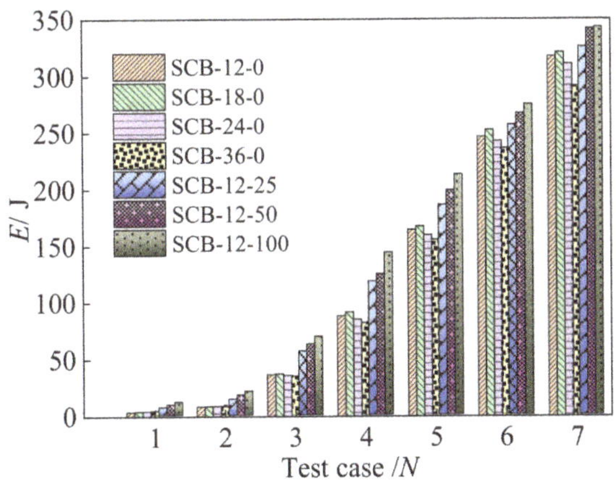

Figure 10. Energy dissipation coefficient of specimens.

3.5. Equivalent Damping Coefficient

The equivalent damping coefficient, ξ_{eq}, is also an important parameter that can be used to evaluate the energy dissipation capacity, which can be calculated as follows [23]:

$$\xi_{eq} = \frac{1}{2\pi} \frac{S_{EBG} + S_{DEG}}{S_{OBA} + S_{ODC}} \tag{2}$$

where S_{EBG} and S_{DEG} represent the areas of the closed geometrical figures EBG and DEG, which are enclosed by the vertical axial force and horizontal displacement of the coordinates, and S_{OBA} and S_{ODC} represent the areas of the triangles OBA and ODC in Figure 11.

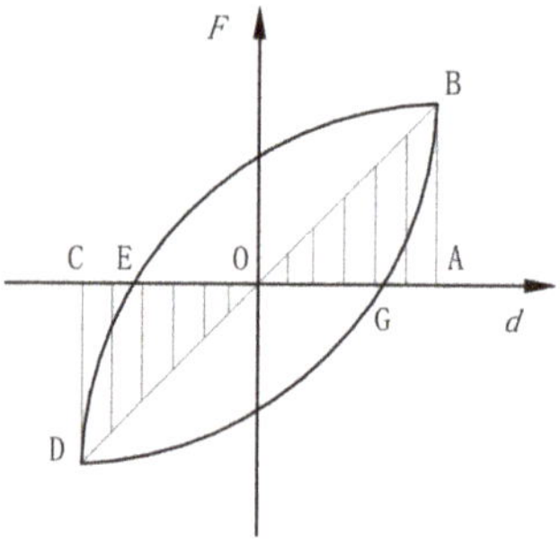

Figure 11. Calculation of the energy dissipation coefficient.

For all of the test specimens, the equivalent damping coefficient during the loading-unloading process is presented in Figure 12. The equivalent damping coefficient of all of the specimens increases as the applied displacement increases. For the specimens with different loading rates, the equivalent damping coefficient of specimen SCB-18-0 is higher than that of specimens SCB-12-0, SCB-24-0, and SCB-36-0; this is mainly caused by the hysteresis area and energy dissipation value. Compared to specimen SCB-12-0, SCB-12-25, SCB-12-50, and SCB-12-100 have a larger maximum axial force F and nearly the same energy dissipation value E_i, resulting in a lower equivalent damping coefficient as the initial strain increases.

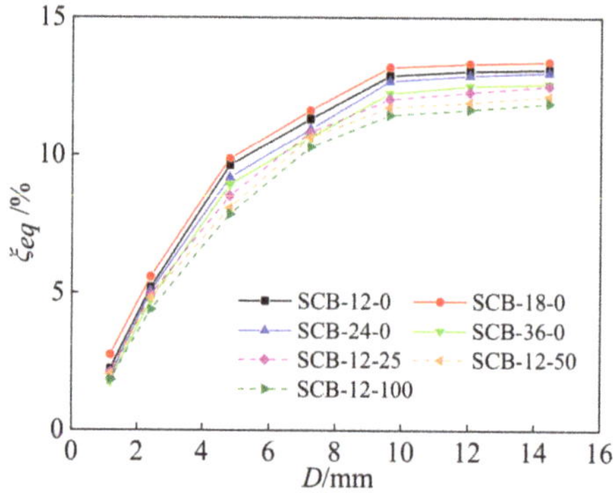

Figure 12. Equivalent damping coefficient of self-centering SMA brace specimens.

3.6. Self-Centering Capacity Ratio

The proposed self-centering SMA brace is mainly composed of a slip component and re-centering component, and consequently, the axial force, F_{SCB}, can be written by:

$$F_{\rm SCB} = F_{\rm SMA} + F_{slip} \tag{3}$$

where F_{SMA} and F_{slip} are the forces of the SMA wires and slip, as shown in Figure 13.

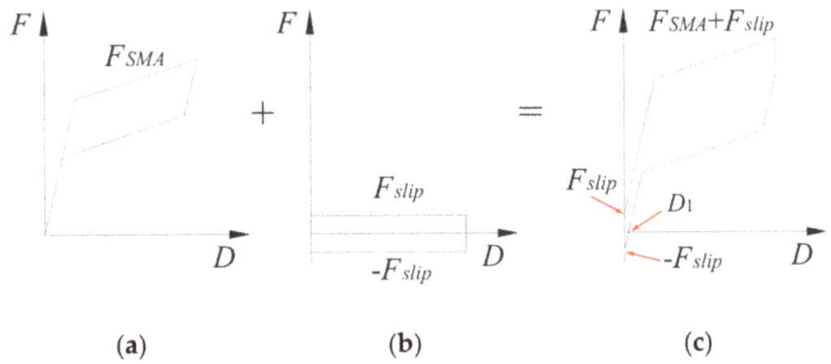

Figure 13. (**a**) SMA, (**b**) slip component, and (**c**) self-centering SMA brace force–displacement curve of self-centering SMA brace.

The self-centering capacity ratio, δ, is an important parameter that can be used to evaluate the re-centering capacity of self-centering SMA braces and can be calculated as follows:

$$\delta = \frac{D - D_1}{D} \quad (4)$$

where D is the maximum applied displacement, and D_1 is the residual displacement.

Based on the test results in Section 3.1, the maximum $F_{u,SCB}$, F_{slip}, F_{SMA}, D, D_1, and δ of each specimen can be concluded and are shown in Table 5. The F_{slip} was obtained by the vertical force of the rectangular loops around the origin point in Figure 7, and the F_{SMA} is equal to the $F_{u,SCB}$ minus the F_{slip}. By increasing the loading rate, the $F_{u,SCB}$ and δ increased, while the F_{slip} and D_1 decreased. In addition, a larger initial strain also resulted in a greater $F_{u,SCB}$ and larger δ at the maximum displacement, and yet both the F_{slip} and D_1 show opposite trends. Moreover, it should be noted that the maximum self-centering capacity ratio is 89.38%, showing that the self-centering SMA brace has an excellent re-centering capacity.

Table 5. Performance indices of self-centering SMA braces.

Specimen	$F_{u,SCB}$/kN	F_{slip}/kN	F_{SMA}/kN	D/mm	D_1/mm	δ/%
SCB-12-0	31.08	1.56	29.52	14.40	1.97	86.32
SCB-18-0	31.49	1.52	29.97	14.40	1.90	86.81
SCB-24-0	31.75	1.46	30.29	14.40	1.87	87.01
SCB-36-0	32.23	1.34	30.89	14.40	1.82	87.36
SCB-12-25	33.21	1.76	31.45	14.40	1.89	86.88
SCB-12-50	36.80	2.15	34.65	14.40	1.69	88.26
SCB-12-100	42.24	2.83	39.39	14.40	1.53	89.38

4. Numerical Results

The SIMULINK toolbox from MATLAB was used to simulate the seismic performance of the self-centering SMA brace, and the numerical and test results will be compared with the same conditions.

4.1. Numerical Model of Self-Centering SMA Brace

In Equation (3), the force of the SMA wires can be expressed by:

$$F_{SMA} = \sigma_{SMA} A_s \quad (5)$$

where σ_{SMA} and A_s are the stress and cross-sectional area of the SMA wires, respectively.

According to the improved Graesser and Cozarelli model by Graesser [24] and Qin [21], the σ_{SMA} can be calculated as follows:

$$\dot{\sigma} = E\left[\dot{\varepsilon} - |\dot{\varepsilon}|\left(\frac{\sigma-\beta}{Y}\right)^{n-1}\left(\frac{\sigma-\beta}{Y}\right)\right] \quad (6)$$

$$\beta = E\alpha\left\{\varepsilon_{in} - \frac{\sigma}{E} + f_T|\varepsilon|^c erf(a\varepsilon)[u(-\varepsilon\dot{\varepsilon})] + f_M[\varepsilon - \varepsilon_{M_f}\mathrm{sgn}(\varepsilon)]^m[u(\varepsilon\dot{\varepsilon})][u(|\varepsilon| - \varepsilon_{M_f})]\right\} \quad (7)$$

where ε, E, and Y are the strain, elastic modulus, and yield stress of the SMA wire, respectively; n is a constant controlling the sharpness of the transition from the elastic state to the phase transformation; β is the one-dimensional back stress; α is equal to $E_y/(E-E_y)$; E_y is the slope of stress–strain curve in the plastic range; ε_{in} is the inelastic strain; f_T, a, and c are the material constant controlling the type and size of the hysteresis, the amount of elastic recovery, and the slope of the unloading stress plateau, respectively; ε_{Mf} is the Martensite finish transformation strain; and f_M and m are the constants controlling the Martensite hardening curve.

The error function *erf* (x), Heaviside function u (x), and sigh function sign (x) can be expressed as [21]:

$$erf(x) = \frac{2}{\sqrt{x}}\int_0^x e^{-t^2}dt \quad (8)$$

$$u(x) = \begin{cases} 1 & (x \geq 0) \\ 0 & (x < 0) \end{cases} \quad (9)$$

$$\mathrm{sgn}(x) = \begin{cases} +1 & x > 0 \\ 0 & x = 0 \\ -1 & x < 0 \end{cases} \quad (10)$$

Therefore, the numerical model of the SMA wires will be revealed in detail using Equations (7) and (8) during the numerical analysis.

To accurately simulate the mechanical properties of the slip component, the Bouc-Wen model [25] is presented and described by:

$$F_{slip} = \lambda kd + (1-\lambda)kD_yZ \quad (11)$$

where d, k, D_y, λ, and Z are the deflection, initial stiffness, yield displacement, ratio of plastic and elastic stiffness, and non-dimensional displacement, respectively.

The first-order non-dimensional displacement equation yields can be expressed by:

$$\dot{Z}D_y = -\gamma\left|\dot{d}\right|Z|Z|^{\eta-1} - \beta\dot{d}|Z|^{\eta} + \theta\dot{d} \quad (12)$$

where γ, β, and θ are the parameters to control the shape and size of the hysteresis curve, and η is a scalar value to govern the smoothness of the transition from the elastic stage to the plastic stage.

In addition, according to the material properties of the SMA wires in Section 2.2.1 and the test results in Section 3, the model parameters of the SMA wire and slip model can be determined and are listed in Table 6.

Table 6. Determined the parameters of the SMA wire and slip model.

SMA Wire		Slip Component
L = 240mm	n = 3	λ = 0.0001, η = 4
A = 43.96 mm^2	ε_{Mf} = 0.04	k = 12000
α = 0.019	f_m = 42500	D_y = 0.1
m = 3	m = 3	γ = 0.5
c = 0.001	a = 240	β = 0.5
f_t = 0.79		θ = 0.9

4.2. Comparison of Test and Numerical Results

4.2.1. Loading Rate

The hysteresis curves of the numerical results were simulated using the improved Graesser model and Bouc–Wen model, which need to combine with the main parameters in Tables 2 and 6. Figure 14 displays the comparison of the test and numerical results under different loading rates. At the applied displacements of 2.40 mm and 4.80 mm for specimens SCB-12-0, SCB-18-0, SCB-24-0, and SCB-36-0, a great difference in the hysteresis curves between the test and numerical results can be observed, which can be explained by the existing errors between the slip bolt and slip hole in the test model. However, both the test and numerical hysteresis curves have almost the same initial stiffness for all of the specimens at the different loading rates. As the applied displacement increases, the hysteresis curves between the test and numerical results for each specimen are very close, and only a small error exists near σ_s^{MA}, which is the same as the error for the SMA wire at the different loading rates [22].

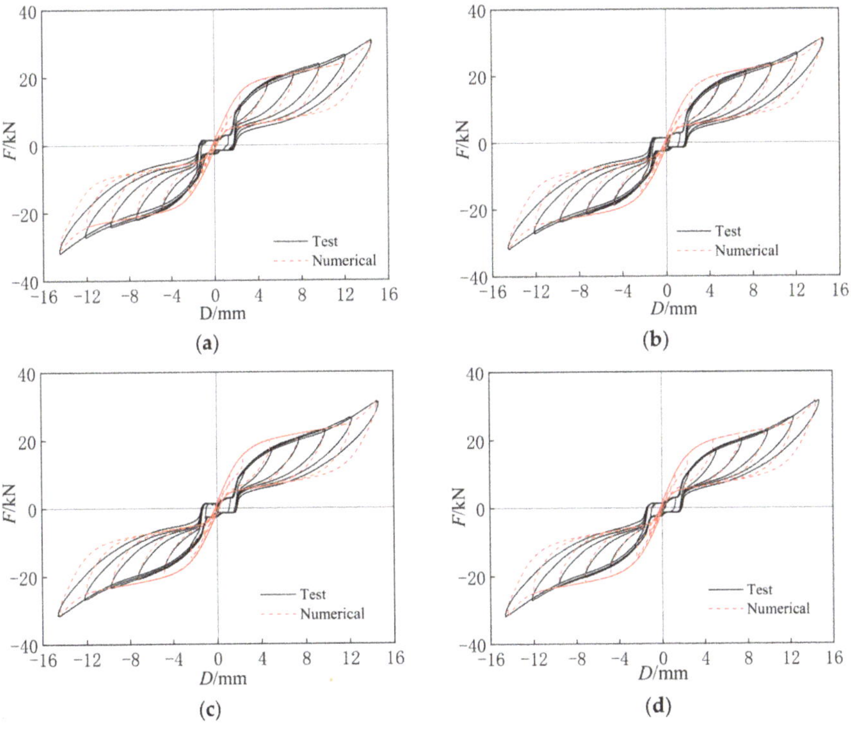

Figure 14. (**a**) SCB-12-0; (**b**) SCB-18-0; (**c**) SCB-24-0; (**d**) SCB-36-0. Comparison between test and numerical hysteresis curves under different loading rates.

For the different loading rates, the secant stiffness and energy dissipation coefficients of the specimens obtained from the test and numerical results with the applied displacement values ranging from 2.40 mm to 14.40 mm are shown in Tables 7 and 8. At the displacements of 2.40 mm and 4.80 mm, the maximum error between the test and numerical secant stiffness is 31.76%. Meanwhile, the maximum error is also 56.63% for the energy dissipation coefficient. The main reason for the error may be due to the difference between the slip bolt and hole in the test brace. As the displacement increases from 7.20 mm to 14.40 mm, the maximum errors in the secant stiffness and energy dissipation coefficient between the test and numerical results are 5.68% and 8.02%, respectively, showing good accuracy. Based on the above analysis, the presented numerical model determined from Equations (3)–(12) can be used to simulate the seismic performance of self-centering SMA braces with different loading rates.

Table 7. Comparison of secant stiffness under different loading rates.

Strain Amplitude/%	Loading Rate											
	0.0012/s			0.0018/s			0.0024/s			0.0036/s		
	Tes.	Num.	Error/%	Tes.	Num.	Error/%	Tes.	Num.	Error/%	Tes.	Num.	Error/%
1	4.87	6.42	31.76	5.28	6.48	22.87	5.40	6.53	20.88	5.70	6.58	15.45
2	3.63	4.09	12.68	3.74	4.17	11.47	4.07	4.18	2.77	4.20	4.19	0.11
3	2.77	2.85	2.66	2.98	3.02	1.37	3.22	3.10	3.63	3.37	3.29	2.32
4	2.39	2.26	5.68	2.47	2.42	2.28	2.60	2.54	2.55	2.73	2.66	2.45
5	2.32	2.24	3.28	2.39	2.30	3.74	2.47	2.42	2.08	2.54	2.49	2.10
6	2.30	2.21	4.23	2.37	2.29	3.65	2.41	2.32	3.39	2.47	2.41	2.22

Table 8. Comparison of energy dissipation coefficient under different loading rates.

Strain Amplitude/%	Loading Rate											
	0.0012/s			0.0018/s			0.0024/s			0.0036/s		
	Tes.	Num.	Error/%	Tes.	Num.	Error/%	Tes.	Num.	Error/%	Tes.	Num.	Error/%
1	8.76	3.94	54.95	9.11	3.98	56.35	9.19	3.99	56.55	9.27	4.02	56.63
2	37.00	42.03	13.59	37.54	43.00	14.56	38.07	43.25	13.62	38.14	43.50	14.05
3	88.33	95.08	7.64	91.83	99.19	8.02	93.43	100.78	7.87	95.31	102.77	7.83
4	164.75	173.86	5.53	167.97	178.94	6.53	170.15	179.28	5.36	172.06	182.42	6.02
5	246.25	257.97	4.76	247.37	259.30	4.82	248.67	261.68	5.23	251.25	266.05	5.89
6	317.37	327.92	3.32	320.81	334.75	4.35	310.73	319.38	2.79	291.13	296.80	1.95

Note. Error = (Numerical − Test)/Test: 'Tes.' and 'Num' denote the test and numerical results, respectively.

4.2.2. Initial Strain

Figure 15 shows the comparison between the test and numerical results under different initial strains at the applied displacements ranging from 2.40 mm to 14.40 mm. At the displacements of 2.40 mm and 4.80 mm for specimens SCB-12-0, SCB-12-25, SCB-12-50, and SCB-12-100, an obvious contrast between the test and numerical results can be observed, and the cause of the contrast is the same as it is with the loading rates. As the displacement increased from 7.20 mm to 14.40 mm, the hysteresis curves of the specimens obtained by the test results all show close agreement with the numerical results.

The secant stiffness and energy dissipation coefficient of the specimens calculated from test and numerical results are shown for the different initial strains in Tables 9 and 10. At the displacement of 2.40 mm, the maximum errors of the secant stiffness and energy dissipation coefficient between the test and numerical results are 54.95% and 31.76%, respectively. As the strain amplitude increases, the numerical results of the secant stiffness and energy dissipation coefficient for all of the specimens become gradually closer to the test results, which have a maximum error of 7.64%, showing great agreement. Therefore, the proposed numerical model can also be used to analyze the seismic performance of the self-centering SMA braces with different initial strains.

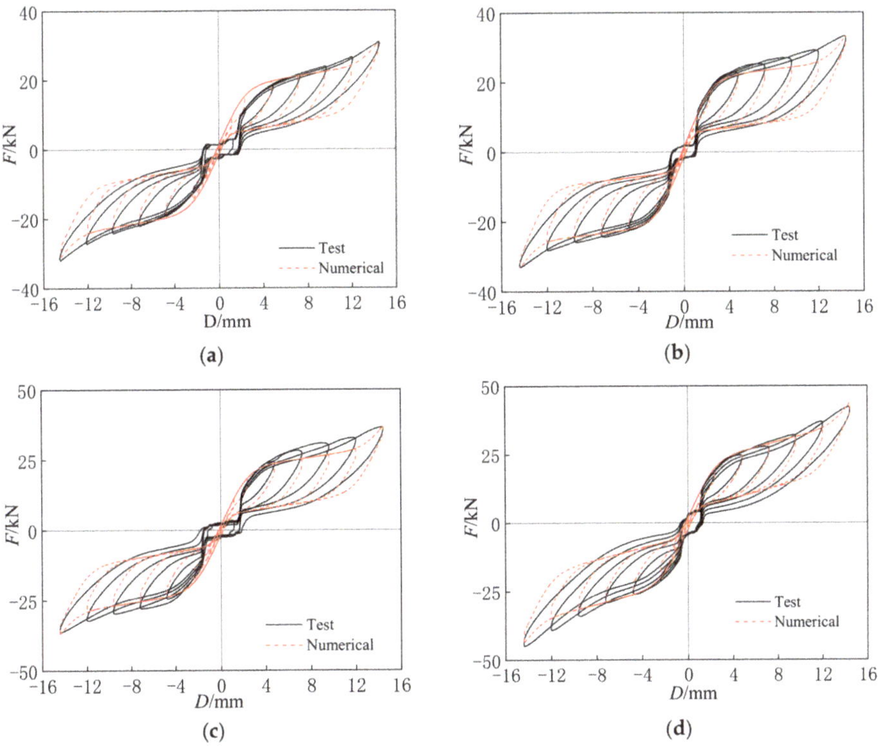

Figure 15. (**a**) SCB-12-0; (**b**) SCB-12-25; (**c**) SCB-12-50; (**d**) SCB-12-100. Comparison between test and numerical hysteresis curves under different initial strains.

Table 9. Comparison of energy dissipation capacity under different initial strains.

Strain Amplitude/%	Initial Strain/%											
	0			0.25			0.5			1		
	Tes.	Num.	Error/%	Tes.	Num.	Error/%	Tes.	Num.	Error/%	Tes.	Num.	Error/%
1	8.76	3.94	54.95	14.25	9.17	35.65	17.96	10.47	41.71	21.29	13.77	35.33
2	37.00	42.03	13.59	56.49	60.06	6.325	62.61	66.37	6.00	69.20	73.22	5.80
3	88.33	95.08	7.64	117.98	124.16	5.23	124.84	130.97	4.91	143.83	150.35	4.53
4	164.75	173.86	5.53	186.16	194.60	4.54	198.96	207.10	4.09	212.85	220.99	3.82
5	246.25	257.97	4.76	256.61	266.45	3.83	267.08	276.25	3.43	275.00	283.34	3.03
6	317.37	327.92	3.32	318.48	328.74	3.22	341.65	350.30	2.53	342.98	351.32	2.43

Table 10. Comparison of secant stiffness under different initial strains.

Strain Amplitude/%	Initial Strain/%											
	0			0.25			0.5			1		
	Tes.	Num.	Error/%	Tes.	Num.	Error/%	Tes.	Num.	Error/%	Tes.	Num.	Error/%
1	4.87	6.42	31.76	7.66	7.31	4.49	7.98	7.69	3.70	8.70	8.28	4.78
2	3.63	4.09	12.68	5.12	4.90	4.36	5.48	5.11	6.83	5.76	5.42	5.83
3	2.77	2.85	2.66	3.73	3.63	2.69	3.94	3.78	3.93	4.26	4.11	3.49
4	2.39	2.26	5.68	2.98	2.85	4.16	3.43	3.28	4.44	3.62	3.47	4.10
5	2.32	2.24	3.28	2.56	2.45	4.11	2.89	2.75	4.71	3.30	3.17	3.90
6	2.30	2.21	4.23	2.42	2.35	3.09	2.66	2.58	3.18	3.11	3.05	2.13

5. Conclusions

In this paper, the cyclic loading test and numerical analysis were studied to carry out an innovative self-centering brace with the effect of the loading rate and initial strain, and the seismic performance of the brace was investigated. The following conclusions can be obtained:

(1) The proposed self-centering SMA brace has an excellent energy dissipation capacity, ductility, and self-centering capacity. The hysteresis curve can be idealized as a flag-shape with small residual deformation, and the self-centering capacity ratio reached 89.38%.

(2) Loading rate: As the loading rate increased, the ultimate axial force and secant stiffness coefficient increased gradually, the residual deformation decreased slightly, and the maximum E_i, and the equivalent damping coefficient at each test case was able to be observed at the loading rate of 0.0018 s^{-1}.

(3) Initial strain: The ultimate axial force, secant stiffness coefficient, and energy dissipation coefficient indeed increased gradually as the initial strain increased from 0 to 0.01, but the maximum equivalent damping coefficient of the self-centering SMA brace appears at the initial strain of 0.

(4) The improved numerical model combined with the Graesser model and Bouc–Wen model can be used to analyze the seismic performance of self-centering SMA braces with different loading rates and initial strains, and the numerical results are consistent with the test results under the same conditions.

Author Contributions: Conceptualization, Y.J., S.H. and S.Z.; methodology, B.Z., S.H. and S.Z.; software, F.T. and S.H.; validation, B.Z., S.Z. and S.H.; formal analysis, Y.J., B.Z., F.T. and S.H.; writing—original draft preparation, B.Z., S.Z. and F.T.; writing—review and editing, Y.J. and S.H.; supervision, Y.J. and S.H.; project administration, Y.J., S.Z. and S.H.; funding acquisition, Y.J., S.Z. and S.H. and W.C. All authors have read and agreed to the published version of the manuscript.

Funding: The authors gratefully acknowledge the research grant provided by the National Nature Science Foundation of China (No. 51908268, 51268044, 51878108).

Institutional Review Board Statement: Not applicable.

Informed Consent Statement: Not applicable.

Data Availability Statement: The data used to support the findings of this study are available from the corresponding author upon request.

Acknowledgments: The authors gratefully acknowledge the support by Guoyang Guan from Engineering Mechanics Experiment Center, Nanchang University during the tests conducted for this study.

Conflicts of Interest: The authors confirm that there are no conflicts of interested associated with this article.

References

1. Ding, Y.; Zhao, C. Cyclic tests for assembled X-shaped buckling restrained brace using two unconnected steel plate braces. *J. Constr. Steel Res.* **2021**, *182*, 106680. [CrossRef]
2. Hu, S.; Zeng, S.; Xiong, J.; Wang, X.; Zhou, Q.; Xiong, X. Seismic Analysis and Evaluation of Y-shaped EBF with an Innovative SSL-SSBC. *Int. J. Steel Struct.* **2020**, *20*, 1026–1039. [CrossRef]
3. Lian, M.; Cheng, Q.; Su, M.; Guan, B.; Zhang, H. Seismic performance of high-strength steel framed-tube structures with web-bolted double-channel shear links. *Struct. Des. Tall Spec. Build.* **2021**, *30*, e1888. [CrossRef]
4. Lian, M.; Li, S.; Li, X.-L. Seismic Story Shear Distribution Based on Inelastic State of Eccentrically Braced Frame with Vertical Link Composite High Strength Steel. *Int. J. Steel Struct.* **2021**, *21*, 1260–1279. [CrossRef]
5. Xie, Q.; Zhou, Z.; Meng, S.-P. Experimental investigation of the hysteretic performance of self-centering buckling-restrained braces with friction fuses. *Eng. Struct.* **2019**, *203*, 109865. [CrossRef]
6. Wang, W.; Fang, C.; Zhao, Y.; Sause, R.; Hu, S.; Ricles, J. Self-centering friction spring dampers for seismic resilience. *Earthq. Eng. Struct. Dyn.* **2019**, *48*, 1045–1065. [CrossRef]
7. Zhang, A.; Ye, Q. Design and testing of prefabricated steel frame with an innovative re-centering energy dissipative brace. *Eng. Struct.* **2019**, *201*, 109791. [CrossRef]
8. Li, J.; Wang, W.; Cao, Z. Self-centering hybrid dampers for improving seismic resilience. *Eng. Struct.* **2021**, *244*, 112829. [CrossRef]
9. Hu, S.-J.; Gu, Q.; Jiang, J.; Song, G.-Q.; Xiong, J.-G. Hysteresis performance and simplified mechanical model of an innovative self-centering SMA brace. *J. Build. Eng.* **2020**, *41*, 66–75. [CrossRef]
10. Lin, C.; Wang, Z.; Yang, X.; Zhou, H. Experimental Study on Temperature Effects on NiTi Shape Memory Alloys under Fatigue Loading. *Materials* **2020**, *13*, 573. [CrossRef]
11. Xue, S.-D.; Wang, L.; Zhuang, P. Design and performance study of a SMA incorporated friction damaper. *World Earthq. Eng.* **2006**, *22*, 1–6. [CrossRef]
12. Qiu, C.-X.; Zhu, S. Performance-based seismic design of self-centering steel frames with SMA-based braces. *Eng. Struct.* **2017**, *130*, 67–82. [CrossRef]
13. Xu, X.; Tu, J.; Cheng, G.; Zheng, J.; Luo, Y. Experimental study on self-centering link beams using post-tensioned steel-SMA composite tendons. *J. Constr. Steel Res.* **2019**, *155*, 121–128. [CrossRef]
14. Fang, C.; Yam, M.C.H.; Lam, C.C.; Xie, L. Cyclic performance of extended end-plate connections equipped with shape memory alloy bolts. *J. Constr. Steel Res.* **2014**, *94*, 122–136. [CrossRef]
15. Fang, C.; Wang, W.; He, C.; Chen, Y. Self-centring behaviour of steel and steel-concrete composite connections equipped with NiTi SMA bolts. *Eng. Struct.* **2017**, *150*, 390–408. [CrossRef]
16. Hu, S.-J.; Gu, Q.; Jiang, G.-Q.; Xiong, J.-G. Experimental study on seismic performance for an innovative self-centering SMA brace. *Eng. Mech.* **2021**, *38*, 109–118. [CrossRef]
17. Li, H.-N.; Liu, M.-M.; Fu, X. An innovative re-centering SMA-lead damper and its application to steel frame structures. *Smart Mater. Struct.* **2018**, *27*, 075029. [CrossRef]
18. Fan, X.; Xu, L.; Li, Z. Seismic performance evaluation of steel frames with pre-pressed spring self-centering braces. *J. Constr. Steel Res.* **2019**, *162*, 105761. [CrossRef]
19. Zhou, H.; Qi, S.; Yao, G.; Zhou, L.; Sun, L.; Xing, F. Damping and frequency of a model cable attached with a pre-tensioned shape memory alloy wire: Experiment and analysis. *Struct. Control Health Monit.* **2017**, *25*, e2106. [CrossRef]
20. Yan, S.; Wang, Q.; Wang, W. Experimental research on mechanical performance for pseudo-elasticity of shape memory alloy. *J. Shenyang Jianzhu Univ.* **2010**, *26*, 458–463.
21. Qian, H.; Li, H.-N.; Song, G.-B. Constitutive model of shape memory alloy based on plastic theory: Experiment and simulation. *J. Funct. Mat.* **2007**, *38*, 1114–1118. [CrossRef]
22. Zhang, B.; Zeng, S.; Tang, F.; Hu, S.; Zhou, Q.; Jia, Y. Experimental and Numerical Analysis of the Mechanical Properties of a Pretreated Shape Memory Alloy Wire in a Self-Centering Steel Brace. *Processes* **2021**, *9*, 80. [CrossRef]
23. *JGJ 101–2015*; National Standard of People's Republic of China. Specification for Test Methods of Seismic Buildings. Architecture Industrial Press of China: Beijing, China, 2015.
24. Graesser, E.J.; Cozzarelli, F.A. Shape-Memory Alloys as New Materials for Aseismic Isolation. *J. Eng. Mech.* **1991**, *117*, 2590–2608. [CrossRef]
25. Triantafyllou, S.P.; Koumousis, V.K. Bouc-Wen Type Hysteretic Plane Stress Element. *J. Eng. Mech.* **2012**, *138*, 235–246. [CrossRef]

materials

MDPI

Article

The Seismic Performance of New Self-Centering Beam-Column Joints of Conventional Island Main Buildings in Nuclear Power Plants

Qiang Pei [1,*], Cong Wu [1], Zhi Cheng [2], Yu Ding [1] and Hang Guo [1]

1 College of Architectural Engineering, Dalian University, Dalian 116622, China; wucforever@163.com (C.W.); wateryu147@163.com (Y.D.); gh897422078@163.com (H.G.)
2 Department of Construction Engineering Management, Southwest Jiaotong University Hope College, Chengdu 610400, China; czh9182021@163.com
* Correspondence: peiqiang@dlu.edu.cn; Tel.: +86-189-4093-2972

Abstract: In order to improve the deformation energy consumption and self-centering ability of reinforced concrete (RC) frame beam-column joints for main buildings of conventional islands in nuclear power plants, a new type of self-centering joint equipped with super-elastic shape memory alloy (SMA) bars and a steel plate as kernel components in the core area of the joint is proposed in this study. Four 1/5-scale frame joints were designed and manufactured, including two contrast joints (a normal reinforced concrete joint and a concrete joint that replaces steel bars with SMA bars) and two new model joints with different SMA reinforcement ratios. Subsequently, the residual deformation, energy dissipation capacity, stiffness degradation and self-centering performance of the novel frame joints were studied through a low-frequency cyclic loading test. Finally, based on the OpenSees finite element software platform, an effective numerical model of the new joint was established and verified. On this basis, varying two main parameters, the SMA reinforcement ratio and the axial compression ratio, a simulation was systematically conducted to demonstrate the effectiveness of the proposed joint in seismic performance. The results show that replacing ordinary steel bars in the beam with SMA bars not only greatly reduces the bearing capacity and stiffness of the joint, but also makes the failure mode of the joint brittle. The construction of a new type of joint with consideration of the SMA reinforcement and the steel plate can improve the bearing capacity, delay the stiffness degradation and improve the ductility and self-centering capability of the joints. Within a certain range, increasing the ratio of the SMA bars can further improve the ultimate bearing capacity and energy dissipation capacity of the new joint. Increasing or decreasing the axial compression ratio of column ends has little effect on the overall seismic performance of new joints.

Keywords: shape memory alloy; self-centering; beam-column joints; seismic performance

Citation: Pei, Q.; Wu, C.; Cheng, Z.; Ding, Y.; Guo, H. The Seismic Performance of New Self-Centering Beam-Column Joints of Conventional Island Main Buildings in Nuclear Power Plants. *Materials* **2022**, *15*, 1704. https://doi.org/10.3390/ma15051704

Academic Editor: Marco Corradi

Received: 27 December 2021
Accepted: 21 February 2022
Published: 24 February 2022

1. Introduction

At present, nuclear power plants have adopted higher seismic design criteria according to the seismic hazard evaluation of the site. For example, the seismic design criteria adopted by Tianwan, Taishan, and Haiyang nuclear power plants in China are 0.2 g, 0.25 g, and 0.3 g, respectively, which are higher than the local seismic level [1]. The safety of domestic and foreign nuclear power plants using this kind of seismic design standard has been verified and affirmed in previous conventional earthquakes. However, for the super-design reference earthquake, the current response measures are mainly to improve the seismic isolation design of nuclear power engineering structures and to analyze and evaluate the nuclear power system through seismic margin assessment and seismic probabilistic risk assessment [2,3]. The above methods cannot be separated from the judgment of empirical data, and the evaluation results have great uncertainty. Especially in the case of two or more unknown super-benchmark accidents, such as accidental damage to buildings adjacent to the nuclear island and secondary disasters caused by beyond-design basis earthquakes, as in the Fukushima nuclear accident in 2011, such a beyond-design basis earthquake may still pose a greater security threat to nuclear power engineering. However, the structural design considering the superposition of multiple disasters, such as earthquakes and tsunamis, is neither economical nor convenient. On the other hand, under the requirements of the current economic sustainable development, the research on seismic engineering has gradually developed from seismic isolation to the direction of recoverable function [4]. In this trend, higher requirements are also put forward for major projects represented by nuclear power projects that are related to the national economy, people's livelihoods and the national economic lifeline, so as to achieve the seismic goal of the function not being interrupted or restored as soon as possible during the earthquake, and the normal use can be achieved without repair or with only slight repair after the earthquake. At the same time, a large number of seismic damage investigation results [5–9] show that beam-column joints, as an important hub for coordinating deformation and transfer load distribution in frame structure systems, are also one of the most seriously damaged parts, especially in related nuclear power engineering frame structure buildings under beyond-design basis earthquake. The traditional seismic design improvement method of concrete beam-column joints is mainly strengthening stirrups or using high strength concrete [10–13]; at the same time, it brings about a substantial increase in the construction cycle and cost and does not break through the performance limitations of traditional building materials. It is difficult to meet the requirements of structural recoverable functions. Concrete-filled steel tubular columns can improve the bearing capacity and seismic performance of the structure to a certain extent, and in recent years, some progress has been made in the experimental and theoretical research of these columns [14–17], but their practical application in nuclear power engineering is still rare. At the same time, the super-elastic shape memory alloy (SMA) has been rapidly developed and applied due to its special material functional properties [18–22]. It also provides a new idea for improving safety in nuclear power frame structure engineering under the action of beyond-design basis earthquakes.

The shape memory alloy (SMA) is a new intelligent material that takes into account sensing and driving functions. When the external force is unloaded, the inverse phase transformation drive can automatically restore the strain up to 8%–10% instantaneously [23]. At present, it has been widely studied and applied in energy dissipation braces [24,25], isolation bearing [26,27], and various dampers [28–30]. Based on these spontaneous and instantaneous recoverable super-elastic characteristics, the improved replacement between SMA bars and ordinary steel bars in conventional concrete beams and columns provides a new research and design idea for improving the mechanical properties of concrete beam-column joints. The research on the seismic performance of RC beam-column joints mainly explores the influence of other structural members, such as wide beams and slabs [31,32], and reinforcement methods, such as FRP and BFRP reinforcement [33–36]. However, there are few experimental studies on the seismic performance of new self-centering concrete beam-column joints based on SMA tendons [37–39], and the design structure of the new joints in the related research reported is relatively larger than that of the traditional concrete joints, which lacks the simplicity and practicability suitable for actual construction.

Based on this, in order to further promote the application of SMA reinforcement in the field of structural engineering and optimize the energy consumption and self-recovery ability of important concrete frame engineering structures, considering that the frame structure edge joints in earthquake damage are often more serious than the internal joints, our research group designed a new type of self-centering and low-damage joint with the conventional frame edge joint of conventional island main buildings in nuclear power plants. Different from the previous related self-centering nodes, in order to enhance the application feasibility, the structural design is closer to the traditional steel binding process. It is proposed to study the failure process, hysteretic characteristics, energy dissipation capacity, stiffness degradation, and self-centering capacity of the new joint through experiment and numerical simulation, so as to provide a certain basis for the practical application of the new joint in important frame structures such as conventional island main buildings in nuclear power plants.

2. Experimental Program

2.1. Test Specimens

Four frame beam-to-column joints with a scale ratio of 1:5 were designed and manufactured, including two self-centering new beam-to-column joint models numbered by PSJD-1 and PSJD-2 (ordinary longitudinal reinforcement plus hybrid joints with different diameters of SMA reinforcement) and two comparative joints (ordinary reinforced concrete beam-to-column joints numbered by PJD-1, SMA reinforced beam-to-column joints numbered by SJD-1). The geometric dimensions of each specimen were the same and were made according to the current concrete design specifications in China. The fixed steel plate required in this experiment not only played the role of connecting shape memory alloy rods, but also played the role of a longitudinal reinforcement elbow. In order to reduce the adverse effect on the anchorage connection end when the joint was damaged and cracked, two steel plates were placed 50 mm away from the outer edge of the beam and the column, respectively. The size reinforcement and specific parameters of the specimen are shown in Figure 1 and Table 1. The steel end plate of the new joint has reserved holes for SMA bars and ordinary steel bars to pass through. The SMA bars and the steel end plate were connected by bolt anchorage, and the ordinary steel bar and the steel end plate were welded. The connection structure is shown in Figure 2.

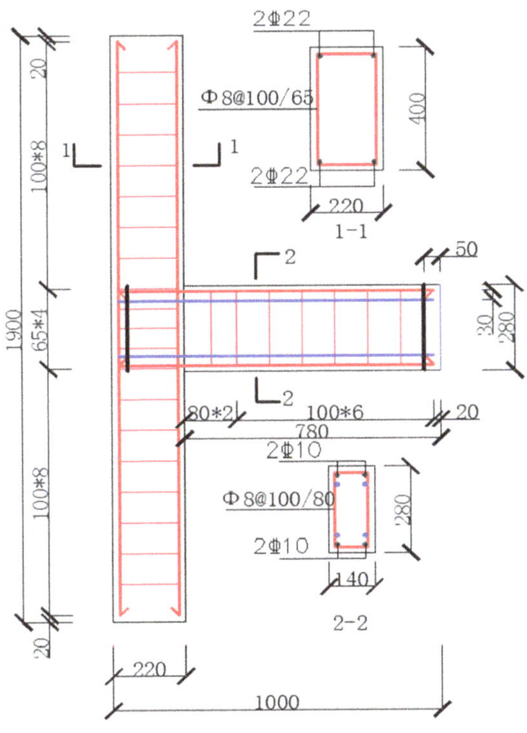

Figure 1. Specimen size and reinforcement details (unit: mm).

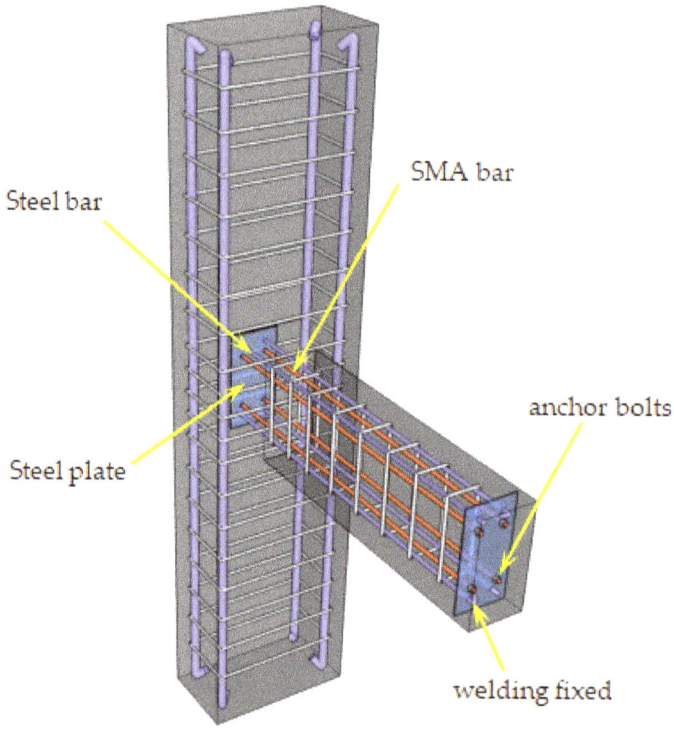

Figure 2. Three dimensional (3D) sketch of the self-centering joint.

Table 1. Basic parameters of specimens.

Specimen Number	Column Reinforcement		Beam Reinforcement		Axial Compression Ratio	Test Purpose
	Longitudinal Reinforcement	Hooped Reinforcement	Longitudinal Reinforcement	SMA Reinforcement Ratio (%)		
PJD-1	4 × C22	A8@65/100	4 × C10	0	0.25	contrast test
SJD-1	4 × C22	A8@65/100	4 × SMA bars (10 mm)	0.800	0.25	contrast test
PSJD-1	4 × C22	A8@65/100	4 × C10 + 4 × SMA bars (8 mm)	0.513	0.25	model test
PSJD-2	4 × C22	A8@65/100	4 × C10 + 4 × SMA bars (10 mm)	0.800	0.25	model test

2.2. Materials

2.2.1. SMA Materials

The Ni–Ti shape memory alloy bar used in this test was customized in Baoji Long Qiangfeng Titanium Industry Co., Ltd (Baoji, China). The chemical composition was approximately as follows: Ni: 54.38%; Ti: 45.575%; others: 0.045%. The length and diameter of the fabricated dumbbell-shaped SMA bar specimens were 130 mm and 10 mm, respectively.

Studies have shown that [37,40] heat treatment can greatly improve the super-elasticity of SMA bars. The heat treatment process used in this experiment was as follows: the SMA specimens were placed in a high-temperature furnace with a constant temperature of 400 °C for 25 min, and the water was taken out immediately after the end. In order to further stabilize the mechanical properties of SMA bars, the specimens after heat treatment were placed in boiling water and ice water for 3 min, and the above cold and hot cycles were repeated five times. The cyclic tensile test of SMA bars was carried out according to the strain amplitude (1%, 2%, ..., 8%). The cyclic loading and unloading at all levels were carried out once, the strain rate was 0.0015 s^{-1}, and the test room temperature was 25 °C. The constant temperature heating furnace for heat treatment and the tensile test loading device are shown in Figure 3. The properties of the Ni–Ti alloy after heat treatment are shown in Table 2.

(a)

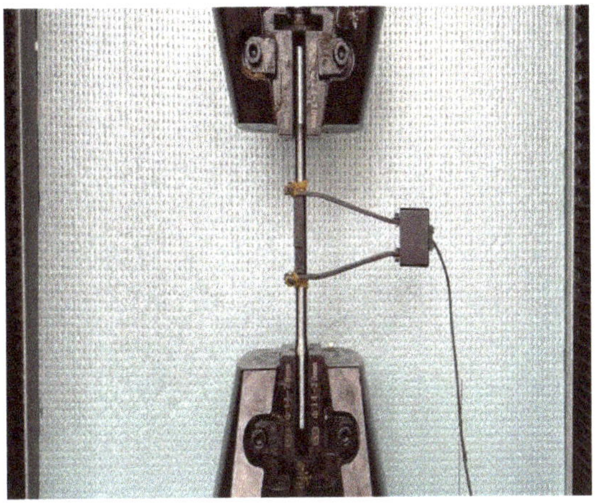

(b)

Figure 3. Heating and loading test device: (**a**) Thermostatic heating furnace; (**b**) SMA rod tensile test device.

Table 2. Material properties of the Ni–Ti alloy.

Material Name	Density ($kg{\cdot}m^{-3}$)	Young's Modulus (GPa)	Tensile Strength (MPa)	Yield Strength (MPa)	Restoration Strain (%)
Ni–Ti	7800	65.4	600	390	4.5

2.2.2. Steel Plate, Reinforced Steel, and Concrete Materials

The steel plate material used in this test were Q235 (ordinary carbon structural steel and had a nominal design yield strength of 235 MPa), and the specific size of the steel plate is shown in Figure 4. The beam column longitudinal steel bar selection model for HRB400 (its standard value of yield strength is 400 MPa), the stirrup selection of HPB300 (similar as HRB400, its standard value of yield strength is 300 MPa), and the mechanical properties of the steel parameters are shown in Table 3. The measured results of the compressive strength of the concrete cube (dimensions: 100 × 100 × 100 mm) test block are shown in Table 4.

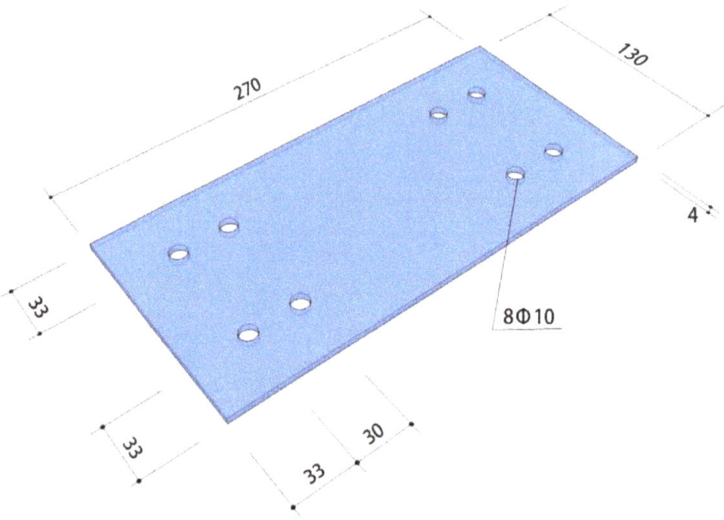

Figure 4. Dimensions of the steel plate (unit: mm).

Table 3. The performance of the reinforcement.

Type of Steel Bar	Steel Bar Diameter d (mm)	Yield Strength f_y (MPa)	Ultimate Strength f_u (MPa)	Young's Modulus (MPa)	Elongation Rate δ (%)
HPB300	6	310.67	460.56	1.90×10^5	19.12
	8	323.98	462.73	1.92×10^5	19.38
HRB400	10	448.56	601.09	2.03×10^5	19.96
	22	438.64	573.86	1.98×10^5	20.60

Table 4. Concrete test block measurement results.

Measurement Items of Concrete Test Block	First Group			Second Heat			End Value
	1	2	3	4	5	6	
Failing load (kN)	543.13	436.56	455.34	485.01	411.73	470.97	
Average value (kN)		445.95			455.90		450.93
Compression strength (MPa)		42.37			43.31		42.84

2.3. Load Equipment

In order to be close to the actual stress state of the joint, the steel box was set at the lower end of the column, the spherical steel hinge was welded on the steel box in the specimen fabrication process, and the spherical hinge with a loading jack was used at the upper end of the column. At the same time, considering the influence of the beam-column self-weight, the specimen was installed by the vertical beam of the column in the horizontal direction, and the lateral support of the specimen was realized by connecting the scaffolding with the reaction wall. The schematic diagram of the test loading device is shown in Figure 5. Figure 6 shows the scene photos of the loading test. The effect of temperature on the performance of the joint is difficult to achieve, so all the tests were carried out at room temperature.

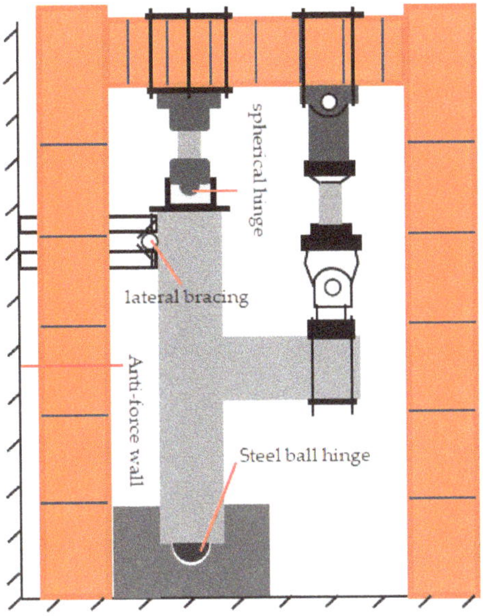

Figure 5. Schematic diagram of the test setup.

Figure 6. Photo of the test setup.

2.4. Layout of Test Points

The resistance strain gauges were affixed to the concrete surface of the longitudinal reinforcement, the SMA reinforcement, and the plastic hinge area of the beam-column member to measure the strain at the corresponding position. The strain gauges used on the steel bar inside the joint specimens were BX120–3AA, and those of the concrete surface were BX120–50AA. The strain gauge arrangement of the specimen is shown in Figure 7.

Considering that the overall size of the component in this test was relatively small, and there was an inevitable slight vibration in the process of low cyclic loading, in order to measure the plastic hinge length of each specimen more accurately, the high-precision non-contact full-field strain measurement system (video image correlate-3d, VIC-3D, Correlated Solutions, Inc., Columbia, SC, USA) based on digital image technology was selected to replace the conventional displacement meter for measurement. The VIC-3D system can measure the full-field three-dimensional displacement and strain of the surface of the object under test and has low environmental requirements. In principle, clear images of the area under the test can be taken indoors and outdoors. The accuracy and feasibility of digital image correlation technology and VIC-3D based on this technology have been proven [41–45].

The measurement range was drawn in the beam-column connection area. Speckles with a diameter not less than 2 mm were arranged in the measurement range, and the correction plate with a side length of 20 mm was selected to correct. The camera was set up in place and focus, and the vertical displacement value Y at any position in the measurement range can be measured by the measurement system. According to the calculation needs, seven measuring points were selected along the beam end in this experiment, and the distance between each measuring point was 60 mm. The measurement range size and the specific location of the selected measuring points are shown in Figure 8. The vertical displacement of each point is recorded as $Y_1, Y_2, \ldots, Y_6, Y_7$. If a plastic hinge is formed between the measuring points marked n ($1 \leq n \leq 7$) from the beam-column junction, the spacing of each measuring point after the measuring point along the beam length should be the initial 60 mm, and the vertical displacement of the next measuring point is denoted as Y_{n+1}, and so on. If $\Delta 1 = Y_{n+1} - Y_n$, $\Delta 2 = Y_{n+2} - Y_n, \ldots, \Delta K = Y_{n+K} - Y_n$ ($2 \leq K \leq 6$), then $\Delta 1$, $\Delta 2$, ΔK should satisfy Equation (1) as follows:

$$\sin\alpha = \frac{\Delta 1}{60} = \frac{\Delta 2}{2 \times 60} = \frac{\Delta 3}{3 \times 60} = \cdots = \frac{\Delta K}{n \times 60} \quad (1)$$

Here, α is an angle which is between the axis of the undeformed beam outside the end point of the plastic hinge and the horizontal direction, as shown in Figure 9. The n value of Equation (1) is the plastic hinge end point. Considering the inevitable measurement error, the absolute value of the sinusoidal difference obtained by Equation (1) is not greater than 1.0×10^{-4}. It can be approximately equal, and the solution schematic diagram is shown in Figure 9. The measured plastic hinge length calculated for each node is shown in Table 5.

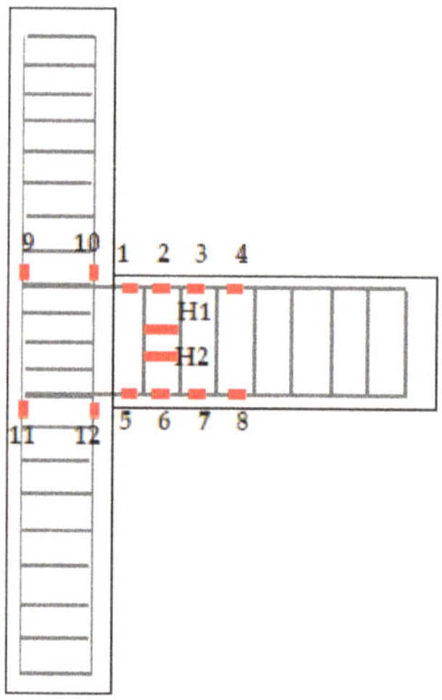

Figure 7. Strain gauge arrangement.

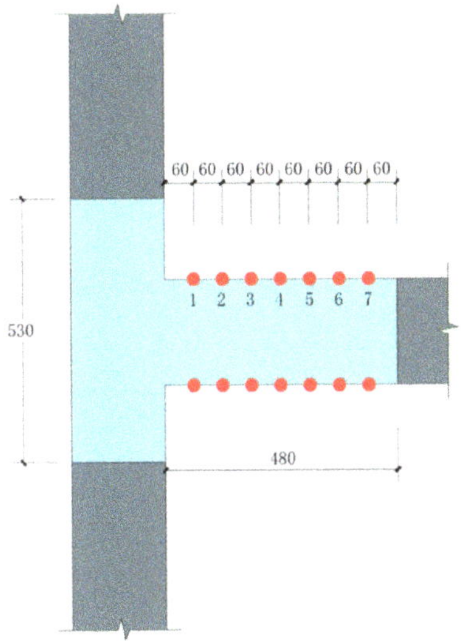

Figure 8. VIC-3D measuring area and the measuring point arrangement.

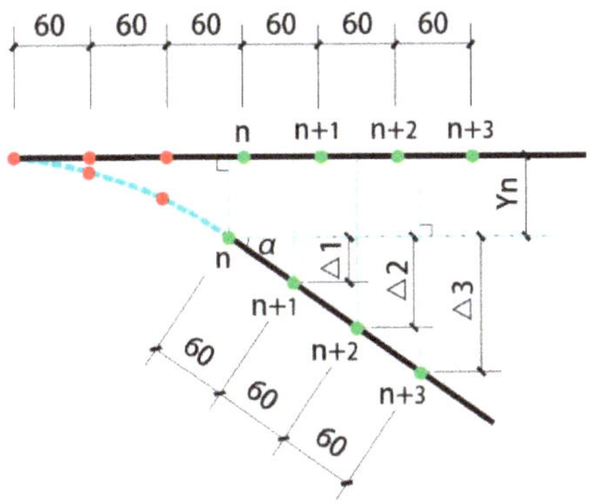

Figure 9. Schematic diagram of the plastic hinge length judgment principle (unit: mm).

Table 5. Measured and calculated value of the plastic hinge length of each node.

Node Number	PJD-1	SJD-1	PSJD-1	PSJD-2
Y1 (mm)	8.13	2.11	9.20	0.71
Y2 (mm)	9.59	3.82	11.37	2.19
Y3 (mm)	11.36	5.52	13.73	3.72
Y4 (mm)	13.15	7.22	15.63	5.26
Y5 (mm)	14.93	8.92	17.74	6.81
Y6 (mm)	16.66	10.63	19.89	8.35
Y7 (mm)	18.39	—	22.05	—
$\sin\alpha$ ($\times 10^{-2}$)	2.90	2.80	3.60	2.50
Plastic hinge end number	5	1	5	4
Plasticity hinge length d (mm)	300	60	300	240

2.5. Loading History

The vertical load was applied to the top of the column before cyclic loading, and the axial force was kept constant after the axial force was relatively stable. The experimental control of the axial compression ratio was 0.25, while the actual vertical load value was 445 kN. The low cyclic loading test of the joint specimen adopted the whole process displacement control method, and the loading rate was 0.2 mm/s. In this experiment, before the specimen yield, the increment of displacement grading loading was 1 mm, and each stage was recycled twice. After the specimen yield, the yield displacement was denoted as Δ, and it was then loaded by an integral multiple of Δ. Each stage was cycled twice. The loading was terminated when the load dropped to 85% of the ultimate load; otherwise, the component was damaged, as shown in Figure 10.

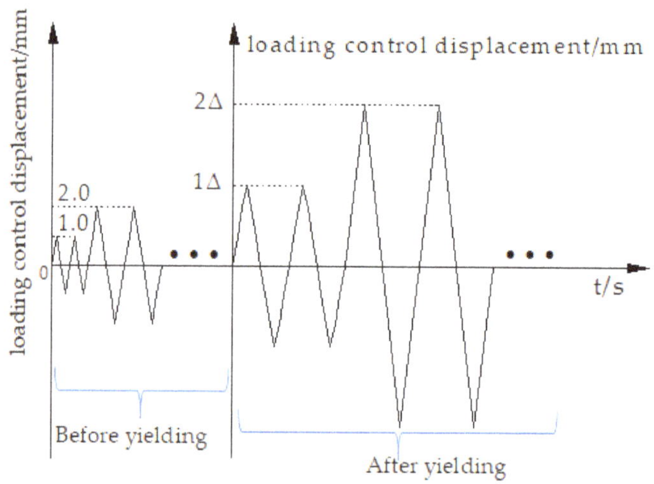

Figure 10. Loading protocol.

3. Results and Analysis

3.1. Test Phenomenon

3.1.1. Comparison Test Piece PJD-1

Specimen PJD-1 was a comparative joint with ordinary steel bars at the beam end. When the loading displacement was 2.0 mm, the first oblique crack appeared in the middle of the beam, with a length of 6.45 cm and a width of 0.10 mm. At the same time, the vertical crack with a length of 4.3 cm and a width of 0.2 mm developed at the root of the beam. The load–displacement curve gradually deviated from the straight line, and the specimen entered the yield stage. When the loading control displacement was one times the yield displacement, the first small crack appears at the bottom of the beam (19 cm away from the core area of the node), and the crack length was 4.16 cm. When the loading control displacement was three times the yield displacement, the vertical cracks at the root of the beam and the oblique cracks in the middle of the beam were successively penetrated from up to down, and were developed in the surroundings, resulting in many small development cracks. At six times the yield displacement, the peak load of the specimen reached 54.72 kN; at 11 times the yield displacement, the load value of the specimen decreased from 50.08 kN to 37.45 kN, and the test ended. At this point, the vertical main crack at the root of the beam was relatively wide, the concrete at the upper and lower sides of the root was obviously crushed, and the concrete of the protective layer was stripped, as shown in Figure 11a.

3.1.2. Comparison Test Piece SJD-1

The SJD specimen was the comparison node, where the longitudinal bars at the beam end were all SMA bars. When the loading displacement was 1.0 mm, the first vertical crack appeared at the root of the specimen, and the length of the crack was about 3.12 cm. When the reverse loading was carried out, the root crack had been quickly penetrated, and the crack width was about 0.3 mm. The load–displacement curve had deviated from the straight line, and the specimen entered the yield stage. When the loading control displacement was four times the yield displacement, the crack width increased, and a small amount of the surface concrete at the root of the beam fell off. At 17 times the yield displacement, a low 'boom' sound was heard, and the load value of the specimen reached 31.66 kN. When the load continued, the load value of the specimen immediately dropped to 16.27 kN, and the test ended. At this point, the vertical main crack at the root of the beam was wider than that at the PJD-1 node, but there was no obvious damage to the upper and lower concrete at the root of the beam, as shown in Figure 11b.

3.1.3. Test Piece PSJD-1 and Test Piece PSJD-2

Both PSJD-1 and PSJD-2 are model test joints with SMA bars and steel endplates. Taking the specimen PSJD-1 as an example, the loading process of this new type of joint was introduced. When the loading displacement was 2.0 mm, the first vertical crack appeared at the root of the beam, with a length of about 9.5 cm and a width of about 0.1 mm. There was no obvious oblique crack in the middle of the beam. When the displacement was controlled to 3 mm, the load–displacement curve begun to deviate from the straight line, the crack width was 0.2 mm, and the crack length reached 10.60 cm; the specimen thus entered the yield stage. When the loading control displacement was three times the yield displacement, the crack width reached 2.0 mm, the residual crack width after unloading was 1.2 mm, and the cracks in the root and middle of the beam were successively penetrated. At seven times the yield displacement, the concrete at the root of the beam showed spalling. At ten times the yield displacement, the load value began to decrease with the low 'boom' sound inside the specimen (which is believed to be the connection fracture between the steel plate and the longitudinal reinforcement), and the final load value of the specimen decreased from 56.11 kN to 46.02 kN. At this point, the crack width of the vertical main crack at the root of the beam was significantly smaller than that of the PJD-1 node, and the compressive area and the stripping volume of the concrete at the upper and lower sides of the root were smaller than those of the PJD-1 node, as shown in Figure 11c.

The first half of the loading process of the PSJD-2 specimen was similar to that of the PSJD-1. When the loading displacement was 2.0 mm, the first vertical crack appeared at the root of the beam, with a length of about 12.0 cm and a width of about 0.1 mm. When the loading displacement was controlled to 4 mm, the root crack width reached 0.5 mm, and the root crack had been penetrated. The first crack appeared in the middle of the beam component, and the crack length was 22.45 cm. The specimen thus entered the yield stage. When the loading control displacement was three times the yield displacement, the core area of the joint specimen began to peel off the concrete block, the root crack width reached 2.0 mm, and the middle crack width reached 1.0 mm. However, when the load was four times the yield displacement, the abnormal sound, similar to that heard in the PSJD-1 experiment, was heard early and originated from inside the specimen, and the load value then immediately decreased from 81.78 kN to 65.80 kN. The specimen thus failed immediately. Due to the relatively early failure of the specimen, the node final crack was smaller than that of the PSJD-1 joint, and there were no obvious signs of concrete compression and spalling, as shown in Figure 11d.

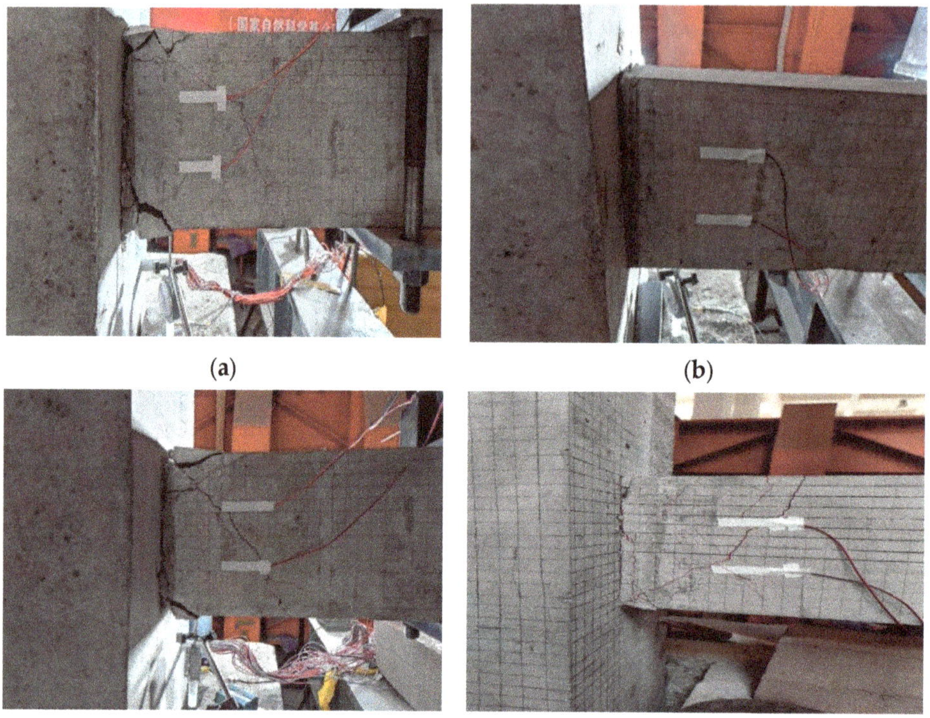

(a) (b) (c) (d)

Figure 11. Failure modes of the specimens: (**a**) PJD-1; (**b**) SJD-1; (**c**) PSJD-1; (**d**) PSJD-2.

3.2. Beam End Load–Displacement Hysteresis Curve

The hysteresis curves of each specimen in this test are shown in Figure 12.

(1) Comparing Figure 12a,c, it can be seen that the pinching phenomenon of the hysteretic curve of the PJD node is obvious. The hysteretic curve of the PSJD-1 node is rounder than that of the PJD-1 node, and the ultimate bearing capacity is increased by about 44%, indicating that the structural form of the built-in SMA reinforcement-steel end plate improves the hysteretic energy dissipation capacity of the node and significantly improves the bearing capacity of the node.

(2) Although the ultimate bearing capacity of the SJD-1 node is lower than that of the PJD-1 node, the residual displacements at all levels before failure are much smaller. This shows that the SMA bar material can significantly improve the self-centering ability of the structure.

(3) Although the PSJD-2 joint failed prematurely in this test, it can be seen, by comparing Figure 12a,c, that the residual deformation of the PJD-1 joint after complete unloading at each loading stage is very large. The hysteresis curve of the PSJD-1 node is obviously 'flag-shaped'. After each loading stage is completely unloaded, the residual deformation of the node is very small, indicating that the new node has a good self-centering ability.

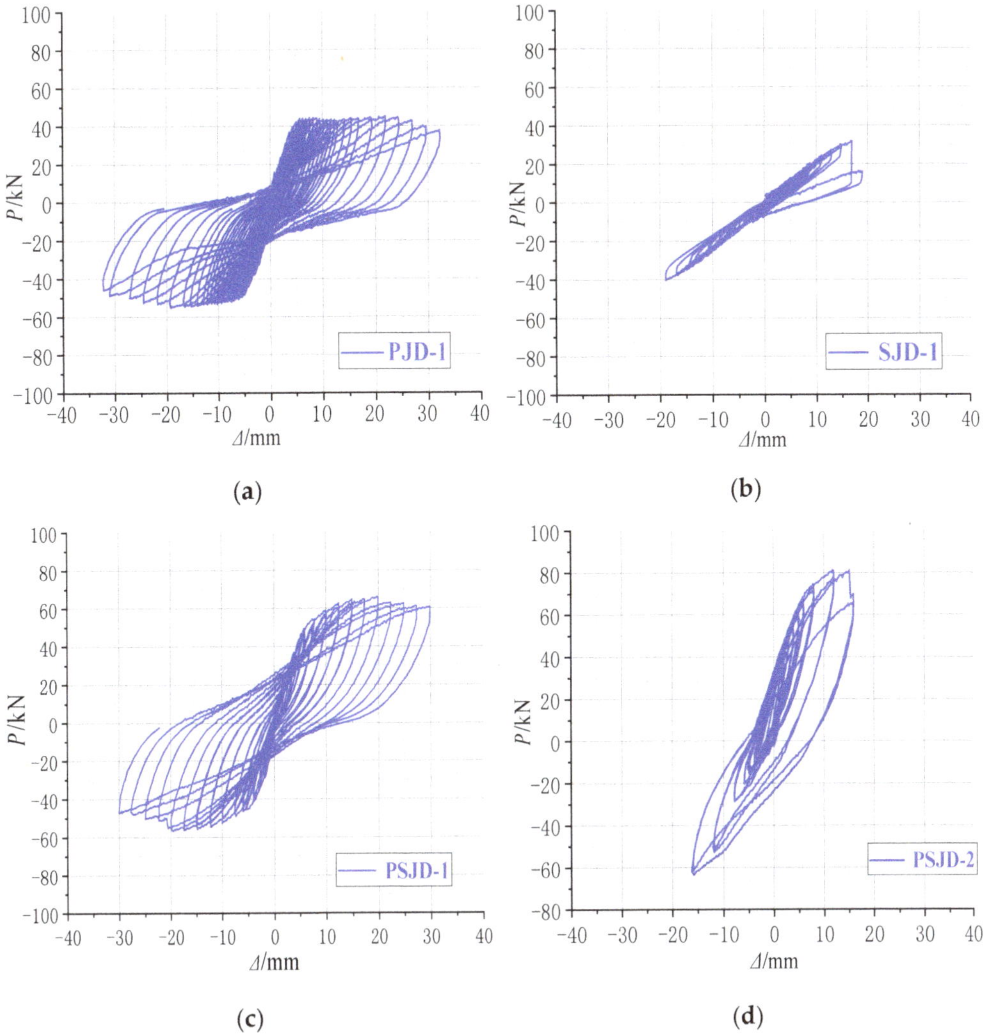

Figure 12. Hysteresis loops of the joint specimens: (**a**) PJD-1; (**b**) SJD-1; (**c**) PSJD-1; (**d**) PSJD-2.

3.3. Skeleton Curve

As an important basis for restoring the force model and nonlinear seismic response analysis, the skeleton curve reflects the deformation, energy consumption, and stiffness degradation of specimens in different stages. The skeleton curve of each node in this test is shown in Figure 13. As the failure of the PSJD-2 specimen is relatively early, this paper mainly analyzes the skeleton curves of the other three specimens.

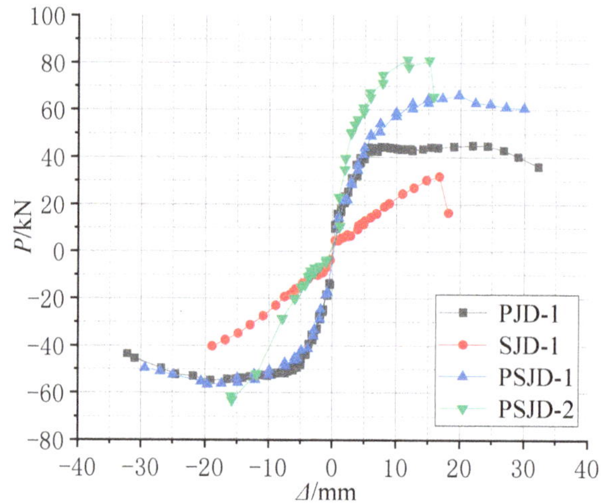

Figure 13. Envelope curves of the specimens.

The following conclusions can be drawn from Figure 13.

(1) The ultimate bearing capacity of the PSJD-1 node and the area surrounded by the skeleton curve and the abscissa axis are larger than those of the PJD-1 node. Although the PSJD-2 node fails earlier, the ultimate bearing capacity before failure and the area surrounded by the abscissa axis are further improved compared to those of the PSJD-1 node. This shows that the ultimate bearing capacity and the energy dissipation capacity of the joint can be significantly improved by adding a super-elastic SMA reinforcement-steel end plate and by increasing the reinforcement ratio of the corresponding SMA reinforcement.

(2) In the initial elastic stage, the skeleton curve slope of the PJD-1 node is much larger than that of the SJD-1 node, indicating that SMA material will reduce the initial stiffness of the component under the same longitudinal reinforcement ratio. This is because the Young's modulus of the SMA bars (65.4 GPa) is much smaller than that of the steel bars (203 GPa).

(3) In the yield stage, the yield displacement value of the PSJD-1 joint is significantly larger than that of the PJD-1 joint, and the bearing capacity of the PJD-1 joint and the SJD-1 joint decreases rapidly after yield. However, the bearing capacity of the new joint PSJD-1 is relatively stable after yield and can withstand relatively larger deformation, indicating that the built-in super-elastic SMA reinforcement-steel end plate can delay the yield of the joint to a certain extent and improve the ductility and damage resistance of the joint.

3.4. Residual Displacement of the Beam End

The residual displacement of the beam end refers to the residual plastic deformation of the beam end after unloading, and its value can reflect the self-centering ability of the component. The beam end loading-residual displacement curve of each joint specimen is shown in Figure 14.

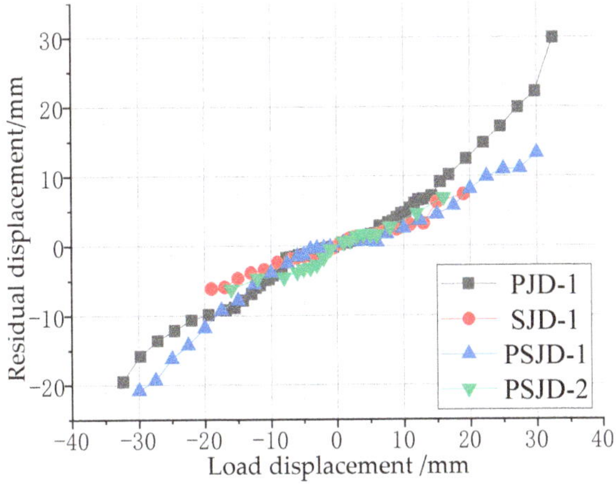

Figure 14. Residual displacement of the beam end after unloading.

It can be seen in the graph that the residual displacement of the three nodes with the built-in super-elastic SMA tendons is less than that of the PJD-1 node under the same loading level. When the peak displacement of the beam end is 30 mm, the recoverable deformation of the PJD-1 joint is 25%, while that of the PSJD-1 joint is 57% (the residual deformation of the beam end is 13 mm); due to the poor anchorage between the SMA bars and the concrete, the failure of the SJD-1 joint occurs earlier (the maximum loading displacement of the beam end is only 20 mm), but the residual displacement of the SJD-1 joint is still less than that of the PJD-1 joint. This shows that, under a certain loading amplitude, the built-in SMA reinforcement can significantly improve the self-centering ability of the node.

3.5. Energy Consumption Capacity

The structure dissipates the ground motion energy mainly through plastic deformation, and the seismic performance of the structure can be reflected by the energy dissipation capacity. The energy dissipation value of each specimen under different load displacement can be obtained by calculating the enclosed area of the hysteresis curve in Figure 12, as shown in Figure 15. Figure 15 shows the energy dissipation curve of the four node specimens.

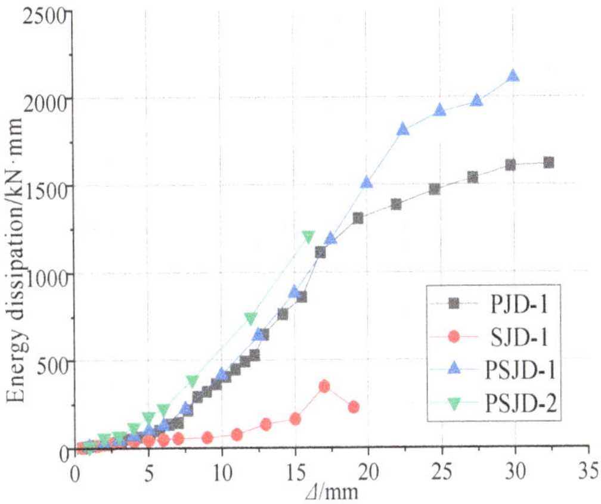

Figure 15. Energy dissipation capacity.

The figure shows the following:

(1) When the specimen is in the elastic stage, the steel bars and SMA bars in the joint do not yield, and the energy dissipation capacity of each joint is very low.
(2) When the specimen is in the plastic stage, the slope of the displacement–energy curve at the beam end of the PSJD-1 node and the PSJD-2 node is greater than that of the PJD-1 node, and the slope of the SJD-1 node is the smallest, indicating that the energy dissipation capacity of the new node is better than that of the ordinary node, while the energy dissipation capacity of the SJD-1 node is the weakest. When the beam end displacement exceeds 20 mm, the energy dissipation capacity of the PJD-1 node tends to be stable, while the energy dissipation of the PSJD-1 node continues to increase. This shows that the built-in SMA rib-steel end plate can effectively improve the energy dissipation capacity of the joint under large displacement deformation.
(3) Comparing the PSJD-1 and PSJD-2 joints, it can be seen that, with an increase in the SMA reinforcement ratio, the energy dissipation capacity of the new joints is further improved.

3.6. Stiffness Degradation

We selected the ring stiffness K as an index to evaluate the stiffness degradation of the beam-column joints. The obtained stiffness–ductility coefficient curve is shown in Figure 16.

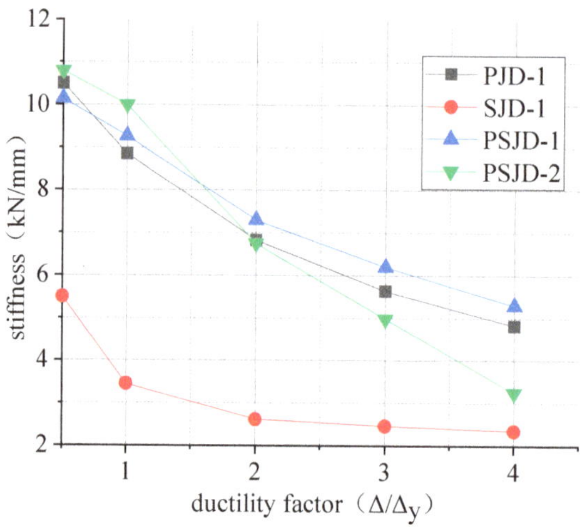

Figure 16. Curves of stiffness degradation.

The following conclusions can be drawn from Figure 16:

(1) Comparing the initial stiffness of the four nodes, it can be seen that the initial stiffnesses of the PSJD-1 node, the PSJD-2 node, and the PJD-1 node show little difference. The initial stiffness of the SJD joints is obviously smaller than that of the PJD-1 joints, because the Young's modulus of the SMA bars is much smaller than that of the steel bars.
(2) When the ductility factor is less than 1, the stiffness degradation rate of the SJD-1 joint is the fastest; this is because the surface of the SMA reinforcement is relatively smooth, and the bonding force between the SMA reinforcement and the concrete is low. After the concrete cracks, the penetrating cracks are quickly formed, and the stiffness decreases rapidly. When the ductility coefficient is greater than 1, the stiffness degradation of the SJD-1 joint is limited; this is because that with a large amount of

concrete cracking out of work, the stiffness ratio increases, and the SMA tendons have excellent super-elasticity.

(3) Comparing the PSJD-1 node, the PSJD-2 node, and the PJD-1 node shows that, when the ductility coefficient is less than 1, the stiffness degradation rates of the PSJD-1 and PSJD-2 nodes are significantly smaller than those of ordinary nodes. When the ductility coefficient is greater than 1, although the premature failure of the internal components of the PSJD-2 node causes the stiffness of the node to rapidly degrade, and thus loses comparability, the stiffness of the PSJD-1 node is greater than that of the ordinary node, and the degradation rate is slower. This shows that the stiffness degradation rate of the node can be delayed with a built-in SMA reinforcement-steel end plate, so that the lateral displacement margin of the structure is larger, which can effectively improve the seismic performance of the structure.

4. Numerical Simulation

4.1. Model Establishment and Verification

4.1.1. Constitutive Model of SMA Material

The self-centering 'double flag' constitutive model developed by Ferdinando Auricchio [46] based on OpenSees is selected, as shown in Figure 17. The constitutive description of the SMA material is relatively complex, and this model simplifies the constitutive curve of the material into three stages: elasticity, phase transformation and hardening. The corresponding rules are as follows:

1. In the elastic stage, when the stress does not exceed σ_y, the loading and unloading processes develop linearly with stiffness k_1.
2. In the phase transformation stage, when the stress value exceeds σ_y, continuous loading develops with k_2 as the stiffness, and this stage ends when the strain reaches ε_b; when unloading in this stage, the stiffness decreases linearly with k_1 and then decreases linearly with k_2.
3. In the hardening stage, when the strain value is such that $\varepsilon_b \le \varepsilon \le \varepsilon_\mu$, the loading and unloading processes develop along a straight line with slope $r \times k_1$. When the unloading strain value is less than ε_b, the continuous unloading is the same as the (1)(2) stage. In general, the strain value does not exceed ε_b.

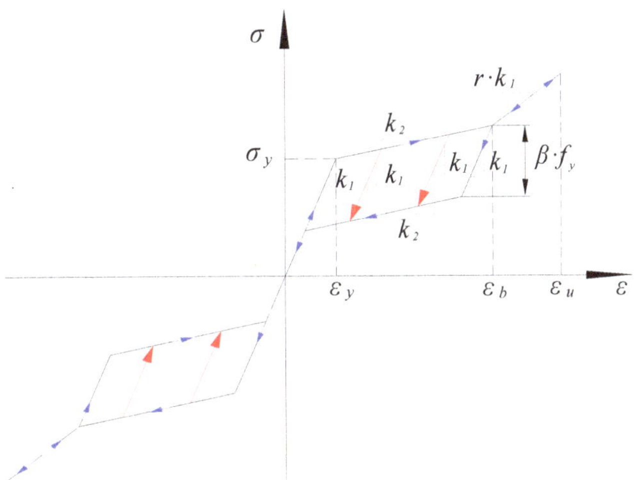

Figure 17. Self-centering constitutive model.

The definition of this constitutive model in OpenSees is implemented with seven parameters. Parameter values are shown in Table 6.

Table 6. SMA simplified constitutive model parameters and values.

Parameter	Physical Significance	Value
k_1	First stiffness (N/mm)	75,000
k_2	Second stiffness (N/mm)	1827
σ_y	Positive phase transition stress (MPa)	400
β	Inverse phase transition stress coefficient	0.80
ε_s	Sliding strain	0.06
ε_b	Hardening strain	0.06
r	Hardening stiffness coefficient	0.39

4.1.2. Constitutive Model of the Steel Bar and the Steel Plate

The constitutive model of the steel bar and steel plate shown in Figure 18 is the Steel 02 model, proposed by Menegotto and Pinto [47], and modified by Filippou [48]. The model not only considers isotropic strain hardening, but also reflects the influence of the Bauschinger effect. It has a high computational efficiency and good numerical stability.

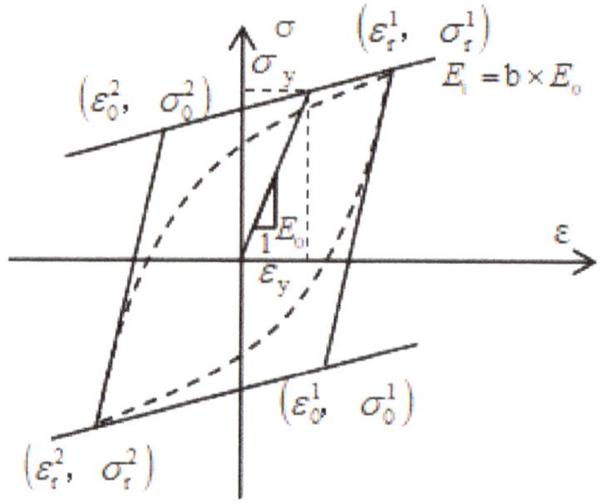

Figure 18. Steel 02 constitutive model.

The constitutive relationship expressions of the Steel 02 model are shown in Equations (2)–(5):

$$\sigma_{eq} = b\varepsilon_{eq} + \frac{(1-b)\varepsilon_{eq}}{\left(1+\varepsilon_{eq}^{R}\right)^{\frac{1}{R}}} \quad (2)$$

$$\sigma_{eq} = \frac{\sigma - \sigma_r}{\sigma_0 - \sigma_r} \quad (3)$$

$$\varepsilon_{eq} = \frac{\varepsilon - \varepsilon_r}{\varepsilon_0 - \varepsilon_r} \quad (4)$$

$$R = R_0 - \frac{a_1\xi}{a_2 + \xi} \quad (5)$$

where σ_{eq} and ε_{eq} represent the normalized stress and strain; σ_0 and ε_0 represent the stress and strain of the rebar in the initial state; σ_r and ε_r represent the stress and strain of the rebar at the yield point; a_1 and a_2 represent the curvature degradation coefficients; R

and R_0 represent the transition curve curvature coefficient and the initial curve curvature coefficient; b and ξ represent the hardening coefficient of reinforcement and the plastic strain in the semi-periodic cycle.

4.1.3. Theoretical Model of Concrete

There are three kinds of concrete constitutive models commonly used in OpenSees: the Concrete 01 model with zero tensile strength, the Concrete 02 model with linear tensile softening, and the Concrete 03 model with nonlinear tensile softening [48]. For the concrete at the joint, its tensile strength can be ignored from a macro perspective. Therefore, for this simulation we selected Concrete 01 as the constitutive model of concrete, as shown in Figure 19.

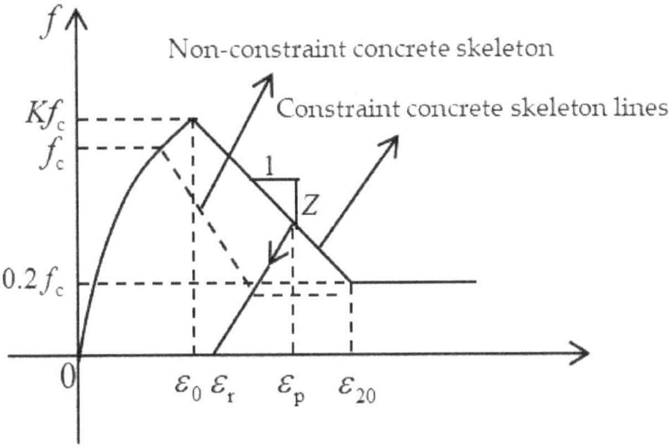

Figure 19. Concrete 01 constitutive model.

In the Concrete 01 model [49], the constitutive relation expression is shown in Equations (6)–(10):

$$\text{when } \varepsilon < \varepsilon_0,\ f = Kf_c\left[\frac{2\varepsilon_c}{\varepsilon_0} - \left(\frac{\varepsilon_c}{\varepsilon_0}\right)^2\right] \quad (6)$$

$$\text{when } \varepsilon_0 \leq \varepsilon \leq \varepsilon_c,\ f = Kf_c[1 - Z(\varepsilon - \varepsilon_0)] \quad (7)$$

$$\text{when}\varepsilon > \varepsilon_c,\ f = 0.2Kf_c \quad (8)$$

$$\varepsilon_0 = 0.002K \quad (9)$$

$$Z = \frac{0.5}{\frac{3+0.29f_c}{145f_c} - 0.002K} \quad (10)$$

where σ_c and ε_c represent the stress and strain of the concrete; f_c is the compressive strength of the concrete cylinder; K is the strain increase coefficient caused by the constraint; Z is the slope of the strain drop segment; ε_0 is the peak pressure and strain of the concrete.

4.1.4. Types of Beam-Column Elements

According to the actual situation of the test, the nonlinear beam-column element (nonlinear beam column) was selected for simulation. The element is characterized by allowing the stiffness to change along the length of the rod. At the same time, the resistance and stiffness matrix of the control section can be determined. In the nonlinear simulation, it has the advantages of fast convergence and high effectiveness [49].

4.1.5. Fiber Model

The fiber model, with a wide application and a high accuracy, was selected for the beam-column section. The fiber model based on the assumption of the plane section divides the section of the component into a certain number of grids. The center of each grid is regarded as a numerical integration point. The longitudinal micro-section of the grid is defined as fiber. The section is divided into several types of fiber bundles, such as confined and non-confined concrete, and steel bars. By calculating the strain stress of each fiber, the stiffness of the whole section is obtained [50].

4.2. Model Validation and Parameter Analysis

4.2.1. Model Validation

According to the test results of material properties, various parameters in the finite element model were set. Based on the actual displacement of each cycle in the test, the displacement-controlled loading simulation analysis of the model was carried out. The hysteresis curves and skeleton curves of the ordinary concrete joint PJD-1 and the new type of joint PSJD-1 with internal SMA reinforcement were compared with the test results, as shown in Figures 20 and 21.

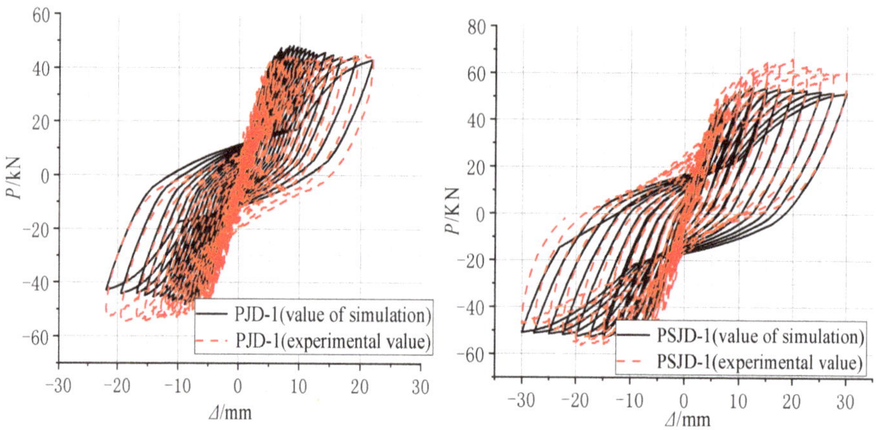

Figure 20. Comparison diagram of numerical simulation and experimental hysteresis curves.

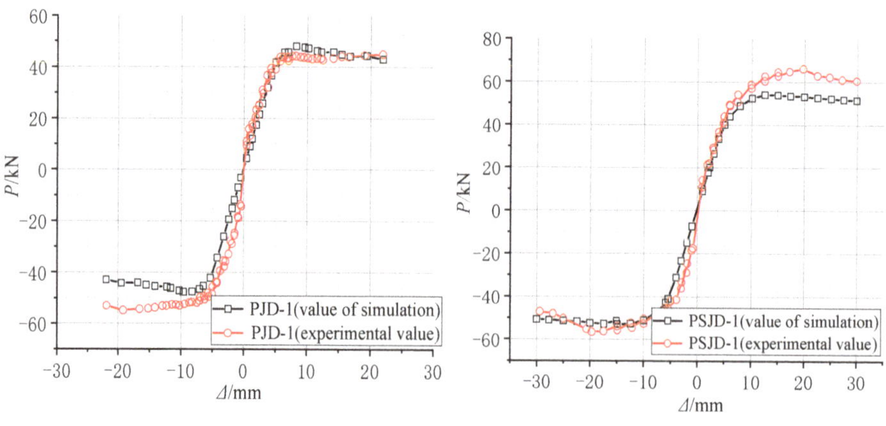

Figure 21. Comparison diagram of numerical simulation and experimental skeleton curves.

Figures 20 and 21 show that the hysteresis curve and the skeleton curve obtained by simulation are basically consistent with the experimental results, and the established model well reflects the self-centering energy dissipation characteristics of the new type of joints. The peak value of forward loading is slightly smaller than that of the test results. Due to the relatively ideal state of the model in the numerical simulation, and many factors such as constraints and measurement on the test site, there are certain differences. However, from the overall perspective, the obtained hysteresis and skeleton curves still well verify the accuracy and effectiveness of the finite element model.

4.2.2. Parameter Analysis

In order to further study the new frame joints to further optimize the design, considering that the PSJD-2 joints failed earlier in the test and failed to fully reflect the influence of SMA reinforcement ratio on the related performance of the new joints, based on the verified finite element analysis model, the parameters such as the reinforcement ratio and the axial compression ratio of the built-in SMA bars in the new joints were simulated and analyzed to further quantify their effects on the mechanical properties, such as hysteretic performance, energy dissipation, and the self-centering ability of the new joints.

Based on the conventional ordinary reinforced concrete frame joints, under the same built-in SMA reinforcement mode, the yield strength of the SMA bars is controlled to be the same, and three new joints (XJD) with different SMA bar diameters were considered, namely, XJD-1 with a diameter of 8 mm, XJD-2 with a diameter of 10 mm, and XJD-3 with a diameter of 12 mm. The diameter of the SMA bars was controlled to be the same (10 mm), and three kinds of SMA bars with different yield strengths were considered to study their effects on the seismic performance of the new joint (XJD): XJD-4 had a yield strength of 300 MPa, XJD-2 had a yield strength of 400 MPa, and XJD-5 had a yield strength of 500 MPa. Specific parameter analysis design is shown in Table 7.

Table 7. Model parameter analysis scheme.

Node Trial Number	SMA Bar Diameter (mm)	Axial Compression Ratio
XJD-1	8	0.25
XJD-2	10	0.25
XJD-3	12	0.25
XJD-4	10	0.3
XJD-5	10	0.35

1. Effect of the SMA Reinforcement Ratio

Figures 22–25 show that, under the condition of suitable reinforcement, the bearing capacity of the new joint also increases significantly with the increase in the diameter of SMA reinforcement. The S-shaped skeleton curve shows that the new joint has good energy dissipation and ductility. With the increase in the diameter of the SMA bars, the overall stiffness increases, but the increase is small, because the Young's modulus of the SMA bars is smaller than that of ordinary steel bars, and the influence of ordinary steel bars on the overall stiffness is small.

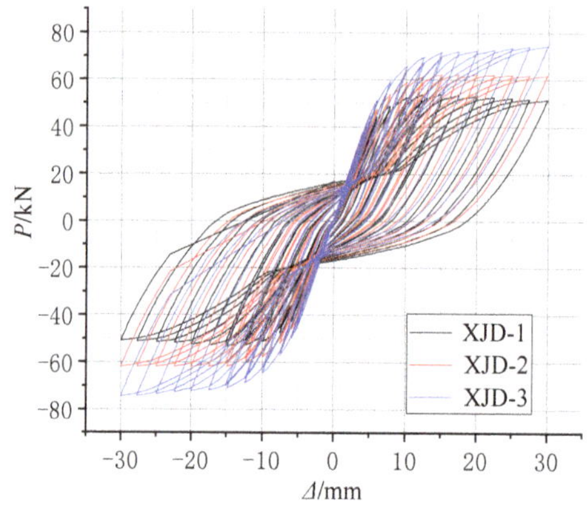

Figure 22. Hysteretic force–displacement curve of different diameters.

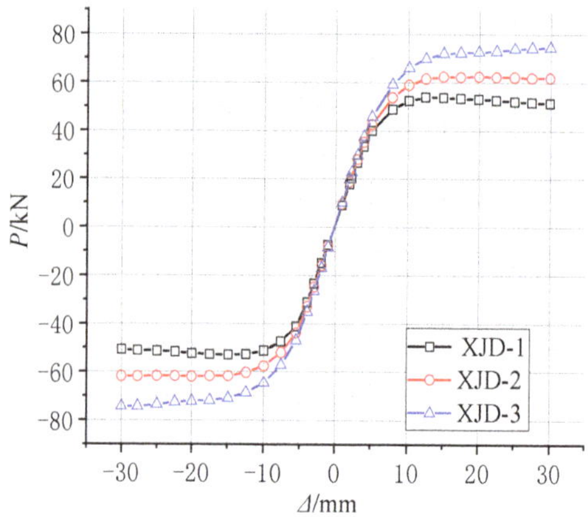

Figure 23. The skeleton curve of different diameters.

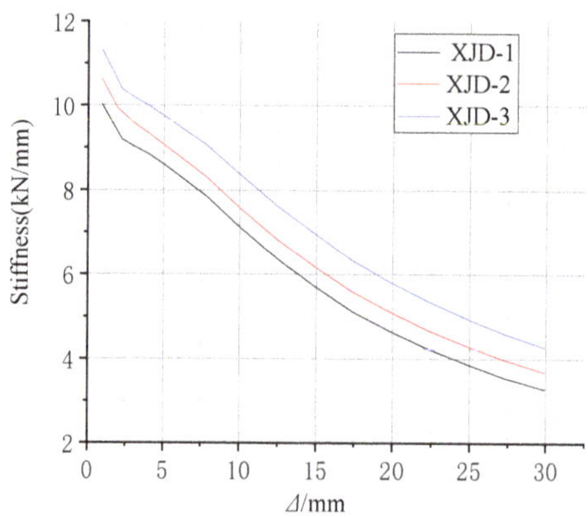

Figure 24. Stiffness degradation curve of different diameters.

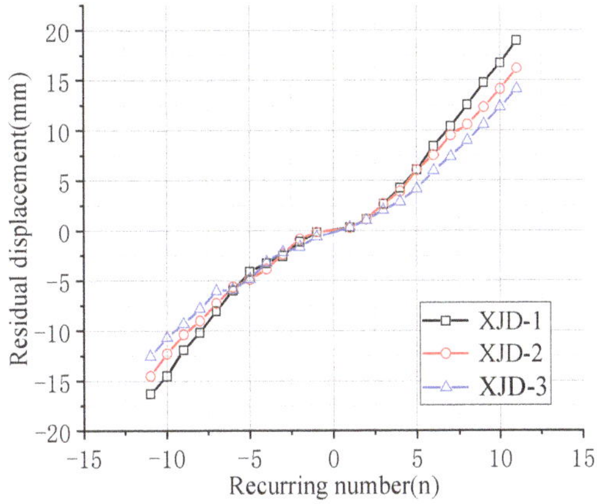

Figure 25. Residual deformation curves with different diameters.

In the initial state, increasing the diameter of SMA reinforcement has little effect on the residual deformation of the new node. With the increase in the number of cycles, the final residual deformation of XJD-3 is reduced by nearly 28% compared with that of XJD-1, indicating that, under the premise of suitable reinforcement, increasing the diameter of SMA reinforcement can significantly improve the self-centering performance of the new node.

2. Effect of Different Axial Pressure Ratio

Figure 26 clearly shows that, when the other parameters of the joint model are exactly the same, within the allowable range, only the axial force applied to the column is changed, and the influence on the seismic performance of the whole joint is almost zero. The main object of the study is the core area of the joint, and the variable parameter object of the study is the beam member; therefore, when the axial compression ratio of the column is changed, there is no obvious effect on the mechanical performance of the whole joint.

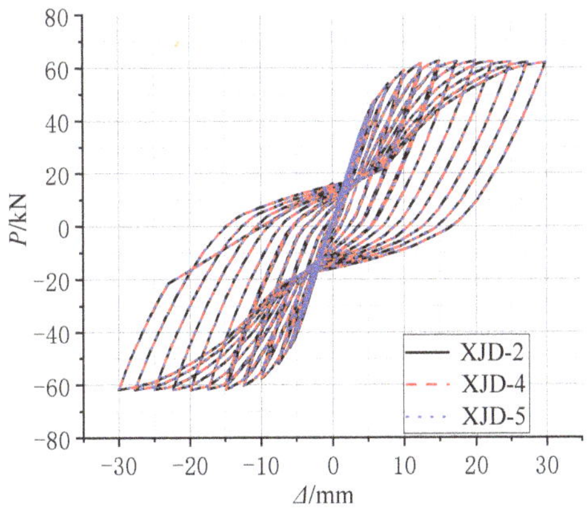

Figure 26. Comparison of hysteresis curves with different axial compression ratios.

5. Conclusions and Discussions

In this research, a new type of seismic-resisting self-centering beam-column joint with built-in SMA reinforcement and a steel end plate was designed and investigated experimentally and numerically based on a conventional beam-column edge joint of conventional island main buildings in a nuclear power plant. In particular, the seismic performance of the joint was validated through a series of low cyclic loading tests on four models. Good agreement was observed in the comparisons between the experimental and numerical results. The main conclusions and discussions are as follows:

(1) The self-centering performance of the joint can be improved by adding super-elastic SMA reinforcement in conventional concrete joints. However, replacing all the longitudinal reinforcement at the beam end with SMA reinforcement not only greatly reduces the bearing capacity and stiffness of the joint, but also makes the failure mode of the joint brittle.
(2) The structural form of the built-in SMA reinforcement-steel end plate can significantly improve the bearing capacity of the joint and improve the cracking damage degree of the joint, so as to improve the post-earthquake reparability of the joint.
(3) The stiffness degradation of the joint can be delayed by using the built-in SMA reinforcement-steel end plate structure, and the joint has good displacement ductility and self-centering energy dissipation performance. When the peak displacement of the beam end is 30 mm, the PSJD-1 joint can recover the deformation up to 57%.
(4) For the new joint constructed with built-in SMA reinforcement and a steel end plate, under the premise of suitable reinforcement, increasing the reinforcement ratio of the SMA reinforcement within a certain range can further improve the bearing capacity and self-centering energy dissipation performance of the joint.
(5) Within a certain range, increasing or reducing the axial compression ratio at the column end has little effect on the overall seismic performance of the new joint.
(6) It is worth noting that the above conclusions are obtained by experiments at room temperature. The tested joints embedded with an Ni–Ti SMA bar are suitable for use in relatively stable environments at room temperature. However, Ni–Ti alloys may not be suitable for outdoor applications because of their extreme sensitivity to temperature. In order to address this limitation, on the one hand, experimental or numerical simulation research on the influence of temperature sensitivity on test results should be carried out. On the other hand, other types of hyperelastic materials, such as monocrystalline materials with a large pseudoelasticity change limit, a high energy dissipation and excellent low-temperature properties, should also be considered for outdoor applications.

Author Contributions: Conceptualization, Q.P. and C.W.; methodology, Q.P. and Z.C.; software, Z.C. and C.W.; validation, Q.P., C.W. and Z.C.; formal analysis, Q.P. and H.G.; investigation, Q.P. and Y.D.; data curation, Y.D. and Q.P.; writing—original draft preparation, C.W. and Z.C.; writing—review and editing, Q.P. and C.W.; supervision, Q.P. and H.G.; project administration, Q.P. and Y.D.; funding acquisition, Q.P., C.W. and Y.D. All authors have read and agreed to the published version of the manuscript.

Funding: This research was funded by the National Nature Science Foundation of China, grant number 51878108, and the Department of Science & Technology Guidance Plan Foundation of Liaoning Province, grant number 2019JH8/10100091.

Institutional Review Board Statement: Not applicable.

Informed Consent Statement: Not applicable.

Data Availability Statement: The data used to support the finding of this study are available from the corresponding author upon request.

Acknowledgments: The authors gratefully acknowledge the support from Yang Liu from the College of Environment and Civil Engineering, Chengdu University of Technology, from Xiaofei Xie and Jiangtao Li from Beijing company Beijing Ruituotech Technology Co., Ltd, during the tests conducted for this study.

Conflicts of Interest: The authors declare no conflict of interest.

Abbreviations

RC	Reinforced concrete
SMA	Shape memory alloy
VIC-3D	Video image correlate-3D
FRP	Fiber-reinforced polymer
BFRP	Basalt fiber-reinforced polymer
PSJD-1	Ordinary longitudinal reinforcement plus hybrid joint with 8mm diameters of SMA reinforcement
PSJD-2	Ordinary longitudinal reinforcement plus hybrid joint with 10mm diameters of SMA reinforcement
PJD-1	Ordinary reinforced concrete beam-column joint
SJD-1	Concrete joint with SMA longitudinal reinforcement in beam
XJD	Ordinary longitudinal reinforcement plus hybrid joints with SMA reinforcement
g	Gravitational acceleration
f_y	Yield Strength
f_u	Ultimate Strength
δ	Elongation Rate

References

1. Qing, M.; Wu, Y.; Zhang, J.; Meng, A.; Zhang, T.; Yang, C.; Liu, F. The development on seismic design and evaluation of CGNPC after Fukushima accident. *Chin. Eng. Sci.* **2013**, *15*, 46–51. [CrossRef]
2. Wang, X.; Lyu, D. Research review of seismic margin assessment methods for nuclear power plants. *China Saf. Sci. J.* **2015**, *25*, 116–122. [CrossRef]
3. Wang, X.; Lyu, D. Review of seismic probability risk assessment of nuclear power plants. *China Civ. Eng. J.* **2016**, *49*, 52–68. [CrossRef]
4. Ying, Z.; Hao, W.; Anqi, G. Earthquake engineering: From earthquake resistance, energy dissipation, and isolation, to resilience. *Eng. Mech.* **2019**, *36*, 1–12. [CrossRef]
5. Liu, W.; Zhang, H.; Huang, Y.; Wang, Y.; Guan, Z. Investigation and discussion of the building seismic damage after Qinghai Maduo Ms7.4 earthquake in 2021. *World Earthq. Eng.* **2021**, *37*, 57–64. [CrossRef]
6. Dai, B.; Tao, Z.; Ye, L.; Gao, Y.; Huang, F. Investigation and analysis of seismic damage to the renovated rural dilapidated residences in Mojiang M5.9 earthquake. *J. Hunan Univ. (Nat. Sci.)* **2021**, *48*, 121–131. [CrossRef]
7. Guo, L.; Liu, J.; Wen, W.; Du, K.; Ding, Y. Investigation and analysis on seismic damage characteristics of building structures in Yangbi M6.4 earthquake in Yunnan province. *World Earthq. Eng.* **2021**, *37*, 64–72. [CrossRef]
8. Tan, H.; Lu, H.; Zhang, G. Damage and analysis of frame structures in wenchuan earthquake. *World Earthq. Eng.* **2016**, *32*, 91–96.
9. An, X.; Li, D. Typical earthquake damage analysis of Ludian earthquake with M_s 6.5. *Build. Struct.* **2020**, *50*, 28–36. [CrossRef]
10. Wei, C.; Wen, Z.; Ju, Y.; Wang, D. Experimental study on seismic behavior and restoring force model of spatial beam-column subassemblies with high strength reinforcements and high strength concrete. *J. Build. Struct.* **2021**, *42*, 145–154. [CrossRef]
11. Rong, X.; Su, T.; Yang, C.; Li, Q. Research on seismic behavior of exterior joints with high-strength reinforced bars and high-toughness concrete. *Ind. Constr.* **2017**, *47*, 26–29. [CrossRef]
12. Zhang, J.; Rong, X.; Wu, H. Experimental research on hysteretic behavior of exterior joints with 600 megapascal reinforced bars. *Ind. Constr.* **2017**, *47*, 30–33. [CrossRef]
13. Yan, C.; Yong, Y.; Jia, J.; Zhang, J.; Liu, S. Study on restoring force model of steel reinforced ultra high strength concrete column and concrete beam joints. *China Civ. Eng. J.* **2014**, *47*, 193–197. [CrossRef]
14. Ahmed, M.; Liang, Q.Q.; Patel, V.I.; Hadi, M.N. Nonlinear analysis of rectangular concrete-filled double steel tubular short columns incorporating local buckling. *Eng. Struct.* **2018**, *175*, 13–26. [CrossRef]
15. Ahmed, M.; Liang, Q.Q.; Patel, V.I.; Hadi, M.N. Numerical analysis of axially loaded circular high strength concrete-filled double steel tubular short columns. *Thin-Walled Struct.* **2019**, *138*, 105–116. [CrossRef]
16. Liang, Q.Q. Nonlinear analysis of circular double-skin concrete-filled steel tubular columns under axial compression. *Eng. Struct.* **2017**, *131*, 639–650. [CrossRef]
17. Lian, M.; Li, S.; Li, X.-L. Seismic Story Shear Distribution Based on Inelastic State of Eccentrically Braced Frame with Vertical Link Composite High Strength Steel. *Int. J. Steel Struct.* **2021**, *21*, 1260–1279. [CrossRef]
18. Shahverdi, M.; Czaderski, C.; Motavalli, M. Iron-based shape memory alloys for prestressed near-surface mounted strengthening of reinforced concrete beams. *Constr. Build. Mater.* **2016**, *112*, 28–38. [CrossRef]
19. Qian, H.; Li, H.; Song, G.; Guo, W. Recentering shape memory alloy passive damper for structural vibration control. *Math. Probl. Eng.* **2013**, *2013*, 963530. [CrossRef]

20. Asgarian, B.; Salari, N.; Saadati, B. Application of Intelligent Passive Devices Based on Shape Memory Alloys in Seismic Control of Structures. *Structure* **2016**, *5*, 161–169. [CrossRef]
21. Rahman Bhuiyan, M.A.; Alam, M.S. Seismic vulnerability assessment of a multi-span continuous highway bridge fitted with shape memory alloy bars and laminated rubber bearings. *Earthq. Spectra* **2012**, *28*, 1379–1406. [CrossRef]
22. Dieng, L.; Helbert, G.; Chirani, S.A.; Lecompte, T.; Pilvin, P. Use of Shape Memory Alloys damper device to mitigate vibration amplitudes of bridge cables. *Eng. Struct.* **2013**, *56*, 1547–1556. [CrossRef]
23. Fang, C.; Wang, W.; Chen, Y.Y. State-of-the-art for application of superelastic shape memory alloy in seismic resistant steel structures. *J. Build. Struct.* **2019**, *40*, 1–12. [CrossRef]
24. Huang, B.; Lao, Y.; Chen, J.; Song, Y. Dynamic Response Analysis of a Frame Structure with Superelastic Nitinol SMA Helical Spring Braces for Vibration Reduction. *J. Aerosp. Eng.* **2018**, *31*, 04018096. [CrossRef]
25. Shu, G.; Li, R. Experiment study on seismic performance of pin-connected steel frame-self-centering-energy-dissipation bracing substructure. *J. Southeast Univ. (Nat. Sci. Ed.)* **2021**, *51*, 187–194. [CrossRef]
26. Wang, J.; Mou, S.; Liu, H. Study on the influencing factors of energy dissipation capacity of SMA-laminated rubber composite isolation bearing. *J. Seismol. Res.* **2017**, *40*, 70–74, 167. [CrossRef]
27. DesRoches, R.; Delemont, M. Seismic retrofit of simply supported bridges using shape memory alloys. *Eng. Struct.* **2002**, *24*, 325–332. [CrossRef]
28. Qian, H.; Li, H.; Song, G. Experimental investigations of building structure with a superelastic shape memory alloy friction damper subject to seismic loads. *Smart Mater. Struct.* **2016**, *25*, 125026. [CrossRef]
29. Mirzai, N.M.; Mansouri, I.; Tezcan, J.; Awoyera, P.O.; Hu, J.W. Estimating optimum parameters of a new SMA damper under different earthquake ground motions. *Structures* **2021**, *33*, 2700–2712. [CrossRef]
30. Hu, S.; Qi, G.; Jiang, G.; Xiong, J. Experimental study on seismic performance for an innovative self-centering SMA brace. *Eng. Mech.* **2021**, *38*, 109–118, 142. [CrossRef]
31. Masi, A.; Santarsiero, G. Seismic Tests on RC Building Exterior Joints with Wide Beams. *Adv. Mater. Res.* **2013**, *787*, 771–777. [CrossRef]
32. Santarsiero, G.; Masi, A. Analysis of slab action on the seismic behavior of external RC beam-column joints. *J. Build. Eng.* **2020**, *32*, 101608. [CrossRef]
33. Realfonzo, R.; Napoli, A.; Pinilla, J.G.R. Cyclic behavior of RC beam-column joints strengthened with FRP systems. *Constr. Build. Mater.* **2014**, *54*, 282–297. [CrossRef]
34. Gand, A.K.; Yeboah, D.; Khorami, M.; Olubanwo, A.O.; Lumor, R. Behaviour of strengthened timber beams using near surface mounted Basalt Fibre Reinforced Polymer (BFRP) rebars. *Eng. Solid Mech.* **2018**, *6*, 341–352. [CrossRef]
35. Gand, A.K.; Sharif, S.; Saidani, M.; Lumor, R.; Fom, P.B.; Yeboah, D.; Ogbologugo, U. Performance of lightweight granulated glass concrete beams reinforced with basalt FRP bars. *Eng. Solid Mech.* **2019**, *7*, 247–262. [CrossRef]
36. Shayanfar, M.A.; Rostamian, M.; Ghanooni-Bagha, M.; Tajban, A.; Nemati, S. Evaluating the plasticity of concrete beam-column connections reinforced with FRP composite rebars. *Eng. Solid Mech.* **2018**, *6*, 331–340. [CrossRef]
37. Qian, H.; Pei, J.; Li, Z.; Kang, L. Experimental study on seismic performance of self-centering beam-column joints reinforced with superelastic SMA and ECC. *China Civ. Eng. J.* **2020**, *53*, 64–73, 80. [CrossRef]
38. Yan, S.; Du, C.; Xiao, Z.; Jiang, X.; Niu, J. Analysis of recoverable deformation behavior of Fe based shape memory alloy concrete frame joints. *J. Seismol. Res.* **2020**, *43*, 491–499, 602. [CrossRef]
39. Oudah, F.; El-Hacha, R. Joint performance in concrete beam-column connections reinforced using SMA smart material. *Eng. Struct.* **2017**, *151*, 745–760. [CrossRef]
40. Dong, J. Experimental research on the self-centering device based on large diameter SMA bar. *Struct. Eng.* **2018**, *34*, 101–108. [CrossRef]
41. Ni, Y.; Sun, L. Experimental study on high-strength concrete under uniaxial compression based on VIC-3D technology. *Concrete* **2014**, *11*, 19–23. [CrossRef]
42. Wang, L.; Hu, J.; Chen, G.; Gong, N. Experimental investigation on interface mechanics of asphalt mixture based on technology of VIC-3D. *J. Funct. Mater.* **2015**, *46*, 3115–3120, 3126. [CrossRef]
43. Tan, Y.; Hou, M.; Zhang, L.; Zhang, K. Studying the strain field distribution of asphalt mixture with the digital speckle correlation method. *Road Mater. Pavement Des.* **2014**, *15*, 90–101. [CrossRef]
44. Wei, K.; Yuan, F.; Dong, Z.; Shao, X.; Wu, G. Precision analysis and civil application of multi-camera digital image correlation method based on encoded targets. *J. Southeast Univ. (Nat. Sci. Ed.)* **2021**, *51*, 219–226. [CrossRef]
45. Zeren, G. *Research on Accuracy Verification Method and Engineering Application of Digital Image Correlation*; University of Science and Technology of China: Hefei, China, 2020. [CrossRef]
46. Auricchio, F.; Taylor, R.L.; Lubliner, J. Shape-memory alloys: Macromodelling and numerical simulations of the superelastic behavior. *Comput. Methods Appl. Mech. Eng.* **1997**, *146*, 281–312. [CrossRef]
47. Menegotto, M.; Pinto, P.E. Method of Analysis for cyclically loaded R.C. plane frames including changes in geometry and nonelastic behavior of elements under combined normal force and bending. Proceeding, Symposium on the Resistance and Ultimate Deformability of Structures Acted on by Well Defined Repeated loads. *Int. Assoc. Bridge Struct. Eng.* **1973**, *36*, 15–22.
48. Spacone, E.; Filippou, F.C.; Taucer, F.F. Fiber Beam-column Model for Nonlinear Analysisof R/C Frames: Part1 Formulation. *Earthq. Eng. Struct. Dyn.* **1996**, *25*, 711–725. [CrossRef]

49. Meng, Z. *Nonlinear Analysis of Circular Steel Tubular Composite Columns with Recycled Aggregate Concrete Based on OpenSees*; Xi'an University of Technology: Xi'an, China, 2020. [CrossRef]
50. Jia, N. *Nonlinear Analysis of Unbonded Partially Prestressed Concrete Beams Based on OpenSees*; Hunan University: Changsha, China, 2016.

materials

MDPI

Article

Experimental Investigation on Bending Behavior of Existing RC Beam Retrofitted with SMA-ECC Composites Materials

Hui Qian, Qingyuan Zhang, Xun Zhang *, Enfeng Deng and Jundong Gao

School of Civil Engineering, Zhengzhou University, Zhengzhou 450001, China; qianhui@zzu.edu.cn (H.Q.); dayuanzi815@126.com (Q.Z.); dengenfeng@zzu.edu.cn (E.D.); gaojundong@zzu.edu.cn (J.G.)
* Correspondence: zhangxun@zzu.edu.cn

Abstract: In order to realize the self-centering, high energy consumption, and high ductility of the existing building structure through strengthening and retrofit of structure, a method of reinforced concrete (RC) beam strengthened by using Shape Memory Alloy (SMA) and Engineered Cementitious Composites (ECC) was proposed. Four kinds of specimens were designed, including one beam strengthened with enlarging section area of steel reinforced concrete, one beam strengthened with enlarging section area of SMA reinforced concrete, beam strengthened with enlarging section area of SMA reinforced ECC, and beam strengthened with enlarging section area of steel reinforced ECC; these specimens were manufactured for the monotonic cycle loading tests study on its bending behavior. The influence on the bearing capacity, energy dissipation performance, and self-recovery capacity for each test specimens with different strengthening materials were investigated, especially the bending behavior of the beams strengthened by SMA reinforced ECC. The results show that, compared with the ordinary reinforced concrete beams, strengthening existing RC beam with enlarging section area of SMA reinforced ECC can improve the self-recovery capacity, ductility, and deformability of the specimens. Finally, a revised design formula for the bending capacity of RC beams, strengthened with enlarging sections of ECC, was proposed by considering the tensile capacity provided by ECC, and the calculated values are in good agreement with the experimental value, indicating that the revised formula can be well applied to the beam strengthening with enlarging section of SMA-ECC Materials.

Keywords: shape memory alloys; engineered cementitious composites; composites materials; self-recovery capacity; bending behavior

Citation: Qian, H.; Zhang, Q.; Zhang, X.; Deng, E.; Gao, J. Experimental Investigation on Bending Behavior of Existing RC Beam Retrofitted with SMA-ECC Composites Materials. *Materials* **2022**, *15*, 12. https://doi.org/10.3390/ma15010012

Academic Editors: Cheng Fang, Canxing Qiu, Yue Zheng and Alessandro P. Fantilli

Received: 24 November 2021
Accepted: 18 December 2021
Published: 21 December 2021

1. Introduction

At present, among many structural forms in China, reinforced concrete (RC) structure is the most widely used structural type. As the service time of the structure increases, the performance of reinforced concrete will be affected by factors, such as adverse environment, aging, concrete carbonization, etc. [1]. These factors will not only affect the mechanical performance of RC structure, but also threaten the personal safety of the users. Therefore, in order to make existing RC structure meet the requirements of using functions and safety, it is great significance to carry out the research on the strengthening and retrofitting of existing RC structures. In recent years, many scholars have proved that the use of high-performance materials and intelligent structural members can better serve the strengthening and retrofitting of existing RC structures [2].

Shape Memory Alloys (SMA) is a new type of intelligent material with shape memory effect, super-elasticity, high damping, and fatigue resistance. If SMA is used as the longitudinal reinforcement of concrete beam, it can provide good self-recovery capacity for concrete beam. However, due to the high price of SMA, it is rarely used in new structures. It can still be used in the strengthening works of some important structures. Many scholars in the world have carried out a series of research on the self-recovery structural system based on SMA. For example, energy dissipation bracings [3,4], dampers [5–8], composite isolation bearings [9–12], energy dissipation coupling beams [13,14], etc. Great progress has been made in such areas. For the seismic performance of structural members, e.g., SMA reinforced beams [15–17], SMA reinforced pier columns [18–20], SMA reinforced shear walls [21,22] and joints [23–25], as well as the structural strengthening and retrofitting technology based on SMA materials [26–28], have been studied.

Engineered Cementitious Composites (ECC) is a kind of high-performance cementitious composite with obvious strain hardening characteristics and good crack control ability [29,30]. Ding et al. [31], Wu et al. [32], Yang et al. [33], and Said et al. [34] have carried out the research on beams, columns, walls, joints, and other components casted with ECC, respectively. These studies indicates that, compared with ordinary concrete, ECC has excellent tensile performance, fine cracking mechanism, and good ductility. It can solve various problems in engineering maintenance and strengthening works, such as improving impermeability, crack resistance, structural durability, and so on. ECC can also improve the bearing capacity and seismic performance of those engineering structural members at the same time.

For the composite structure of SMA reinforced ECC, scholars have studied beams [35,36], pier columns [37,38], shear walls [39], joints [40], and other structural members. This research indicates that the combination with SMA and ECC can insure both ECC and SMA in use with their optimal capacity respectively, and thereby satisfy the structural demands.

Enlarged section method is a traditional strengthening method of concrete structure. It is a strengthening method to improve the bearing capacity of original members by increasing section area and reinforcement area. This method can significantly improve the mechanical performance of members because of the increase of member section. However, the component size becomes larger after strengthening, which may affect the serviceability of the structure. Therefore, the premise of this strengthening method is that it does not affect the serviceability of the structure. At present, the strengthening method of concrete structure pasted with FRP has also been widely studied [41–43]. Its advantage is that the strength and durability of structural members can be improved without increasing the self-weight of the structure and the member section. However, the fire resistance of FRP is poor, and the fire prevention treatment further increases the cost of strengthening works.

In summary, the durability of concrete can be improved significantly by the superelasticity of SMA and high toughness and fine cracking mechanism of ECC. Therefore, this paper proposes to use the enlarge section area of SMA reinforced ECC to strengthen the existing RC beams. Four types of strengthened beams were designed and fabricated. Through monotonic cycle loading tests, the influences on the bearing capacity, energy dissipation performance, and self-recovery capacity of the test beams with different strengthening materials are investigated, especially the bending behavior of the beams strengthened by SMA reinforced ECC.

2. Test Overview

2.1. Specimen Design

Due to the limitation of the loading capacity of the testing device, the section of the specimen needs to be controlled below then 130 × 130 mm. At the same time, in order to meet the requirements of the minimum thickness of concrete/ECC cover of enlarged section, the height of enlarged section must meet the minimum requirements of 30 mm. Based on the above principles, the member section is determined as: the existing beam length is 1000 mm, and the original beam section is a rectangle of width × height = 120 mm × 80 mm before the strengthening, and the upper and lower reinforcements are 2 HRB355 steel bars with diameter of 6 mm, and the stirrup is HPB300 steel bars with diameter of 6 mm and spacing of 100 mm. Strengthening is carried out after the existing beam has been fully cured. The strengthening method is: firstly, chisel off the 10 mm protective layer at the bottom of the test piece and then roughen the bottom surface; finally, the enlarged section will be poured at the beam bottom by secondary pouring, as shown in Figure 1. The cross section of the beam after strengthening is 120 mm × 110 mm, the enlarged section at the bottom of beam is Δh = 40 mm (including the chiseled 10 mm protective layer). The specifications of the test specimens are shown in Figure 1.

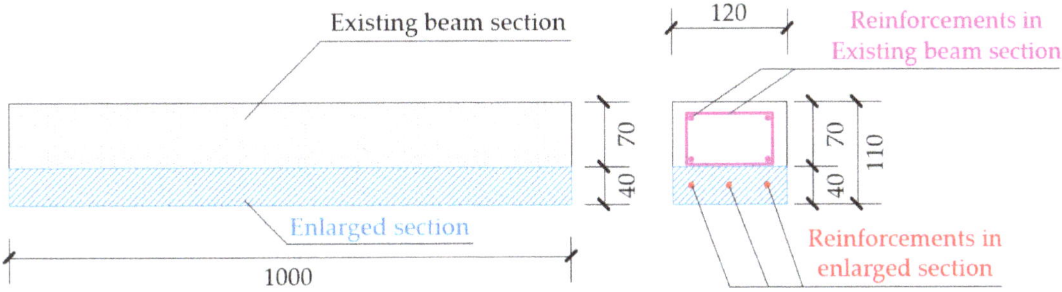

Figure 1. Specifications of the test specimens.

A total of 6 specimens with the same shape and size were produced in this test, of which SJ-1 is strengthened with steel reinforced concrete, SJ-2 is strengthened with SMA reinforced concrete, SJ-3 is strengthened with SMA reinforced ECC, and SJ-4 is strengthened with steel reinforced ECC. SJ-5 and SJ-6 are two spare test pieces, which are designed as the same as SJ-1 and SJ-3, respectively. The reinforcement ratio of enlarged section is designed according to the principle of the same total tensile bearing capacity of reinforcements in the enlarged section. The design parameters of specimens are shown in Table 1.

Table 1. The design parameters of specimens.

Serial Number of Specimen	Strengthening Material	Section Size (mm)	Beam Length (mm)	Reinforcement	Reinforcement Diameter
SJ-1	Steel-concrete	120 × 110	1000	2 HRB355 steel bars	6 mm
SJ-2	SMA-concrete	120 × 110	1000	3 SMA bars	5.5 mm
SJ-3	SMA-ECC	120 × 110	1000	3 SMA bars	5.5 mm
SJ-4	Steel-ECC	120 × 110	1000	2 HRB355 steel bars	6 mm
SJ-5	Steel-concrete	120 × 110	1000	2 HRB355 steel bars	5.5 mm
SJ-6	SMA-ECC	120 × 110	1000	3 SMA bars	5.5 mm

2.2. *Material Test of Specimens*

2.2.1. Shape Memory Alloy

In the material test, the test specimen of SMA bar has a diameter of 5.5 mm, a length of 250 mm, and a gauge length of 150 mm, as shown in Figure 2. The composition of SMA is Ti-56.35at%Ni, and the completion temperature of reverse martensitic transformation (A_f) is −10 °C. After the test piece is processed into an annealed state, it will be heat-treated. The heat treatment process of SMA bar is at 400 °C for 30 min, followed by water quenching [44].

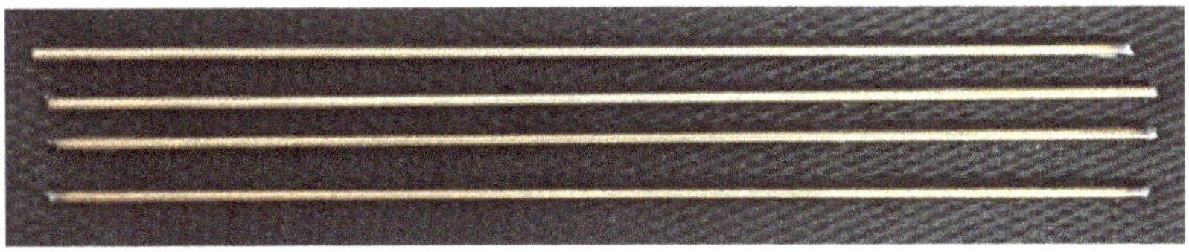

Figure 2. NiTi SMA bars with the diameter of 5.5 mm.

The test device adopts the CMT (Crane Motor Traction) electro-hydraulic servo universal material testing machine controlled by a microcomputer, as shown in Figure 3. In order to ensure the stability of the material performance of SMA, the SMA bar should be treated under thermal-cooling cycle treatment before the material test. The thermal-cooling cycle treatment method requires that the SMA bars should be placed in boiling water (100 °C) for 5 min, and then taken out and placed in cold water for 5 min. This treatment method was performed alternately five times before the test. Finally, the test specimens should be taken out from boiling water and cooled naturally at room temperature. The material tensile tests were performed on SMA bars with increasing strain amplitudes as 1%, 2%, 3%, 4%, 5%, 6%, etc. The material tensile tests data are shown in Figure 4.

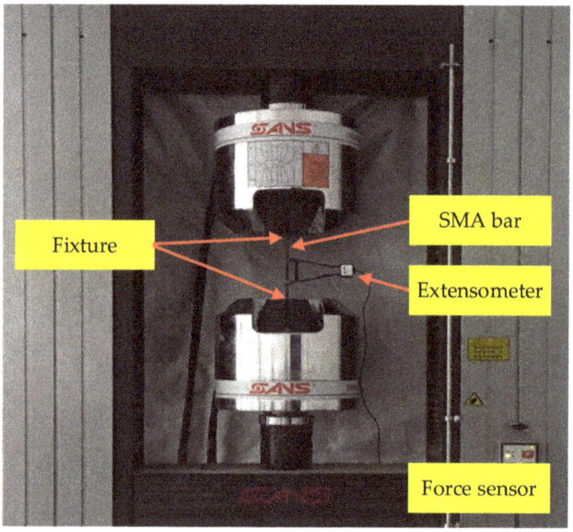

Figure 3. CMT electro-hydraulic servo universal material testing machine.

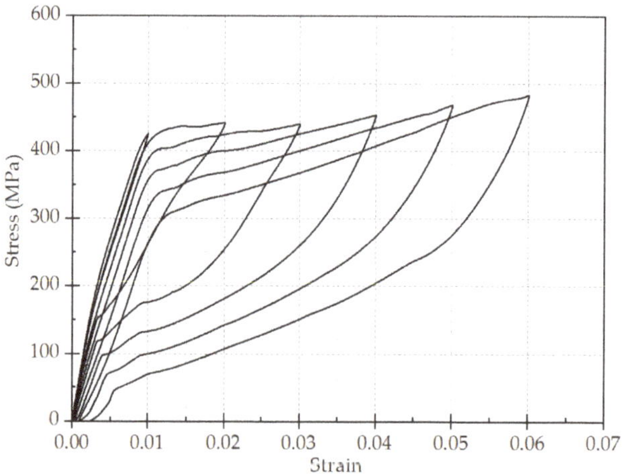

Figure 4. Stress–strain relationship curve of NiTi SMA bars under cyclic tensile load.

Through the analysis of the material tensile test results, it can be concluded that:

(1) With the increase of strain amplitude, the phase transformation stress of the hyperelastic SMA bar gradually decreases, the recovery stress gradually increases, and finally tends to be stable with the decrease of strain amplitude. Therefore, the SMA bar is stretched under circulation of loading and unloading is conducive to the stability of its material properties before it is used.

(2) The phase transformation stress and recovery stress tend to be stable with the increase of loading cycle; the residual strain gradually increases during the loading process, and the variation range becomes smaller and smaller. Therefore, SMA bars are stretched under circulation of loading and unloading before use, which is also conducive to improve the super-elasticity of SMA.
(3) As the strain amplitude increases, the residual strain of SMA gradually increases, and the maximum residual strain is only 0.003, indicating that the SMA bars used in the material tests have good recovery ability. With the increasing loading cycle, the increasing rate of residual strain gradually slows down, and the residual strain tends to be stable.

In summary, monotonic cycle loading can make the mechanical properties of SMA more stable, in order to ensure that the material properties of SMA can be significantly displayed in the tests.

2.2.2. ECC and Ordinary Concrete

This test uses concrete with a strength grade of C30, and ECC adopts high-strength PVA (polyvinyl alcohol) fiber-reinforced cement mortar. Its components include cement, water, fly ash, fine sand, PVA fiber, and admixtures, which are configured according to the mix proportion given in Table 2. Among them, the content of PVA fiber is 2% by volume, the specification of PVA fiber is A 0.02 × 8, and the tensile strength is 1400 MPa. Tensile tests are performed on 3 ECC test specimens, and the test results are given in Table 3. The average tensile strength of the test pieces was 3.87 MPa, which indicates that ECC has good ductility. ECC tensile stress–strain curve is shown in Figure 5.

Table 2. Mix proportion of ECC.

Element	Cement	Water	Fly Ash	Fine Sand	Admixture	PVA Fiber
Proportion	1	1.43	1.43	0.86	0.18	2%

Table 3. Tensile test results of ECC specimens.

Specimen Number	First Group	Second Group	The Third Group	Average Value
Tensile strength (MPa)	4.26	3.89	3.46	3.87

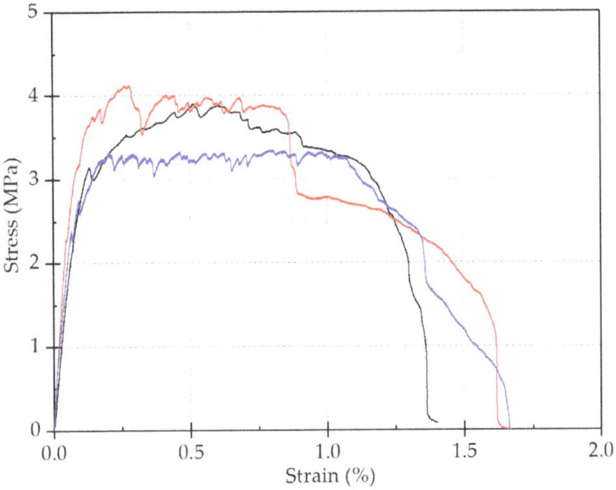

Figure 5. ECC tensile stress–strain curve.

2.3. Monotonic Cycle Loading Test

2.3.1. Test Device and Loading System

This test loading device adopts the CMT microcomputer controlled electro-hydraulic servo universal testing machine in the Structure Laboratory of Zhengzhou University, and the measuring range of force sensor is 2000 kN. The test adopts four-point bending loading, and the force were applied to the trisection point of the beam by the distributing beam. All test data are automatically collected by the test software of Power Test V3.4. The loading device is shown in Figure 6. The measuring points of strain of reinforcements are arranged in the middle of the span.

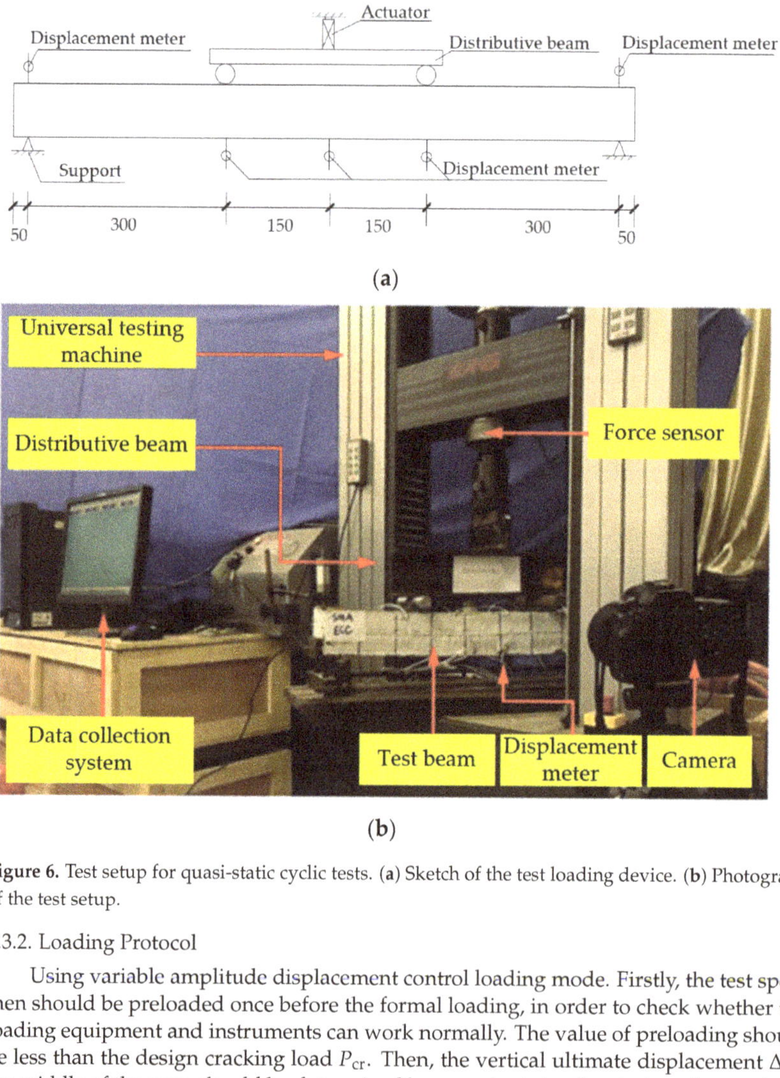

Figure 6. Test setup for quasi-static cyclic tests. (**a**) Sketch of the test loading device. (**b**) Photograph of the test setup.

2.3.2. Loading Protocol

Using variable amplitude displacement control loading mode. Firstly, the test specimen should be preloaded once before the formal loading, in order to check whether the loading equipment and instruments can work normally. The value of preloading should be less than the design cracking load P_{cr}. Then, the vertical ultimate displacement Δ in the middle of the span should be determined by monotonic loading test on the specimen of SJ-5. The results of monotonic loading test on the specimen of SJ-5 shows that the Δ is about 15 mm. Therefore, in the formal loading, the displacement of the initial loading cycle is 1.5 mm, and then the displacement of each cycle increases by 1.5 mm progressively. The monotonic cycle loading protocol is shown in Figure 7.

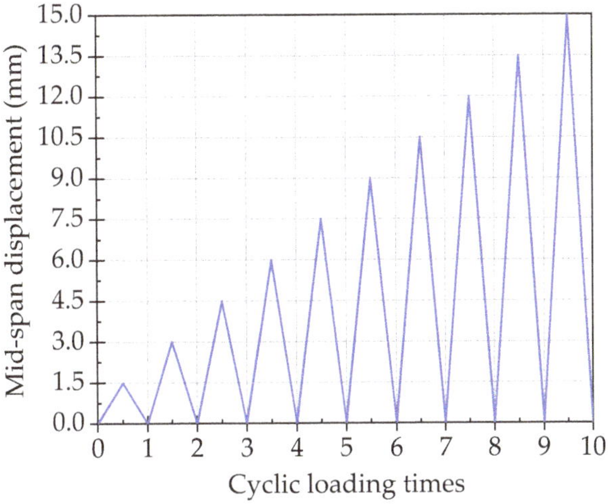

Figure 7. Monotonic cycle loading protocol.

3. Test Results and Analysis

3.1. Failure Process

(1) SJ-1 (Strengthened by Steel reinforced concrete)

Six cracks were observed in the pure bending section of the specimen. As the load increased, the width of the cracks increased, and finally part of the concrete was crushed. After unloading, the number of cracks and the crack widths remain basically unchanged, and the ultimate bearing capacity was 32.16 kN. The failure mode of the specimen after unloading is shown in Figure 8a.

Figure 8. Failure modes of specimens. (**a**) SJ-1; (**b**) SJ-2; (**c**) SJ-3; (**d**) SJ-4.

(2) SJ-2 (Strengthened by SMA reinforced concrete)

The failure mode of specimen SJ-2 was similar to that of SJ-1, and the number of vertical cracks observed by SJ-2 is slightly reduced. After unloading, the maximum crack width at the beam bottom decreased, some small cracks were closed, and the ultimate bearing capacity was 27.17 kN. The failure mode of the specimen after unloading is shown in Figure 8b.

(3) SJ-3 (Strengthened by SMA reinforced ECC)

Due to the characteristics of ECC, there was no obvious main crack in the strengthened section at the bottom of the beam until the end of loading. In the existing beam section, there were still main cracks that appeared. With the increase of loading cycle, the number of micro-cracks in the strengthened section of the specimen increased, but the width of these cracks increased slowly. Finally, more than 70 micro-cracks were counted in the pure bending section of the strengthened section. The crack distribution of specimen SJ-3 was shown in the Figure 9. After unloading, most of micro-cracks were closed, only 3 cracks were still observed, and the ultimate bearing capacity of the specimen was 25.31 kN. The failure mode of the specimen after unloading is shown in Figure 8c.

Figure 9. The crack distribution of SJ-3.

(4) SJ-4 (Strengthened by steel reinforced ECC)

ECC material was also used in specimen SJ-4, so the development of cracks during loading was basically similar to SJ-3, mainly a large number of fine micro-cracks. With the increase of the load, the number of micro-cracks in the strengthened section of the specimen increased obviously. Finally, more than 40 micro-cracks were counted in the pure bending section of the strengthened section. The width of the cracks was small, and it appeared as obvious multiple micro-cracking when it failed. After unloading, the crack width was basically unchanged, and the ultimate bearing capacity was 30.16 KN. The failure mode of the specimen after unloading is shown in Figure 8d.

It can be clearly seen from Figures 8 and 9 that there are no obvious cracks at the connection interface between the enlarged section and existing beam section of all specimens, which indicates that the bonding performance between the enlarged section and existing beam section is reliable. After strengthening, the enlarged section and existing beam section are commonly worked together and deformed harmoniously.

3.2. Load–Displacement Curves

The load–displacement curves for all the specimens are shown in Figure 10. By comprehensively comparing the load–displacement curves of the four specimens, it can be seen that:

(1) Comparing the number of loading cycles of the load–displacement curves of the four specimen, it can be found that the number of loading cycles of beams strengthened by ECC is more than that of beams strengthened by concrete, indicating that strengthening with ECC can significantly improve the ductility of specimens. Among all tested members, the ductility of SJ-3 is the best, followed by SJ-4, and the ductility of SJ-1 and SJ-2 are the worst.

(2) The ultimate bearing capacities of SJ-1 and SJ-4 are higher than that of the other two specimens, because both SJ-1 and SJ-4 are reinforced by steel bars. The total tensile

capacity of steel reinforcements is slightly greater than that of SMA bars, and the bond strength of ribbed steel bar in concrete or ECC is better than that of the SMA bar.

(3) The residual deformation of SJ-2 and SJ-3 after unloading is smaller than the other two specimens. However, the self-recovery performances of SJ-2 and SJ-3 are still not obvious, and the super-elasticity of SMA is not significantly displayed while strengthening beams.

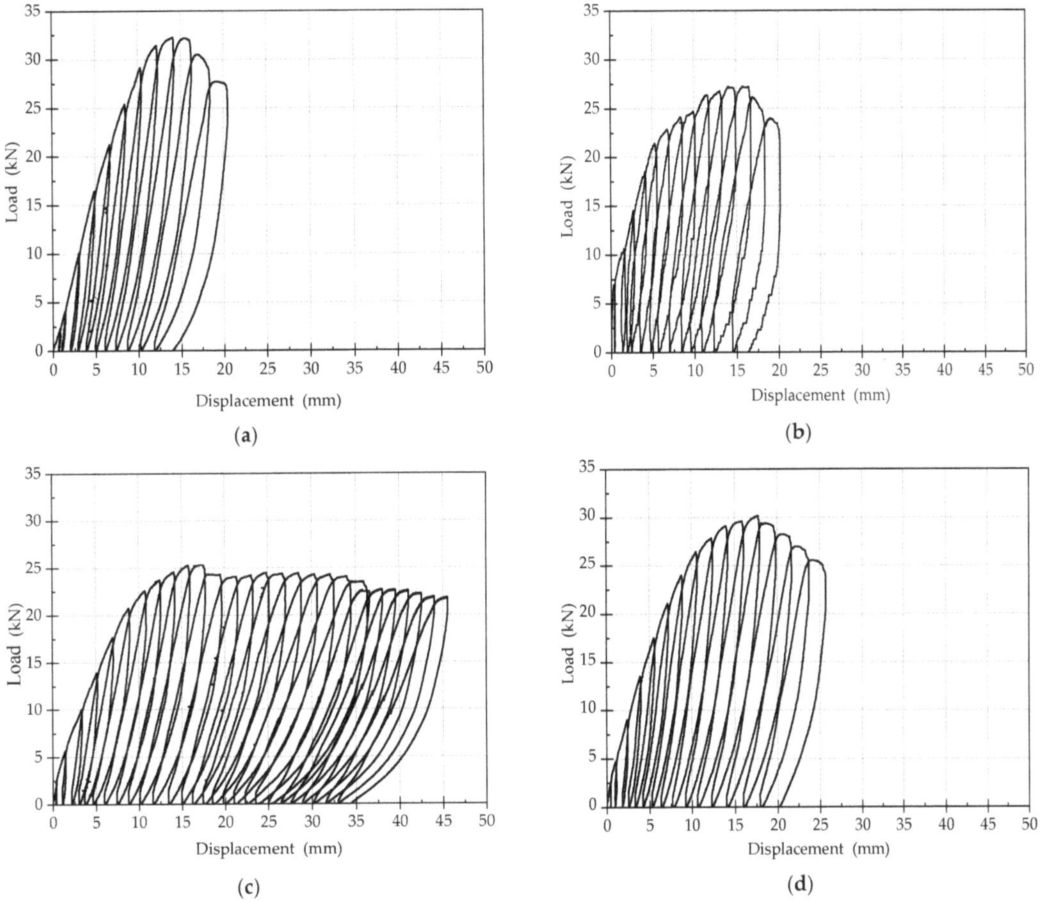

Figure 10. Load displacement curves (**a**) SJ-1; (**b**) SJ-2; (**c**) SJ-3; (**d**) SJ-4.

3.3. Skeleton Curves

The skeleton curves of 4 specimens are shown in Figure 11. By comprehensively comparing those skeleton curves, it can be seen that:

(1) All the test processes of the four specimens have gone through three stages, respectively, which are elastic stage, elastic-plastic stage, and failure stage.

(2) Replacing the steel bars in the concrete enlarged section by SMA bars with equivalent strength, the ultimate bearing capacity of the beam reduces by 15.5%. If the concrete is replaced by ECC, the ultimate bearing capacity decreases by about 6%. However, the number of loading cycles of ECC specimen is significantly more that of the concrete specimen, which shows that ECC can improve the ductility of specimens.

(3) After reaching the ultimate bearing capacity, the bearing capacity of SJ-3 decreases significantly slower than that of the other three members, followed by SJ-4, and SJ-1 decreases the fastest. It indicates that the ductility of SJ-3 and SJ-4 is significantly improved while being strengthened by ECC.

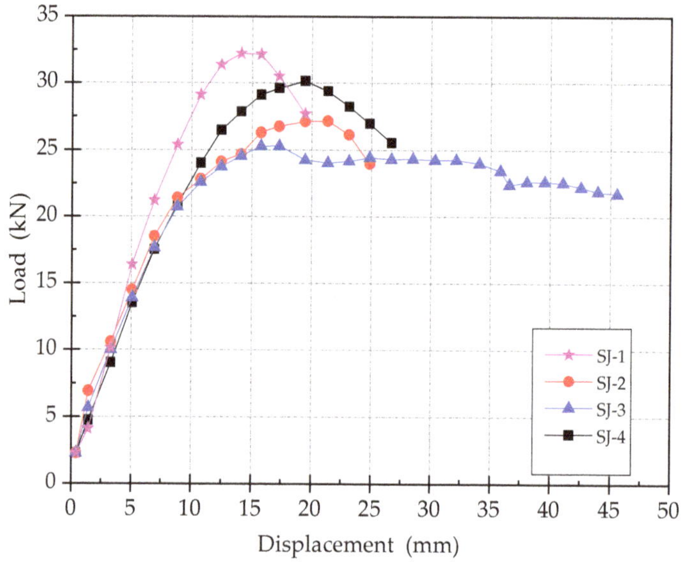

Figure 11. Skeleton curves.

3.4. Maximum Crack Width

The curves of maximum crack width distribution during loading and unloading are shown in Figure 12. The following conclusions can be drawn from the comparison between those curves:

(1) As the number of loading cycle increases, the cracks continue to develop and the crack width becomes wider. After unloading, the values of maximum crack width for all specimens reduce, but the reductions are quite different.

(2) By comparing the maximum crack width before and after unloading, it can be seen that the reduction of the maximum crack width of SJ-1 and SJ-4 after unloading is very small and can be basically ignored. The maximum crack width of SJ-2 and SJ-3 decreases obviously after unloading, the decreasing rates are 19.2% and 31.8%, which indicates that the self-recovery performances of specimens can be improved while strengthening with SMA. Due to the use of plain SMA bars as the reinforcement, the bond strength between SMA bars and concrete/ECC is small. Therefore, the super-elasticity of SMA cannot be fully utilized in the deformation process of the specimen. That is the reason why the specimens of SJ-2 and SJ-3 can only be partially recovered after unloading.

(3) By comparing the values of maximum crack width for all specimens, the maximum crack widths of SJ-3 and SJ-4 are much smaller than the other 2 specimens, which are less than 500μm before the 16th loading cycle. Due to its good ductility, the specimen of SJ-3 can be continuously loaded until the vertical displacement in mid-span reaches 42 mm (the 28th loading cycle). The maximum crack width of SJ-3 at this time is only 1078 μm, which is still less than the maximum crack widths of SJ-1 and SJ-2. It proves that using ECC as reinforcement layer can effectively control the development of crack width in the tensile zone of the beam section.

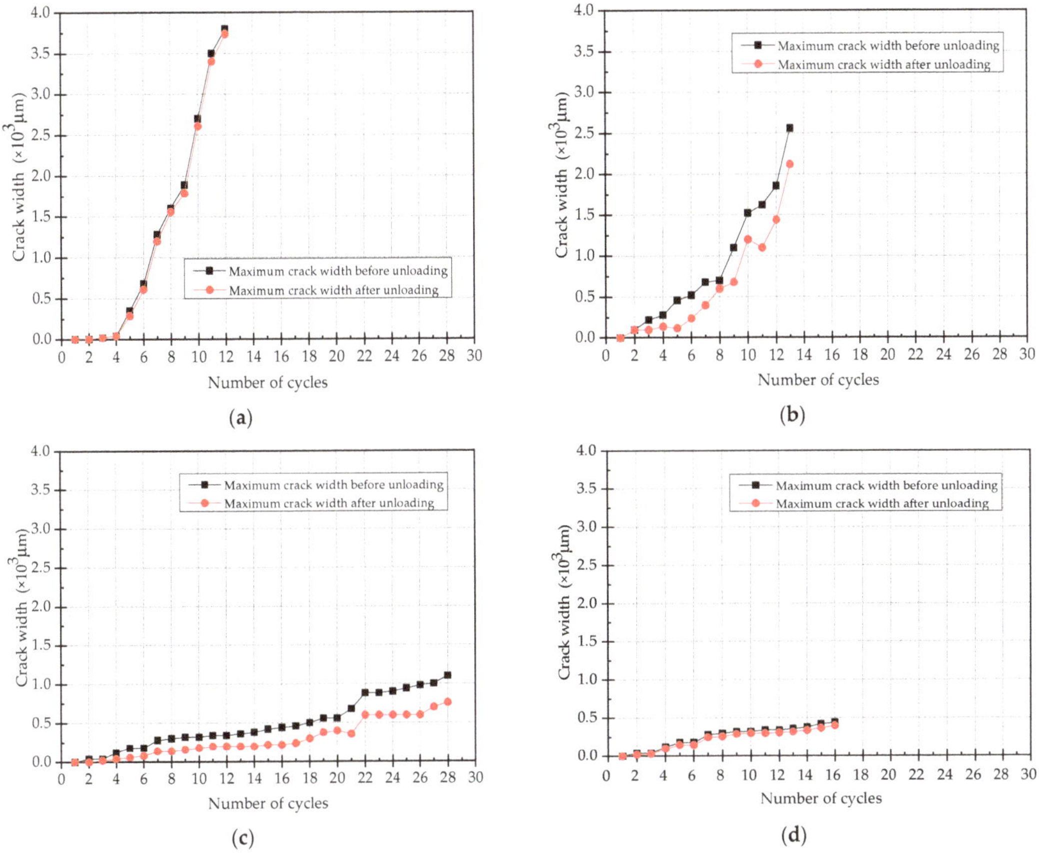

Figure 12. The curves of maximum fracture width under loading and unloading. (**a**) SJ-1; (**b**) SJ-2; (**c**) SJ-3; (**d**) SJ-4.

3.5. Number of Cracks

The curves of number of cracks for each specimen is shown in Figure 13. By comparing the results of data analysis, it can be found that:

(1) For specimens strengthened with ECC, the number of cracks in the tensile area of beam section is significantly more than that of concrete specimens. With the increase of loading cycles, ECC strengthened specimens will quickly produce new cracks, but the crack width does not increase significantly. However, after the cracking of concrete specimens, the crack width increases with the increasing load in order to form obvious main cracks, and the number of cracks does not increase significantly in the later cycles of loading.

(2) By comparing the curves of number of cracks of SJ-2 and SJ-3, it can be seen that the number of cracks of SJ-2 does not reduce significantly after unloading, but the number of cracks of SJ-2 decreases significantly after unloading. It indicates that the development of fine cracks is conducive to the shape memory effect and superelasticity of SMA. The self-recovery performance of beams can be better realized by strengthening beams with SMA reinforced ECC layer.

(3) By comparing the curves of number of cracks of SJ-1 and SJ-2, two curves are basically the same, and the number of cracks after unloading does not decrease. This is because

the bonding performance between SMA and concrete is poor, so the super-elasticity of SMA is not effective under this situation.

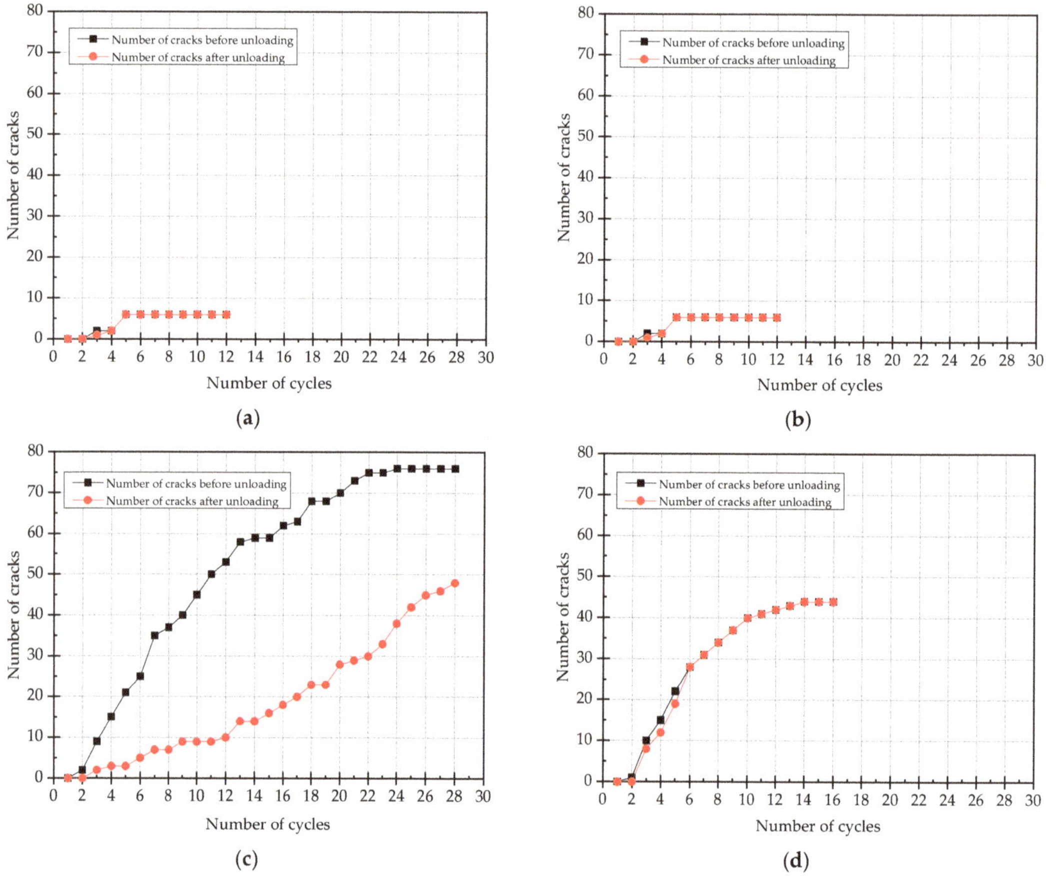

Figure 13. The curves of number of cracks under loading and unloading. (**a**) SJ-1; (**b**) SJ-2; (**c**) SJ-3; (**d**) SJ-4.

3.6. Mid-Span Deflection

The curves of mid-span deflection for all the specimens are shown in Figure 14. By comparing these curves, it can be seen that as the loading cycle increases, the mid-span deflection of the specimen increases linearly. After unloading, the mid-span deflections of the four specimens recover, while the self-recovery performance of SJ-3 is obviously the best, followed by SJ-2 and SJ-4, and the value of recovered mid-span deflection for SJ-1 is the minimum. The maximum recovery rates for all components are 28.4% for SJ-1, 42.7% for SJ-2, 26.1% for SJ-3, and 27.1% for SJ-4, respectively. It indicates that: (1) The mid-span deflection of the strengthened beam can be actively recovered by use of shape memory effect and super-elasticity of SMA; (2) The self-recovered value of mid span deflection of beams strengthened with ECC is obviously better than that of concrete beams, which proves that the failure mode of fine cracks of ECC can provide good conditions for self-recovery of the strengthened beams after unloading; (3) The recovery value of mid span deflection of ECC members is large, but due to the low stiffness of ECC members, the recovery rate is less than that of concrete members; (4) The recovery value of mid span deflection of ECC members is large, but due to the low stiffness of ECC members, the recovery rate is less than that of concrete members

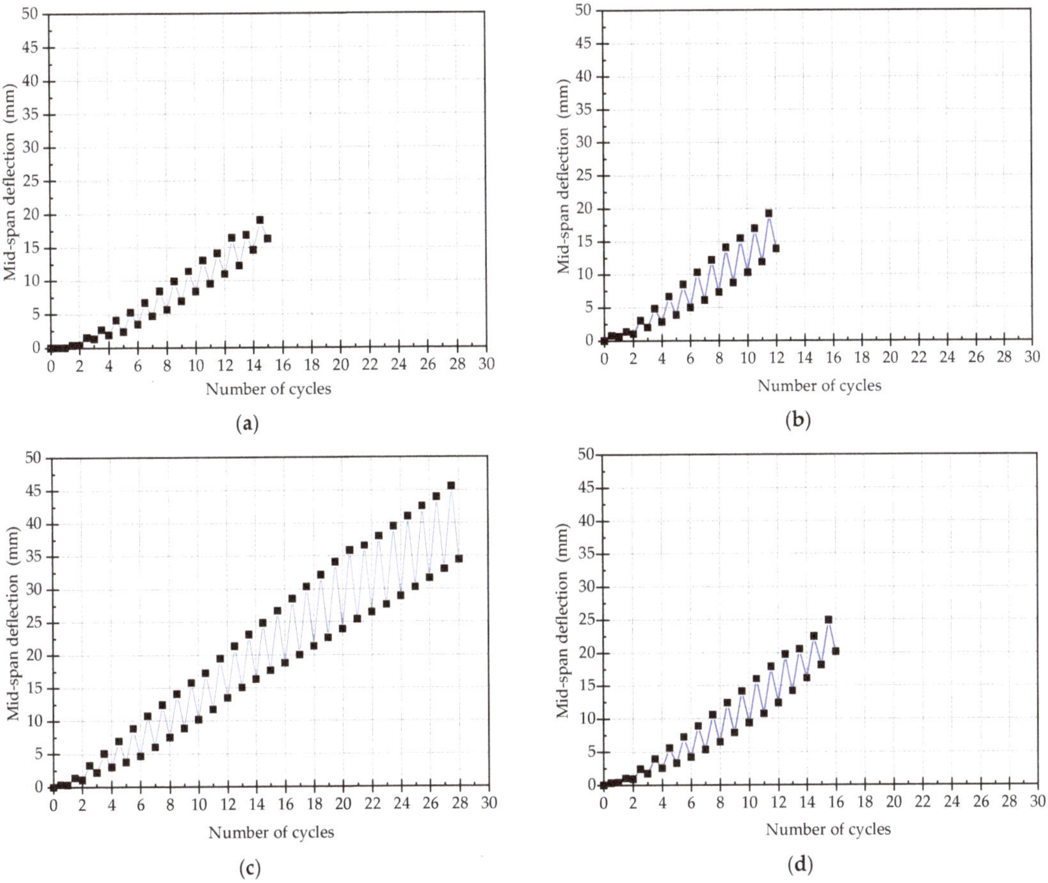

Figure 14. The curves of mid-span deflection. (**a**) SJ-1; (**b**) SJ-2; (**c**) SJ-3; (**d**) SJ-4.

3.7. Energy Consumption Capacity

The energy consumption capacity of the specimen can be determined by the area enveloped by the load–displacement curve of each level of loading. The curves of energy consumption capacity for all the specimens are shown in Figure 15. It can be seen that the energy consumption capacity of the specimen SJ-3 is the highest. Before the 22nd cycle, the energy dissipation capacity is mainly borne by ECC, which is in the strain-hardening stage. At the 23rd cycle, the ECC layer began to failure, the energy consumption capacity is mainly borne by SMA at this moment, and it is significantly reduced. Before the specimen is completely failed, the values of energy consumption for all specimens are basically the same. This is because the steel bars inside the original beam section are still retained after strengthening of the beams. With the reinforcement of the enlarged section, all the reinforcements at the bottom section of the beam do not yield obviously during the tests. Therefore, there is no obvious difference in the energy consumption capacity between each specimen.

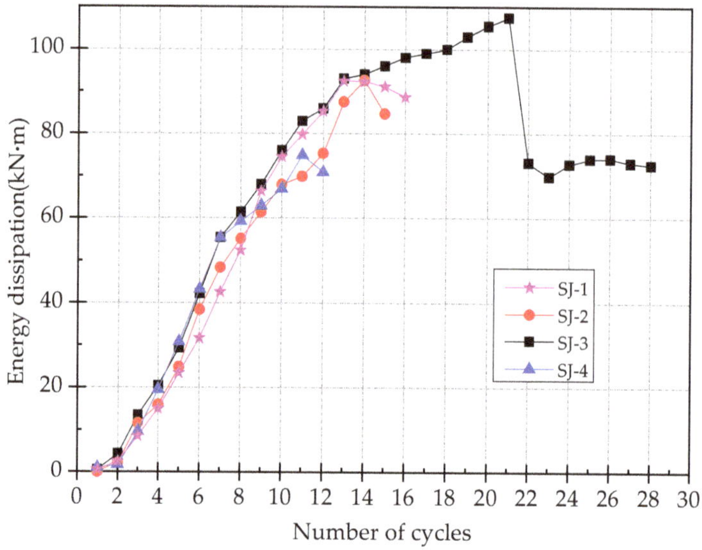

Figure 15. The curves of energy consumption capacity.

3.8. Mechanical Performance of Reinforcements

Figure 16 shows the load–strain relationship between the reinforcement in the enlarged section and the steel bar in original section. It can be concluded from the analysis of those results that: (1) The strain development of the original reinforcements is basically similar, which shows that all the strengthening methods of the specimens can give full play to the material properties of the original reinforcements, and the mechanical performances of the strengthened specimens are good. (2) The strain development of reinforcements in the enlarged section is also relatively stable in the elastic stage. With the increase of load, the reinforcement of SJ-1 yields first, followed by SJ-2 and SJ-4, and the reinforcement of SJ-3 yields last. This shows that ECC in the enlarged section can bear part of the tensile force, which makes the reinforcements in the tensile area yield later so as to improve its recovery ability.

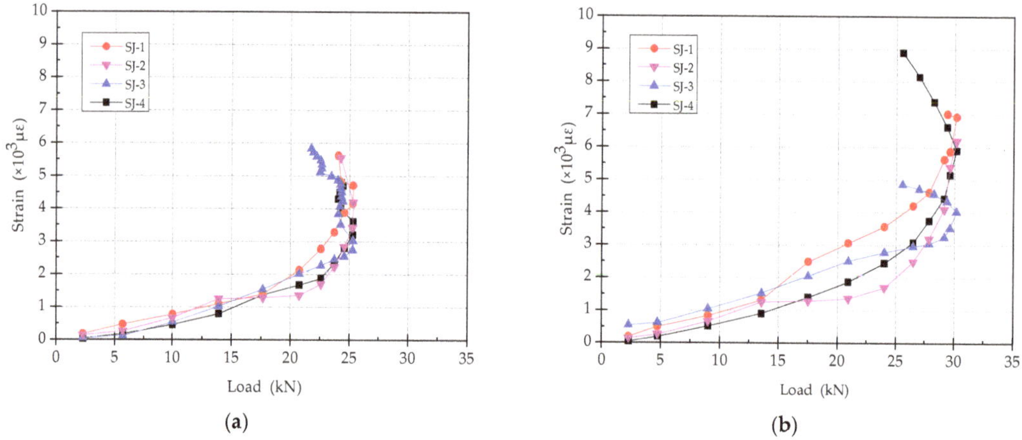

Figure 16. The load–strain relationship between the reinforcement. (**a**) The steel bar in existing beam section; (**b**) The reinforcements in enlarged section.

4. Flexural Capacity Formula

4.1. Basic Formula

According to the "Code for design of strengthening concrete structure" (GB50367-2013) [45], the flexural bearing capacity of the existing beam section should be determined according to Formula (1):

$$M \leq \alpha_s f_y A_s (h_0 - \frac{x}{2}) + f_{y0} A_{s0} (h_{01} - \frac{x}{2}) + f'_{y0} A'_{s0} (\frac{x}{2} - a') \quad (1)$$

where:

M—Design value of bending moment after strengthening of member (kN·m)
α_s—Strength utilization factor of reinforcements in enlarged section, taken as $\alpha_s = 0.9$
f_y—Design value of tensile strength of reinforcements in enlarged section (N/mm^2)
A_s—The cross-sectional area of the reinforcements in enlarged section (mm^2), as shown in Figure 17.

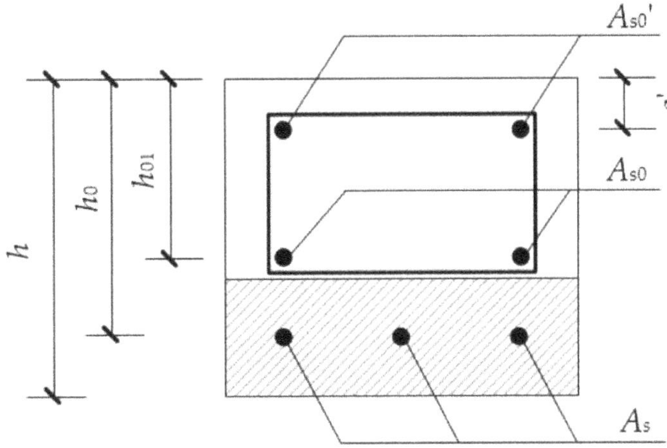

Figure 17. Dimensions of strengthened beam.

h_0, h_{01}—Effective height of section after strengthening and before strengthening (mm), as shown in Figure 17.
x—Height of compression zone of the section concrete (mm)
f_{y0}, f'_{y0}—Design value of tensile and compressive strength of steel bars in existing structure member (N/mm^2)
A_{s0}, A'_{s0}—The cross-sectional area of the tensile reinforcements and compressive reinforcements (mm^2), as shown in Figure 17.
a'—The distance from the resultant force point of longitudinal compressive reinforcements to the edge of compression zone of the beam (mm), as shown in Figure 17.

4.2. Flexural Capacity of ECC Reinforced Beams

Through the verification of the test results, it is found that when ECC is used to strengthen the tensile zone of the beam, due to the excellent tensile performance of ECC, the contribution of ECC to the bending capacity of the strengthened beam must be considered. Therefore, formula 1 needs to be revised. The revised formula is shown as Formulas (2) and (3).

$$M \leq \alpha_s f_y A_s (h_0 - \frac{x}{2}) + f_{y0} A_{s0} (h_{01} - \frac{x}{2}) + f'_{y0} A'_{s0} (\frac{x}{2} - a') + F_{t,ecc} \times (h - \frac{x}{2} - \frac{h_1}{2}) \quad (2)$$

$$F_{t,ecc} = f_{t,ecc} \times b \times h_1 \quad (3)$$

where:
$f_{t,ecc}$—Equivalent strength of ECC (N/mm^2), Use value of σ_{tc} in the tensile stress–strain relationship curve of the ECC shown in Figure 18 [46]
b—Section width (mm)
h_1—The height of enlarged section strengthened by ECC (mm), h_1 = 40 mm in these tests

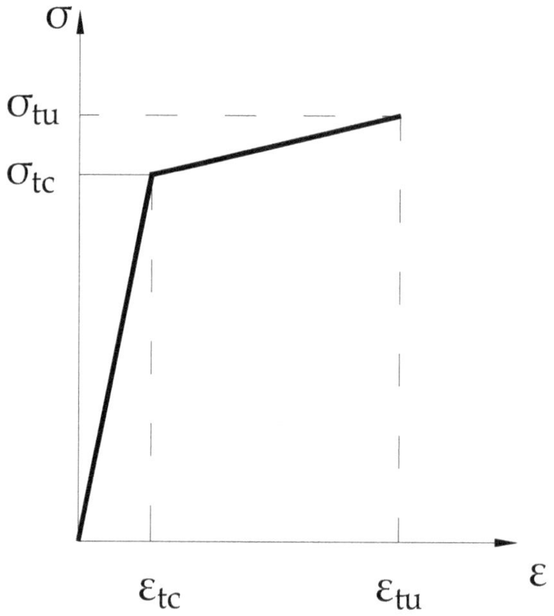

Figure 18. ECC tensile stress–strain curve.

4.3. *Verification of the Revised Formula*

By using the parameters of material properties in Tables 4 and 5, the theoretical value of flexural bearing capacity of each specimen can be calculated through the revised formula.

Table 4. Yield strength of longitudinal reinforcement.

Material	SMA	Tensile Longitudinal Bar	Compressed Longitudinal Bar
Yield Strength (MPa)	296.17	397.17	397.17

Table 5. Concrete strength.

Material	Concrete	ECC
Compressive strength (MPa)	17.48	18,021
Tensile strength (MPa)	-	5.1

The theoretical values and the experimental value of flexural bearing capacity for all 4 specimens are shown in Table 6, where M_{cu} is the calculated theoretical value of the bearing capacity of the beam strengthening with increasing section area, M_{tu} is the experimental value of the bearing capacity of the beam strengthening with increasing section area, which can be determined by the loading when the reinforcements are yielded, and M_{cu}/M_{tu} is the ratio of the theoretical value to the experimental value.

Table 6. The theoretical values and the experimental values of bending capacity for the strengthened beams.

Specimen Number	Reinforcement Material	M_{cu} (kN·m)	M_{tu} (kN·m)	M_{cu}/M_{tu}
SJ-1	Steel-Concrete	2.75	2.96	0.92
SJ-2	SMA-Concrete	2.51	2.76	0.91
SJ-3	SMA-ECC	2.51	2.78	0.90
SJ-4	Steel-ECC	2.75	2.93	0.94

It can be seen from Table 6 that the bending capacity calculated by Formula (2) for all the specimens are in good agreement with the test value, and the errors are all within 10%, indicating that the accuracy of the revised formula can be guaranteed. The values of M_{cu} are always less than the value of M_{tu}, representing that the theoretical values calculated by Formula (2) are much safer compared with the actual value, and the revised formula can be well applied to beam strengthening with increasing section of ECC.

5. Conclusions

(1) The effects of heat treatment, strain amplitude, and number of loading cycle on the mechanical properties of SMA were studied. The results show that the mechanical properties of SMA can be improved by heat treatment significantly; the stability of mechanical properties of SMA can be significantly improved by increasing the strain amplitude and loading cycle.

(2) The reinforced concrete beam strengthened with increasing section of ECC have good toughness, the cracking characteristics of the strengthened beam is the fine cracks when it fails. The strengthened beam can continue to bear the load beyond the ultimate bearing capacity, and the bearing capacity decreases slowly, indicating that the beam strengthening with increasing section of ECC has good energy dissipation capacity.

(3) The total tensile capacity of steel reinforcements is slightly greater than that of SMA bars, and the bond strength of ribbed steel bar in concrete or ECC is better than that of SMA bar. Therefore, the bearing capacity of specimens strengthened with steel is better than that of SMA bar.

(4) The crack width, number of cracks, and recovery performance of concrete beams strengthened with SMA bars are better than those of ordinary reinforced concrete beams. In order to give full play to shape memory effect and super-elasticity of SMA, the bond strength between SMA and concrete/ECC should be improved. The effect of temperature on the material properties of SMA cannot be ignored.

(5) The combination of SMA and ECC gives full play to their own respective advantages, respectively. ECC provides good toughness and cracking characteristics, and SMA provides excellent recovery ability. These two materials working together can significantly improve the reliability of the structure.

(6) Based on the design formula of bending capacity recommended by the design code and considering the tensile capacity provided by ECC in the strengthened section, a revised design formula for the bending bearing capacity of RC beams strengthened with increasing section of ECC is proposed. The revised design formula are well demonstrated by the test results, indicating that the revised formula can be well applied to the beam strengthening with increasing section of ECC.

The purpose of this paper is to reveal the bending capacity, failure mode, and self-recovery capacity of concrete beams strengthened with SMA/ECC enlarged section based on full-scale beam specimens with small dimensions. In the follow-up research, the influence of different design parameters, such as section size, flexure reinforcement ratio, and material strength, etc., on the flexural performance of concrete beams strengthened with SMA/ECC enlarged section will be further analyzed theoretically and experimentally to improve the design principle of the strengthening method and provide a theoretical basis for design of strengthening works.

Author Contributions: Conceptualization, H.Q. and X.Z.; methodology, Q.Z. and E.D.; software, Q.Z., E.D. and J.G.; validation, X.Z. and J.G.; formal analysis, Q.Z.; investigation, Q.Z. and X.Z.; resources, E.D.; data curation, Q.Z.; writing—original draft preparation, X.Z.; writing—review and editing, H.Q.; visualization, J.G.; supervision, X.Z.; project administration, H.Q.; funding acquisition, X.Z. All authors have read and agreed to the published version of the manuscript.

Funding: This research was funded by National Natural Science Foundation of China, Grant number 51987631, 51478438; The Key Research Projects of Henan Higher Education Institutions, Grant number 20A560002.

Conflicts of Interest: The authors declare no conflict of interest.

References

1. Pan, Y. Durability Analysis of Concrete Structure. *Appl. Mech. Mater.* **2013**, *2382*, 312. [CrossRef]
2. Tabrizikahou, A.; Kuczma, M.; Nowotarski, P.; Kwiatek, M.; Javanmardi, A. Sustainability of Civil Structures through the Application of Smart Materials: A Review. *Materials* **2021**, *14*, 4824. [CrossRef] [PubMed]
3. Qiu, C.X.; Zhu, S.Y. Shake table test and numerical study of self-centering steel frame with SMA braces. *Earthq. Eng. Struct. Dyn.* **2017**, *46*, 117–137. [CrossRef]
4. Qiu, C.X.; Zhu, S.Y. Performance-based seismic design of self-centering steel frames with SMA-based braces. *Eng. Struct.* **2017**, *130*, 67–82. [CrossRef]
5. Mirzai, N.M.; Mansouri, I.; Tezcan, J.; Awoyera, P.O.; Hu, J.W. Estimating optimum parameters of a new SMA damper under different earthquake ground motions. *Structures* **2021**, *33*, 2700–2712. [CrossRef]
6. Mirzai, N.M.; Cho, H.M.; Hu, J.W. Experimental study of new axial recentering dampers equipped with shape memory alloy plates. *Struct. Control. Health Monit.* **2020**, *28*, e2680. [CrossRef]
7. Wang, W.; Tang, Y.C.; Xu, W.K. Experimental and FEA Studies of a New Type SMA Viscous Damper. *Appl. Mech. Mater.* **2013**, *2545*, 353–356. [CrossRef]
8. Wang, B.; Zhu, S. Super-elastic SMA U-shaped dampers with self-centering functions. *Smart Mater. Struct.* **2018**, *27*, 55003. [CrossRef]
9. Wilde, K.; Gardoni, P.; Fujino, Y. Base isolation system with shape memory alloy device for elevated highway bridges. *Eng. Struct.* **2000**, *22*, 222–229. [CrossRef]
10. Jiang, S. The Earthquake Response Analysis of Re-Centring Isolation Structure with Shape Memory Alloys. Master's Thesis, Xi'an University of Architecture and Technology, Xi'an, China, 2004. (In Chinese).
11. Narjabadifam, P.; Noori, M.; Cardone, D.; Eradat, R.; Kiani, M. Shape memory alloy (SMA)-based Superelasticity-assisted Slider (SSS): An engineering solution for practical aseismic isolation with advanced materials. *Smart Struct. Syst.* **2020**, *26*, 89–102.
12. Shinozuka, M.; Chaudhuri, S.R.; Mishra, S.K. Shape-Memory-Alloy supplemented Lead Rubber Bearing (SMA-LRB) for seismic isolation. *Probabilistic Eng. Mech.* **2015**, *41*, 34–45. [CrossRef]
13. Qian, H.; Xu, J.; Zhang, X.; Deng, E.; Liu, Y.; Fan, J. Seismic performance of shear wall structure with self-centering energy-dissipating coupling beam. *J. Civ. Environ. Eng.* **2021**, *43*, 9–15. (In Chinese)
14. Li, S.B.; Liang, Q.G.; Li, J.P. Finite Element Analysis for the Crack Self-Repairing Behavior of the Concrete Beam Embedded Shape Memory Alloy. *Appl. Mech. Mater.* **2014**, *2972*, 507. [CrossRef]
15. Deng, Z.; Li, Q.; Sun, H. Behavior of concrete beam with embedded shape memory alloy wires. *Eng. Struct.* **2006**, *28*, 1691–1697. [CrossRef]
16. Elbahy, Y.I.; Youssef, M.A. Flexural behaviour of superelastic shape memory alloy reinforced concrete beams during loading and unloading stages. *Eng. Struct.* **2019**, *181*, 246–259. [CrossRef]
17. Azadpour, F.; Maghsoudi, A.A. Experimental and analytical investigation of continuous RC beams strengthened by SMA strands under cyclic loading. *Constr. Build. Mater.* **2020**, *239*, 117730. [CrossRef]
18. Saiidi, M.S.; Wang, H. Exploratory study of seismic response of concrete columns with shape memory alloys reinforcement. *ACI Struct. J.* **2006**, *103*, 436–443.
19. Billah, A.H.M.M.; Alam, M.S. Probabilistic seismic risk assessment of concrete bridge piers reinforced with different types of shape memory alloys. *Eng. Struct.* **2018**, *162*, 97–108. [CrossRef]
20. Zheng, Y.; Dong, Y. Performance-based assessment of bridges with steel-SMA reinforced piers in a life-cycle context by numerical approach. *Bull. Earthq. Eng.* **2019**, *17*, 1667–1688. [CrossRef]
21. Wang, B.; Zhu, S. Seismic behavior of self-centering reinforced concrete wall enabled by superelastic shape memory alloy bars. *Bull. Earthq. Eng.* **2018**, *16*, 479–502. [CrossRef]
22. Cortes-Puentes, L.; Zaidi, M.; Palermo, D.; Dragomirescu, E. Cyclic loading testing of repaired SMA and steel reinforced concrete shear walls. *Eng. Struct.* **2018**, *168*, 128–141. [CrossRef]

23. Youssef, M.A.; Alam, M.S.; Nehdi, M. Experimental Investigation on the Seismic Behavior of Beam-Column Joints Reinforced with Superelastic Shape Memory Alloys. *J. Earthq. Eng.* **2008**, *12*, 1205–1222. [CrossRef]
24. Nahar, M.; Islam, K.; Billah, A.M. Seismic collapse safety assessment of concrete beam-column joints reinforced with different types of shape memory alloy rebars. *J. Build. Eng.* **2020**, *29*, 101106. [CrossRef]
25. Azariani, H.R.; Esfahani, M.R.; Shariatmadar, H. Behavior of exterior concrete beam-column joints reinforced with Shape Memory Alloy (SMA) bars. *Steel Compos. Struct. Int. J.* **2018**, *28*, 83–98.
26. Nguyen, H.D.; Choi, E.; Nguyen, S.N.; Pham, T.K. Performance of self-centering devices containing superelastic SMA bars and their application via finite element analysis. *Eng. Struct.* **2021**, *237*, 112113. [CrossRef]
27. Gholampour, A.; Ozbakkaloglu, T. Understanding the compressive behavior of shape memory alloy (SMA)-confined normal- and high-strength concrete. *Compos. Struct.* **2018**, *202*, 943–953. [CrossRef]
28. Qian, H.; Ren, Z.; Xiong, J.; Kang, L. Experimental investigation on seismic performance of RC pier columns confined with shape memory alloy wires. *China Civ. Eng. J.* **2020**, *53*, 265–271. (In Chinese)
29. Zhang, J.; Leung, C.K.; Cheung, Y.N. Flexural performance of layered ECC-concrete composite beam. *Compos. Sci. Technol.* **2006**, *66*, 1501–1512. [CrossRef]
30. Yuan, F.; Pan, J.; Leung CK, Y. Flexural behaviors of ECC and concrete/ECC composite beams reinforced with basalt fiber-reinforced polymer. *J. Compos. Constr.* **2013**, *17*, 591–602. [CrossRef]
31. Ding, Y.; Yu, K.; Yu, J.; Xu, S.L. Structural behaviors of ultra-high performance engineered cementitious composites (UHP-ECC) beams subjected to ending-experimental study. *Constr. Build. Mater.* **2018**, *177*, 102–115. [CrossRef]
32. Wu, C.; Pan, Z.F.; Su, R.K.L.; Leung, C.K.; Meng, S. Seismic behavior of steel reinforced ECC columns under constant axial loading and reversed cyclic lateral loading. *Mater. Struct.* **2017**, *50*, 78–92. [CrossRef]
33. Yang, J.; Liang, S.T.; Zhu, X.J.; Wu, D.Y.; Wang, G.Y. Research on the Design Method of a Scale Model of ECC Reinforced Concrete Precast Shear Wall Structure. *Appl. Mech. Mater.* **2016**, *858*, 29–33. [CrossRef]
34. Said, S.H.; Abdul Razak, H. Structural behavior of RC engineered cementitious composite (ECC) exterior beam–column joints under reversed cyclic loading. *Constr. Build. Mater.* **2016**, *107*, 226–234. [CrossRef]
35. Li, X.; Li, M.; Song, G. Energy-dissipating and self-repairing SMA-ECC composite material system. *Smart Mater. Struct.* **2015**, *24*, 025024. [CrossRef]
36. Liu, X. The Characterization and Control of Shape Memory Alloy Cables for Reinforcement of Engineered Cementitious Composite Beams. Master's Thesis, University of Houston, Houston, TX, USA, 2013.
37. Wang, H. A Study of RC Columns with Shape Memory Alloy and Engineered Cementitious Composites. Master's Thesis, University of Nevada Reno, Reno, NV, USA, 2005.
38. Hosseini, F.; Gencturk, B.; Lahpour, S.; Gil, D.I. An experimental investigation of innovative bridge columns with engineered cementitious composites and Cu-Al-Mn super-elastic alloys. *Smart Mater. Struct.* **2015**, *24*, 085029. [CrossRef]
39. Rojas, M.A.S. Behavior of Engineered Cementitious Composite Repaired Superelastic-Shape Memory Alloy Reinforced Shear Walls. Master's Thesis, YORK University, Toronto, ON, Canada, 2020.
40. Khan, M.S. Seismic performance of deficient RC frames retrofitted with SMA-reinforced ECC column jacketing. *Innov. Infrastruct. Solut.* **2021**, *6*, 157. [CrossRef]
41. Aksoylu, C.; Yazman, Ş.; Özkılıç, Y.O.; Gemi, L.; Arslan, M.H. Experimental analysis of reinforced concrete shear deficient beams with circular web openings strengthened by CFRP composite. *Compos. Struct.* **2020**, *249*, 112561. [CrossRef]
42. Özkılıç, Y.O.; Yazman, Ş.; Aksoylu, C.; Arslan, M.H.; Gemi, L. Numerical investigation of the parameters influencing the behavior of dapped end prefabricated concrete purlins with and without CFRP strengthening. *Constr. Build. Mater.* **2021**, *275*, 122173. [CrossRef]
43. Aksoylu, C. Experimental analysis of shear deficient reinforced concrete beams strengthened by glass fiber strip composites and mechanical stitches. *Steel Compos. Struct.* **2021**, *40*, 267–285.
44. Wang, W.; Shao, H. Experimental Investigation on Mechanical Properties of Shape Memory Alloy Bars in Different Sizes. *Struct. Eng.* **2014**, *6*, 168–174. (In Chinese)
45. GB50367-2013. *Design Code for Reinforcement of Concrete Structures*; China Architecture Industry Press: Beijing, China, 2015. (In Chinese)
46. Li, Q. Theoretical Analysis and Experimental Investigation on Bending Performances of Reinforced Ultra High Toughness Cementitious Composites. Ph.D. Thesis, Dalian University of Technology, Dalian, China, 2009. (In Chinese)

materials

Article

Seismic Response Mitigation of a Television Transmission Tower by Shape Memory Alloy Dampers

Jingbo Wu [1], Bo Chen [1,*], Lunhai Zhi [2] and Xinxin Song [1]

[1] Key Laboratory of Roadway Bridge and Structural Engineering, Wuhan University of Technology, Wuhan 430070, China; wujingbo0618@whut.edu.cn (J.W.); songxinxinfx@163.com (X.S.)
[2] School of Civil Engineering, Hefei University of Technology, Hefei 230009, China; zhilunhai1979@163.com
* Correspondence: cebchen@whut.edu.cn

Abstract: High-rise television transmission towers are of low damping and may vibrate excessively when subjected to strong earthquakes. Various dynamic absorbers and dampers are proposed to protect television transmission towers from excessive vibrations and damages. Up to now, the seismic damage reduction in television towers, using SMA dampers under seismic excitations, has not been conducted. To this end, the response reduction in a flexible television tower, disturbed by earthquakes using SMA dampers, is conducted in this study. A two-dimensional dynamic model is developed for dynamic computation at first. The mathematical model of an SMA damper is proposed, and the equations of motion of the tower, without and with, are established, respectively. The structural dynamic responses are examined in the time and the frequency domain, respectively. The effects of damper stiffness, service temperature, hysteresis loops, and earthquake intensity on control efficacy are investigated in detail. In addition, the power spectrum density curves, of dynamic responses and the energy responses, are compared to provide deep insights into the developed control approach. The control performance of SMA dampers is compared with that of widely-used friction dampers. The analytical observations indicate that SMA dampers with optimal parameters can substantially reduce the vibrations of TV transmission towers under seismic excitations.

Keywords: television transmission tower; seismic excitation; shape memory alloy damper; parametric study; vibration control

Citation: Wu, J.; Chen, B.; Zhi, L.; Song, X. Seismic Response Mitigation of a Television Transmission Tower by Shape Memory Alloy Dampers. *Materials* **2021**, *14*, 6987. https://doi.org/10.3390/ma14226987

Academic Editor: Antonio Mattia Grande

Received: 19 August 2021
Accepted: 9 November 2021
Published: 18 November 2021

1. Introduction

To be important high-rise infrastructures, television (TV) transmission towers are widely used across the world for TV transmission and sightseeing. These TV transmission towers, with low damping, may vibrate excessively in severe environments, inducing possible damage and even failure [1,2]. For instance, a real steel TV transmission tower, with a height of 136m, in China collapsed under strong winds in 2012 [3]. The collapse of transmission towers subjected to strong earthquakes in China is also reported [4]. The simplest approach to reduce vibration of TV transmission towers is to increase inherent structural ductility, although it is often uneconomical. Thus, various control devices, such as dynamic absorbers and energy-dissipating dampers, are developed to protect TV transmission towers from excessive vibrations and damages [5–7]. The dynamic absorbers are firstly applied in the vibration control of high-rise towers. The wind-induced responses of high-rise towers are studied by Hitchcock et al. [8] and Balendra et al. [9], using tuned liquid dampers and active tuned liquid column dampers. Similar studies are carried out using tuned mass dampers (TMD), such as Yang et al. [10] for a steel tower, Wu and Yang [11] for the Nanjing TV tower, Ghorbani-Tanha et al. [12] for the Milad Tower, and Lu et al. [13] for Shanghai World Financial Center Tower. An obvious defect of dynamic absorbers for vibration control of TV transmission towers is the large space requirement, which often conflicts with the structural function in operation.

As an alternative approach, friction dampers and fluid dampers have been utilized in the structural vibration control of truss towers for many years [14–17]. Chen et al. [18] conducted the seismic response mitigation of TV transmission towers by using friction dampers. Similar control approaches have been developed by fluid dampers due to their simple configuration and low cost. For instance, Zhang and Li [19] investigated the responses, of a flexible tower under earthquakes, by fluid viscous dampers. Recently, smart materials are widely used in the field of vibration engineering to fabricate various smart dampers for structural vibration control. Xu et al. [20] carried out the semi-active control of a wind-excited tower by piezoelectric friction dampers, associated with a local feedback control algorithm. Chen et al. [21] conducted the semi-active control of a flexible tower subjected to winds using magnetorheological dampers.

A typical smart material, shape memory alloy (SMA) has been recently gaining great popularity due to the properties of high strength, super-elasticity, energy dissipation, better fatigue, and corrosion resistance [22–26]. SMA can exhibit excellent super-elasticity, and this feature is explored to fabricate SMA energy dissipation systems, including SMA bracings, SMA dampers, and connection elements, for vibration control. Zheng et al. [27] proposed a self-adaptive Ni-Ti SMA-cable-based frictional sliding bearing to mitigate pounding and unseating damage of bridges subjected to earthquakes. Liu et al. [28] investigated the seismic response control of a steel frame using base isolation systems with SMA springs.

However, the seismic response suppression, of TV transmission towers with SMA dampers, has not yet been studied. In this regard, the seismic behavior of TV transmission towers with SMA dampers is examined in this study. The two-dimensional (2D) lumped mass model is developed based on the three-dimensional (3D) finite element (FE) model for structural dynamic computation. The model of an SMA damper is proposed, and the equations of motion of the tower, without and with control, are established, respectively. Three schemes are studied and compared to determine the rational location of SMA dampers. The effects of damper stiffness, service temperature, hysteresis loops, and earthquake intensity on control efficacy are examined through a detailed parametric study. In addition, the power spectrum density (PSD) curves of dynamic responses are examined to explore the vibration properties in the frequency domain. The control performance of SMA dampers is compared with that of widely-used friction dampers. The energy responses of the uncontrolled and controlled towers are also compared to provide deep insights into the developed control approach. The made observations indicate that SMA dampers with optimal parameters are beneficial to the vibration control of TV transmission towers under seismic excitations.

2. Seismic Responses of a TV Transmission Tower

The 3D FE model of the TV transmission tower is firstly constructed with the aid of the commercial package ABAQUS, as displayed in Figure 1a. The configuration of the TV transmission tower is quite complicated, and the number of the dynamic degree-of-freedoms (DOFs) of the 3D FE model is quite large. Thus, the dynamic response computation is time-consuming, and the displacement-force relationship of SMA dampers is nonlinear. The dynamic analysis of the tower-damper system needs a lot of iterations at each time step. Thus, the response computation and parametric study on control performance are quite tedious and time-consuming for a complicated 3D FE model [18,20]. To this end, 2D lump models, with satisfactory accuracy, are commonly utilized to simplify dynamic computation and parametric study [18,20], as shown in Figure 1b. Thirty nodal floors are determined, and each floor is assumed to have only one dynamic DOF. The stiffness matrix is determined based on the flexibility matrix of the original 3D FE model.

During the construction of the 2D lumped dynamic model of the TV transmission tower, the structural masses are concentrated at limited nodal floors, and the mass matrix of the reduced 2D dynamic model is established using a lumped mass matrix. The stiffness matrix of the 2D lumped model can be determined based on the flexibility matrix of the 3D FE model of the tower. Apply the same horizontal unit force at the thirty node floors one by one and compute the flexibility coefficient of each floor. Then, form the flexibility matrix of the tower and inverse the flexibility matrix to obtain the stiffness matrix of the 2D model.

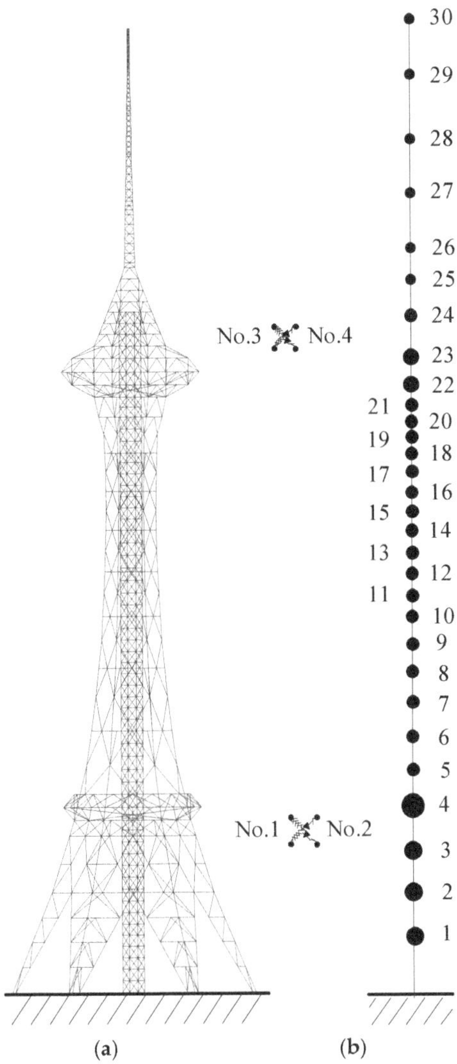

Figure 1. Analytical model of a TV transmission tower. (**a**) 3D FE model; (**b**) 2D dynamic model.

3. Model of SMA Damper

The model of SMA material can be described based on thermodynamic theory, and the commonly accepted constitutive model is:

$$\sigma - \sigma_0 = E(\xi)\varepsilon - E(\xi_0)\varepsilon_0 + \Omega(\xi)\xi_S - \Omega(\xi_0)\xi_{S0} + \Theta(T - T_0) \quad (1)$$

where σ and ε are the stress and strain of the SMA wire, respectively; σ_0 and ε_0 are the stress and strain of the SMA wire at the initial state, respectively; ξ is the volume percentage of the martensite; ξ_0 is the martensite content of the SMA wire at initial state; T_0 is the temperature of the SMA wire at initial state; ξ_{S0} is the initial volume content of the martensite; ξ_S is the volume content of the martensite; E is the Young's modulus of SMA material; Ω is the phase transition tensor; Θ is the thermo-elastic tensor of SMA material; T is the temperature at a reference state.

The Young's modulus of an SMA wire, related to the martensite content $E(\xi)$, is given by

$$E(\xi) = \xi E_M + (1-\xi)E_A \quad (2)$$

where E_A and E_M are Young's moduli at the austenite and martensite phases, respectively.

The inherited properties of SMA wires, such as self-centering, better fatigue, and corrosion resistance, are useful in the design of energy-dissipating dampers. An SMA damper with re-centering capacity is designed according to the super-elasticity and high damping characteristics of the SMA wire, as shown in Figure 2. The SMA damper consists of the outer tube, inner tube, SMA wires, and two circular plates. The SMA wire is designed always in tension when the damper works. The SMA wire can dissipate vibrant energy, during its reciprocal movement, to suppress the structural vibration.

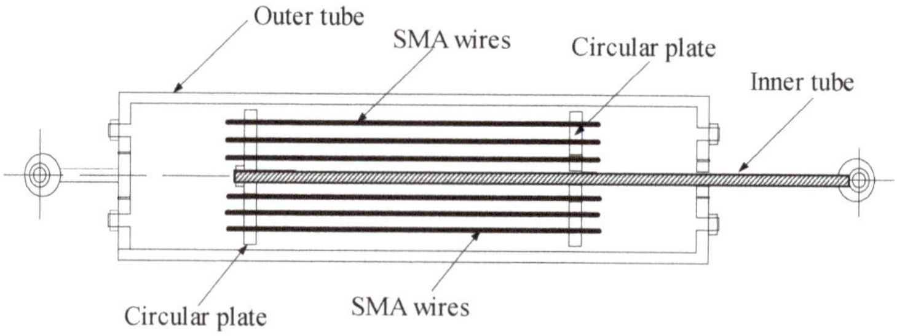

Figure 2. Schematic configuration of SMA damper.

The hysteretic model of the Ni-Ti SMA material is displayed in Figure 3. In the figure, M_f and M_s denote the martensite finish temperature and martensite starts temperature, respectively. A_f and A_s denote the austenite finish temperature and austenite start temperature, respectively; σ_{M_s} and ε_{M_s} represent the critical stress and strain of the Ni-Ti SMA material at M_s, respectively; σ_{M_f} and ε_{M_f} are the critical stress and strain at M_f, respectively; σ_{A_s} and ε_{A_s} are the critical stress and strain at A_s, respectively; σ_{A_f} and ε_{A_f} are the critical stress and strain at A_f, respectively; ε_L is the maximum residual strain.

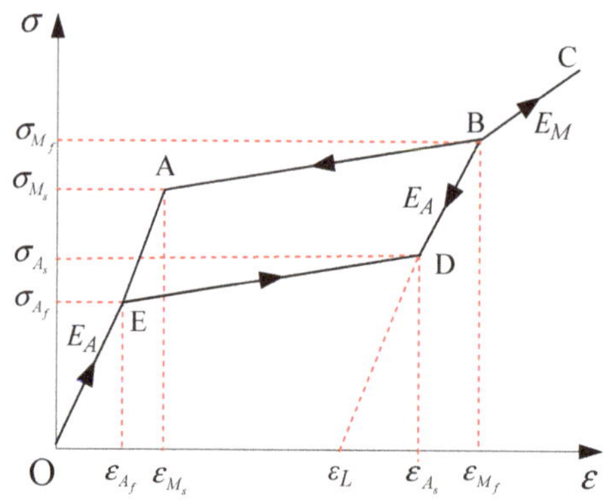

Figure 3. Flag-shaped hysteretic model of an SMA wire.

When the working temperature of the SMA material is given, the mechanical property of the SMA material will be known [29–31]. As illustrated in Figure 3, the relationships of strain and stress, on different paths associated with Ni-Ti SMA material, can be expressed as follows:

(1) Paths O-A and E-O are elastic (full austenite):

$$\sigma(t) = E_A \varepsilon(t) \tag{3}$$

The control force of the SMA damper $u(t)$ is:

$$u(t) = \frac{E_A A}{l_s}(x^d(t) - l_s) \tag{4}$$

where $x^d(t)$ is the length of the damper after deformation; A is the cross-sectional area of SMA wires; l_s is the initial length of the SMA wires.

(2) Path A-B (forward transformation):

$$\sigma(t) = \sigma_{M_s} + \frac{\sigma_{M_f} - \sigma_{M_s}}{\varepsilon_{M_f} - \varepsilon_{M_s}}(\varepsilon(t) - \varepsilon_{M_s}) \tag{5}$$

$$u(t) = \sigma(t)A \tag{6}$$

(3) Path B-C is elastic (full martensite):

$$\sigma(t) = \sigma_{M_f} + E_M(\varepsilon(t) - \varepsilon_{M_f}) \tag{7}$$

$$u(t) = \frac{E_M A}{l_s}(x^d(t) - l_s) \tag{8}$$

(4) Path B-D is elastic (full martensite)

$$\sigma(t) = \sigma_{M_f} + E_M(\varepsilon(t) - \varepsilon_{M_f}) \tag{9}$$

$$u(t) = \sigma_{M_f} A + E_M\left[(x^d(t) - l_s) - \varepsilon_{M_f}\right]A \tag{10}$$

(5) Path D-E (reverse transformation):

$$\sigma(t) = \sigma_{M_s} + \frac{\sigma_{M_s} - \sigma_{A_f}}{\varepsilon_{A_s} - \varepsilon_{A_f}}(\varepsilon(t) - \varepsilon_{A_s}) \tag{11}$$

$$u(t) = \sigma(t)A \tag{12}$$

in which $\sigma(t)$ and $\varepsilon(t)$ are the stress and the strain of SMA wires, respectively.

4. Dynamic Analysis of the Controlled Tower

The equations of motion of the controlled tower with SMA dampers is given by:

$$\mathbf{M}\ddot{\mathbf{x}}(t) + \mathbf{C}\dot{\mathbf{x}}(t) + \mathbf{K}\mathbf{x}(t) = -\mathbf{M}\mathbf{I}_g\ddot{x}_g(t) + \mathbf{H}\mathbf{u}(t) \tag{13}$$

$$\mathbf{u}(t) = \begin{bmatrix} u_1 & u_2 & \cdots & u_m \end{bmatrix} \tag{14}$$

where $\mathbf{x}(t)$, $\dot{\mathbf{x}}(t)$, and $\ddot{\mathbf{x}}(t)$ are the displacement, velocity, and acceleration responses of the TV transmission tower, respectively; $\mathbf{M}$, and $\mathbf{K}$ are the mass and stiffness matrices of the tower, respectively; $\mathbf{C}$ is the Rayleigh damping matrix; $\mathbf{I}_g$ is the influence vector for ground acceleration; $\ddot{x}_g(t)$ is the seismic acceleration; $\mathbf{H}$ is the position matrix of control force; $\mathbf{u}(t)$ is the control force of SMA dampers; m is the number of SMA dampers.

The incremental equations of the TV transmission tower are given by:

$$\mathbf{M}\Delta\ddot{\mathbf{x}}(t) + \mathbf{C}\Delta\dot{\mathbf{x}}(t) + \mathbf{K}\Delta\mathbf{x}(t) = -\mathbf{M}\mathbf{I}_g\Delta\ddot{x}_g(t) + \mathbf{H}\Delta\mathbf{u}(t) \tag{15}$$

where $\Delta\mathbf{x}(t)$, $\Delta\dot{\mathbf{x}}(t)$, and $\Delta\ddot{\mathbf{x}}(t)$ are the displacement increment, velocity increment, and acceleration increment, respectively; $\Delta\ddot{x}_g(t)$ is the increment of the ground acceleration.

The above incremental equation is solved by the Newmark-β method with a time step of Δt. It is noted that the increment vector of SMA dampers $\Delta \mathbf{u}(t)$ can only be determined under the known $\Delta \mathbf{x}(t)$ and $\Delta \dot{\mathbf{x}}(t)$ in the numerical computation. However, the displacement increment $\Delta \mathbf{x}(t)$ and velocity increment $\Delta \dot{\mathbf{x}}(t)$ is commonly determined based on $\Delta \mathbf{u}(t)$. Thus, a dynamic iteration process is developed to ensure computational convergence. The iteration process proceeds until $\Delta \mathbf{x}(t)$ satisfies the following equation:

$$abs\left[\left(\Delta \mathbf{x}(t)^{(i+1)} - \Delta \mathbf{x}(t)^{(i)}\right)/\Delta \mathbf{x}(t)^{(i)}\right] < \kappa \tag{16}$$

where κ is a prescribed error limitation related to an actual TV transmission tower.

Energy response is an important index to describe the vibration of TV transmission towers. The energy representation of the example tower is formed by integrating Equation (15). The absolute energy equation of the uncontrolled tower is given by:

$$E_{KE} + E_{DE} + E_{SE} = E_E \tag{17}$$

Similarly, the energy equation for the tower with SMA dampers is:

$$E_{KE} + E_{DE} + E_{SE} + E_{PE} = E_E \tag{18}$$

where E_E is the total input energy from earthquakes; E_{KE} and E_{SE} are kinetic energy and strain energy; E_{DE} and E_{PE} are the energy dissipated by structural damping and SMA dampers, respectively.

5. Case Study

5.1. Analytical Parameters

A real TV tower, with a height of 340 m, is used to examine the feasibility of the proposed control method, as displayed in Figure 1. Two turrets are developed for sightseeing. The 3D model of the TV tower is established by ABAQUS, and then, a 2D dynamic model is constructed by a developed computer program. The validity of the 2D model is examined by comparing structural dynamic characteristics, as listed in Table 1. The fundamental frequency of the 3D and 2D models are 0.129 Hz and 0.128 Hz, respectively, with an error of 0.621%. It is also found that the average error of the first eight natural frequencies is 3.641%. Thus, the dynamic characteristics of the 2D dynamic model are close to that of the original 3D model, and the validity of the 2D model can be guaranteed. To evaluate the effectiveness of the proposed control method, three far-field and two near-field historical records are selected [32–35]: (1) Taft earthquake on 21 July 1952; (2) El Centro earthquake on 18 May 1940; (3) Kobe earthquake on 17 January 1995; (4) Hachinohe earthquake on 16 May 1968; (5) Northridge earthquake on 17 January 1994, respectively. The original peak ground accelerations (PGAs) of the five seismic records are 1.559, 3.417, 8.178, 2.250, and 8.2676 m/s^2, as shown in Figure 4. All the earthquakes are scaled to have the same PGA of 4.0 m/s^2 for comparison. The time step and damping ratios in the dynamic analysis are 0.02 s and 0.01, respectively.

Table 1. Comparison of dynamic properties of the 3D and 2D models.

Freq.no	3D Finite Element Model	2D Dynamic Model	Error
1	0.129	0.128	0.621%
2	0.416	0.386	7.211%
3	0.729	0.781	−7.133%
4	1.171	1.163	0.683%
5	1.514	1.456	3.831%
6	2.221	2.142	3.556%
7	2.578	2.569	0.349%
8	2.632	2.783	−5.737%

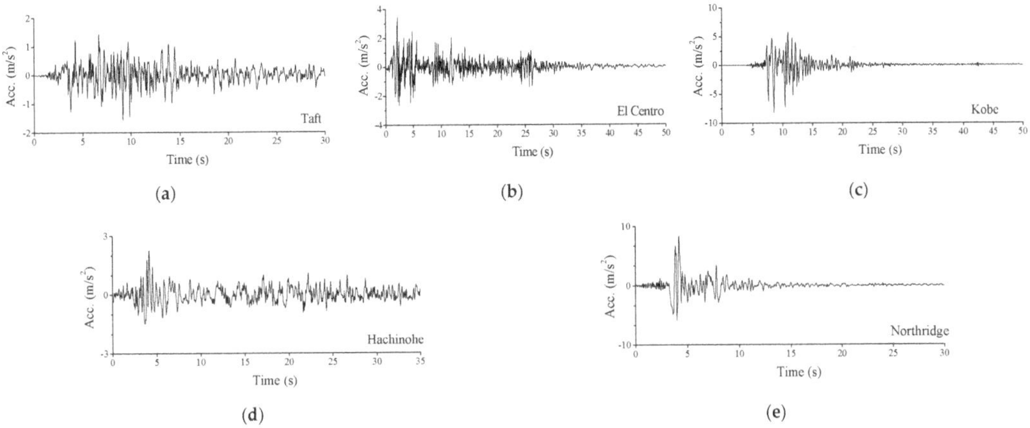

Figure 4. Time histories of four historical seismic records. (**a**) Taft; (**b**) El Centro; (**c**) Kobe; (**d**) Hachinohe; (**e**) Northridge.

Obvious whiplash effects of the mast are observed based on analytical results. Owing to the small geometric size of the mast, sixty SMA dampers are evenly divided into four groups and installed around the two turrets. Thirty SMA dampers (group nos. 1 and 2) are installed at the low turret, and the other two groups of SMA dampers are at the upper turret. An SMA damper with an axial brace connects spherical nodes in the axial direction of a member. The physical parameters of SMA dampers are listed in Table 2. The Young's modulus of the SMA damper brace is 2.12×10^8 kN/m, and the cross-sectional area of the SMA damper is 50 cm^2.

Table 2. Physical parameters of SMA dampers.

Parameter	Value	Parameter	Value	Parameter	Value
M_f	−46 °C	C_M	10 MPa/°C	σ_s^{cr}	120 MPa
A_s	−18.5 °C	C_A	15.8 MPa/°C	σ_f^{cr}	190 MPa
M_s	−37.4 °C	D_A	75,000 MPa	ε_L	0.079
A_f	−6 °C	D_M	29,300 MPa		

5.2. Control Performance Comparison

The vibration reduction factor (VRF) is used in this study to evaluate the control efficacy of SMA dampers:

$$\mathrm{VRF} = \frac{R_{nc} - R_{co}}{R_{nc}} \tag{19}$$

where R_{nc} and R_{co} are the peak response without and with SMA dampers, respectively.

The curves in Figure 5 indicate that the peak responses are remarkably mitigated using SMA dampers. It is found that the displacement VRFs of the tower top, upper turret, and low turret are 42.739%, 9.420%, and 37.856%, respectively. Similarly, the counterparts of the acceleration are 28.937%, 9.533%, and 26.405%, respectively. In addition, the displacement and acceleration VRFs are slightly better than those of velocity and responses. As a result, satisfactory control can be achieved using SMA dampers, particularly for the whiplash effects of the mast. The dynamic responses of the tower top and the upper turret, with and without SMA dampers under the Taft earthquake, are displayed in Figure 6. Owing to the additional damping produced by SMA dampers, the vibrations of the controlled tower damped out more quickly than that of the original tower. The application of SMA dampers can mitigate the seismic responses of the TV transmission tower to a great extent. In particular, the control efficacy on whiplash effects of the mast is better than that of the other parts of the tower.

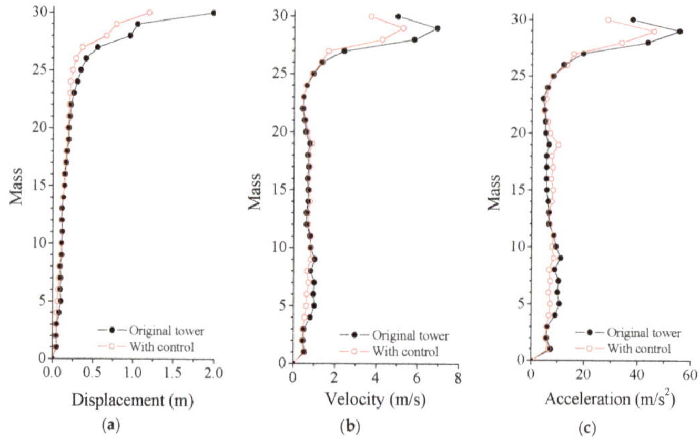

Figure 5. Maximum responses of the tower without/with control. (**a**) Displacement; (**b**) Velocity; (**c**) Acceleration.

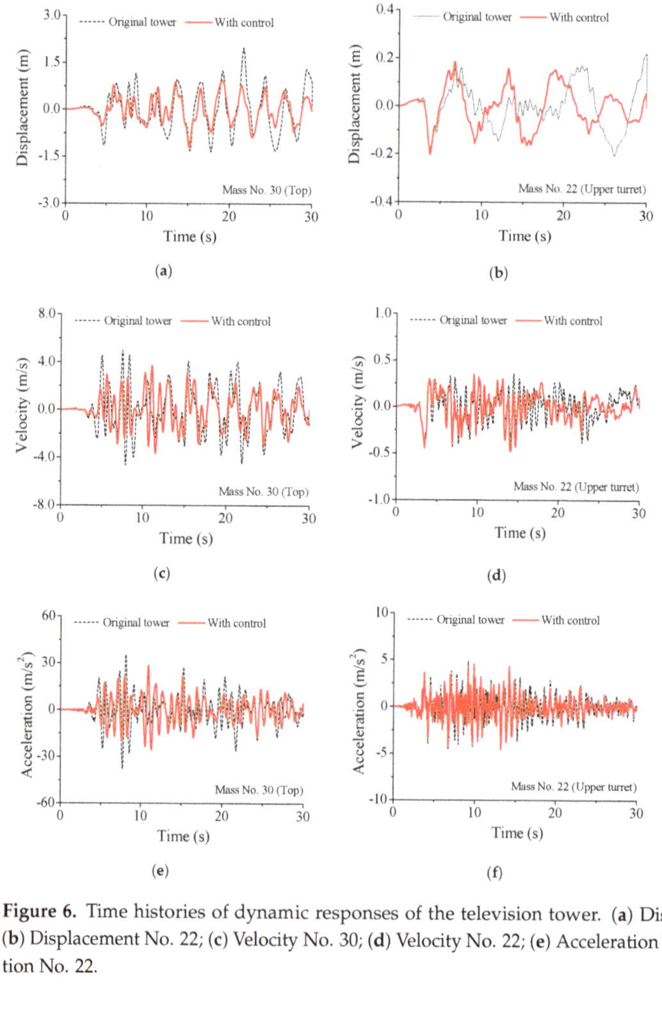

Figure 6. Time histories of dynamic responses of the television tower. (**a**) Displacement No. 30; (**b**) Displacement No. 22; (**c**) Velocity No. 30; (**d**) Velocity No. 22; (**e**) Acceleration No. 30; (**f**) Acceleration No. 22.

Three damper installation schemes are considered to compare the effects of damper position on control efficacy, as listed in Table 3. Scheme 1 is mentioned above, in which sixty dampers are around the two turrets. For scheme 2, all the SMA dampers are installed around the low turret (Mass No. 4). For scheme 3, all the SMA are installed around the upper turret.

Table 3. Damper installation schemes.

Scheme 1	Scheme 2	Scheme 3
30 dampers around the low turret (Mass No. 4) 30 dampers around the upper turret (Mass No. 24)	60 dampers around the low turret (Mass No. 4)	60 dampers around the upper turret (Mass No. 24)

Figure 7 shows the structural peak responses of different control schemes. For scheme 2, with all the dampers around the low turret, the control efficacy of displacement responses is much better than that of velocity and acceleration responses. Similar observations can be made for scheme 3, whose control performance is slightly inferior to that of scheme 2. The overall control efficacy of scheme 1 is better than that of schemes 2 and 3. Thus, the optimal positions for SMA dampers are both two turrets instead of a single turret. It is also observed that the control efficacy of displacement of all the schemes is slightly better than that of velocity and acceleration responses. Therefore, scheme 1 is adopted for the parametric study in the following sections.

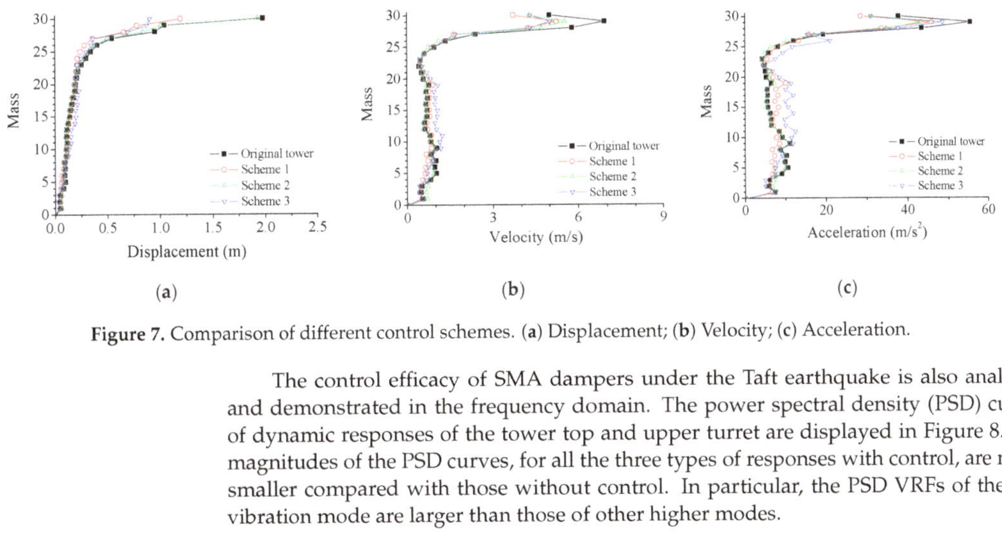

Figure 7. Comparison of different control schemes. (**a**) Displacement; (**b**) Velocity; (**c**) Acceleration.

The control efficacy of SMA dampers under the Taft earthquake is also analyzed and demonstrated in the frequency domain. The power spectral density (PSD) curves of dynamic responses of the tower top and upper turret are displayed in Figure 8. The magnitudes of the PSD curves, for all the three types of responses with control, are much smaller compared with those without control. In particular, the PSD VRFs of the first vibration mode are larger than those of other higher modes.

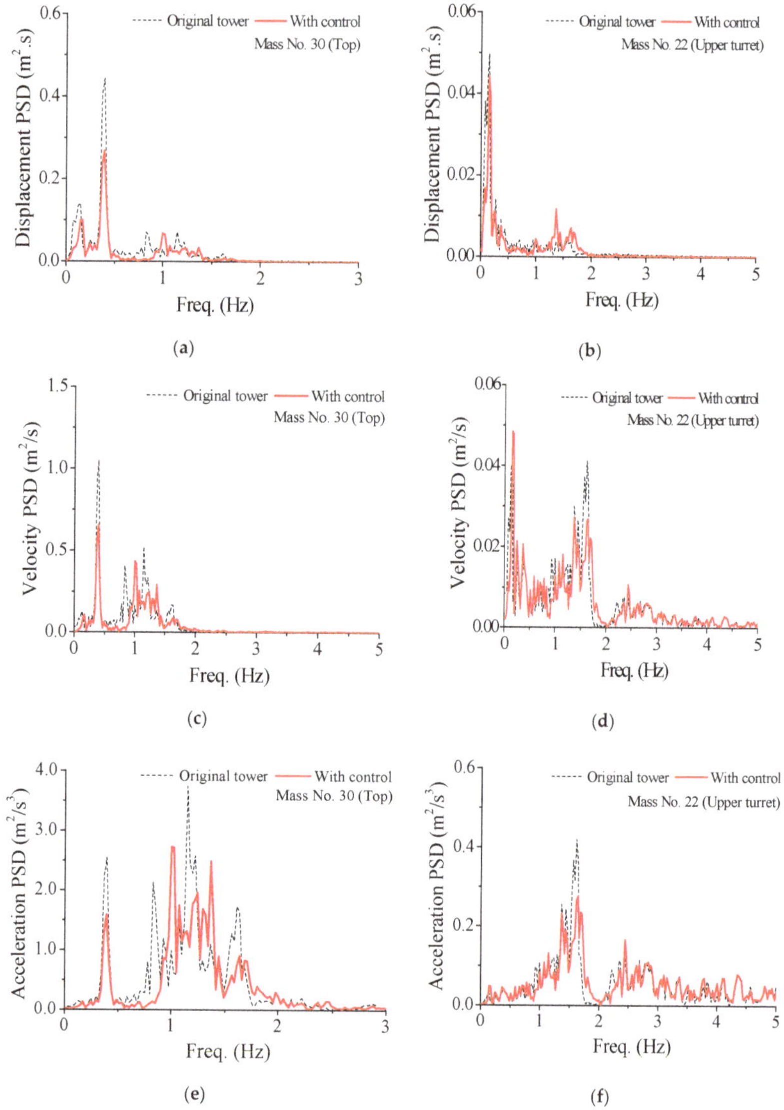

Figure 8. Comparison of PSD curves of response time histories. (**a**) Displacement PSD No. 30; (**b**) Displacement PSD No. 22; (**c**) Velocity PSD No. 30; (**d**) Velocity PSD No. 22; (**e**) Acceleration PSD No. 30; (**f**) Acceleration PSD No. 22.

6. Parametric Study on Control Performance

6.1. Stiffness Effects of SMA Damper

The stiffness coefficient of an SMA damper is given by:

$$\mathrm{SC} = \frac{k_d^{SMA}}{k_0^{SMA}} \tag{20}$$

where k_d^{SMA} is the stiffness value in computation; k_0^{SMA} is the initial stiffness of an SMA damper.

The stiffness effects of SMA dampers on the peak responses under the Taft earthquake are investigated, and the results are displayed in Figure 9. The peak displacement and velocity responses of the mast are gradually reduced with an increasing SC until approximately 1.0. A further increase in stiffness of SMA dampers does not generate further significant response reduction when SC is larger than 1.0. The peak acceleration of the mast reduces at first and then increases with the increasing SC, which means there exist optimal SC values for structural peak responses. It is not beneficial to set a very large SC value to avoid unnecessary cost waste. Similar results can be achieved from the dynamic responses of the tower body and turrets. Thus, an optimal SC value for all the responses does not exist, and optimal SC values for three kinds of peak responses, at different places, are different to some extent. The optimum SC value for the example tower is selected as 1.0 considering the overall control efficacy of the tower.

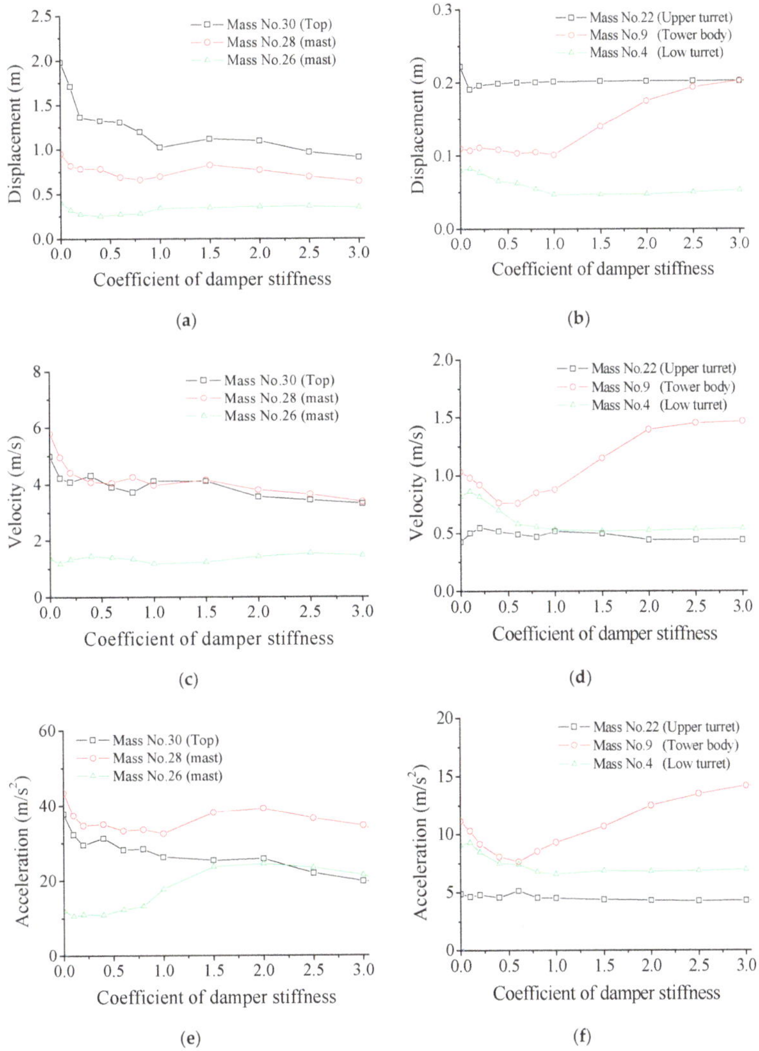

Figure 9. Effects of damper stiffness on peak responses. (**a**) Displacement at mast; (**b**) Displacement at tower; (**c**) Velocity at mast; (**d**) Velocity at tower; (**e**) Acceleration at mast; (**f**) Acceleration at tower.

6.2. *Properties of Hysteresis Loops*

The hysteresis loop is an important index to describe the energy-dissipating capacity of an SMA damper under earthquakes. The variation of a hysteresis loop with damper stiffness is displayed in Figure 10. A small SC value (SC = 0.6) reflects the small enclosed area of the hysteresis loop. Thus, the SMA damper is easy to extend while the energy-dissipating capacity is poor. Increasing the damper stiffness (SC = 1.0 or 2.0) can improve the shape of the hysteresis loops and the energy-absorbing capacity. After the SC value reaches 2.0, a further increase (SC = 3.0) in damper stiffness may increase the peak control force but reduce the enclosed area of the hysteresis loop. Therefore, an optimal SC value of an SMA damper can also be determined based on the configuration of hysteresis loops.

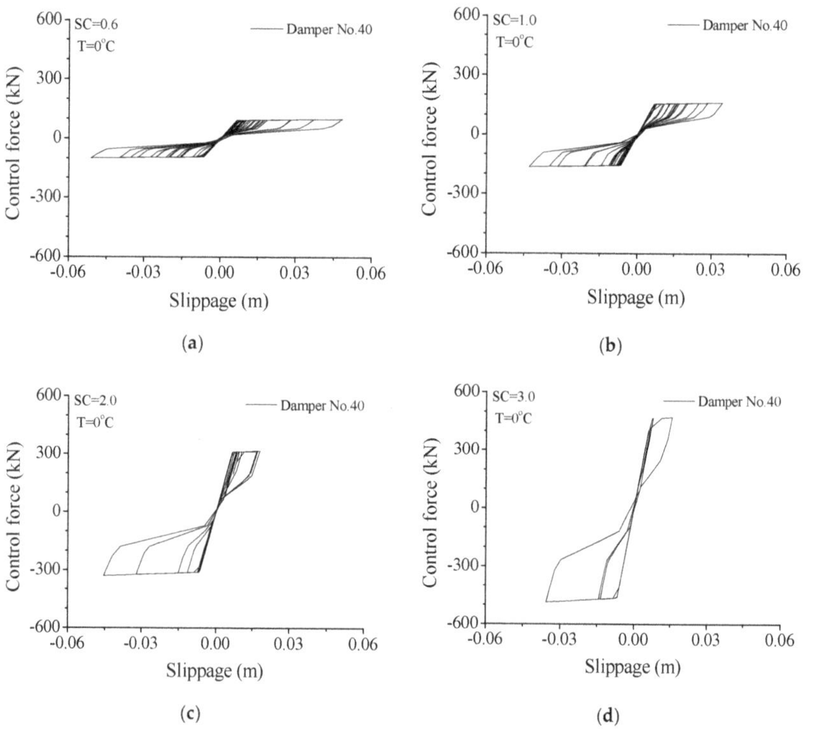

Figure 10. Variation of hysteresis loop with damper stiffness. (**a**) SC = 0.6; (**b**) SC = 1.0; (**c**) SC = 2.0; (**d**) SC = 3.0.

6.3. *Effect of Service Temperature*

The variation of damper force and deformation with service temperature is displayed in Figure 11. The peak forces of damper No. 40 are 163.71, 193.82, 224.87, and 284.08 kN, respectively, at 0, 10, 20, and 40 degrees Celsius. With the increase in service temperature, the peak damper force gradually increases, while the damper deformation decreases to some extent. Similar observations can be made for the other SMA dampers. The effects of service temperature on damper performance can also be found from the variation of damper hysteresis loops. The increasing service temperature may induce the increment in damper force and reduction in damper deformation. The enclosed areas of damper hysteresis loops reduce to some extent. Thus, a very large service temperature changes the damper states, and the damper may behave as a steel brace does and lose its energy-dissipating capacity.

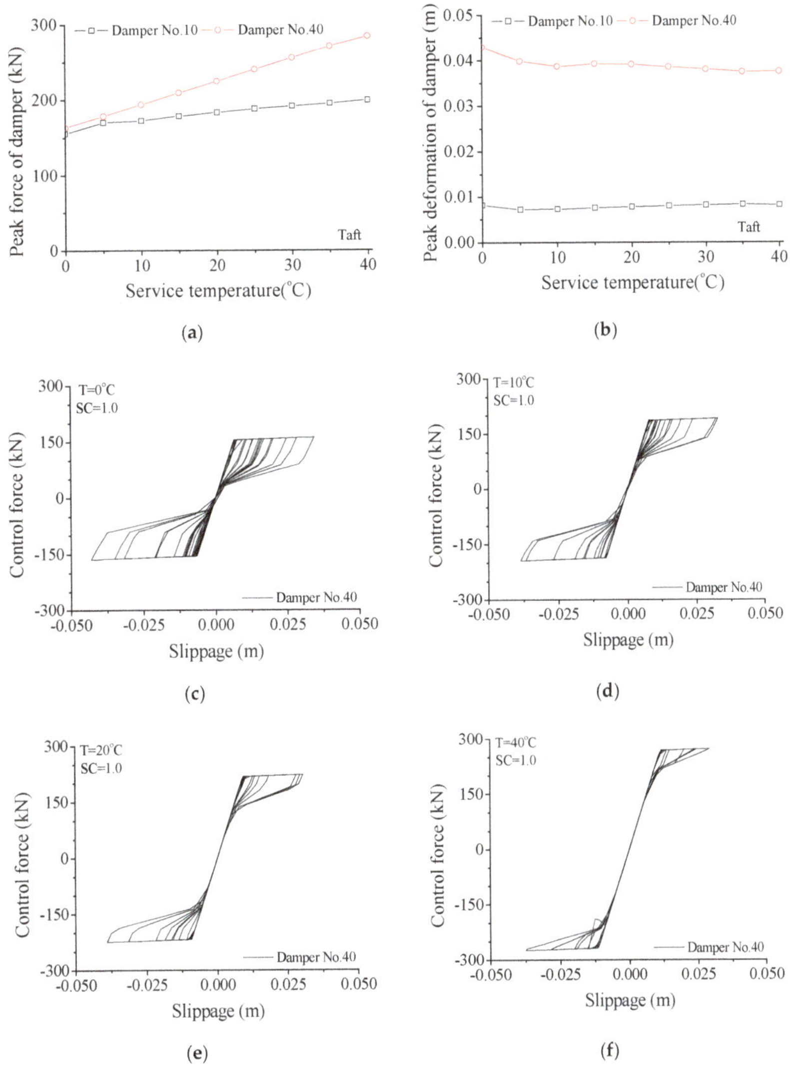

Figure 11. Variation of damper force and deformation with service temperature. (**a**) Peak force; (**b**) Peak deformation; (**c**) T = 0 °C; (**d**) T = 10 °C; (**e**) T = 20 °C; (**f**) T = 40 °C.

The effects of service temperature on maximum responses of the TV transmission tower are studied, as shown in Figure 12. The responses of the tower top, mast, tower body, and two turrets are computed, respectively, under the common temperature range between 0 to 40 degrees Celsius. The displacement responses of the TV tower are almost stable with increasing service temperature, while the velocity and acceleration responses slightly increase. It is noted that a relatively large service temperature may affect the shape of damper hysteresis loops, as displayed in Figure 11, while the control performance is stable. This is because the peak forces of SMA dampers increase automatically with increasing temperature.

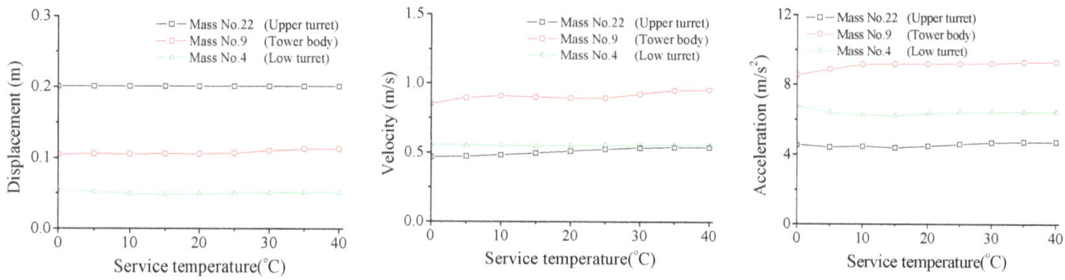

(**a**) (**b**) (**c**)

Figure 12. Effects of service temperature on maximum responses of the tower. (**a**) Displacement; (**b**) Velocity; (**c**) Acceleration.

6.4. Effects of Earthquake Intensity

The effects of PGA on control performance are investigated in this section. Figure 13 displays the variation of peak control force of SMA dampers with PGA. The deformation of SMA dampers almost increases linearly with the increasing PGA under constant damper stiffness. The peak control force of the SMA damper gradually increases with the increasing earthquake intensity. This means that the performance of SMA dampers presents certain robustness under external excitations.

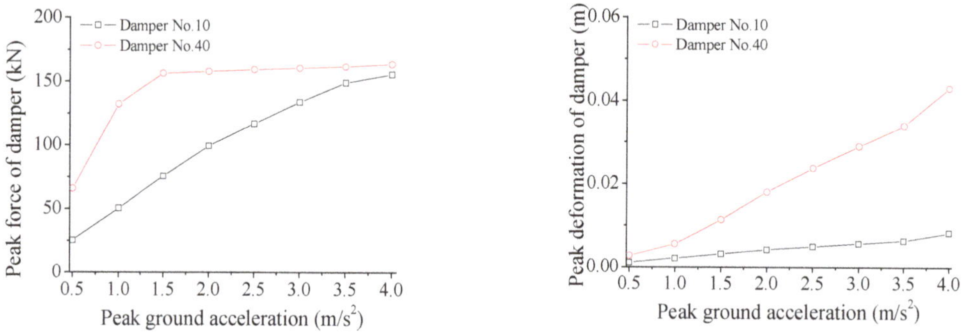

(**a**) (**b**)

Figure 13. Variation of peak control force of SMA dampers with PGA. (**a**) Peak force; (**b**) Peak deformation.

As displayed in Figure 14, the displacement VRF of the tower top gradually increases, with the increment in PGA, while the velocity and acceleration VRFs almost keep constant. All three types of VRFs of the mast are stable with increasing earthquake intensity. Similar observations can be made for the upper turret and low turret. Thus, it is noted that the overall VRFs of the TV transmission tower are stable when subjected to different seismic excitations. The control performance by SMA dampers is robust and versatile for the earthquake-disturbed tower.

6.5. Efficacy Subjected Other Earthquakes

The seismic responses under the other three earthquakes are also computed and compared with the same PGA of 4.0 m/s^2. The response VRFs of the example tower under different earthquakes are listed in Table 4. Three far-field and two near-field historical records are selected to evaluate the performance of the proposed control method based on SMA dampers [32–35]. The SMA dampers can reduce structural seismic responses while the control efficacy, under different seismic inputs, is different. In addition, the overall VRFs of the tower top and mast are much better than those of the two turrets. Thus, the whiplash effects of the example tower can be substantially reduced. In reality, the different frequency components of the example earthquakes in the frequency domain induce the difference in structural dynamic responses. Therefore, control performances under different earthquakes cannot maintain optimality at all times.

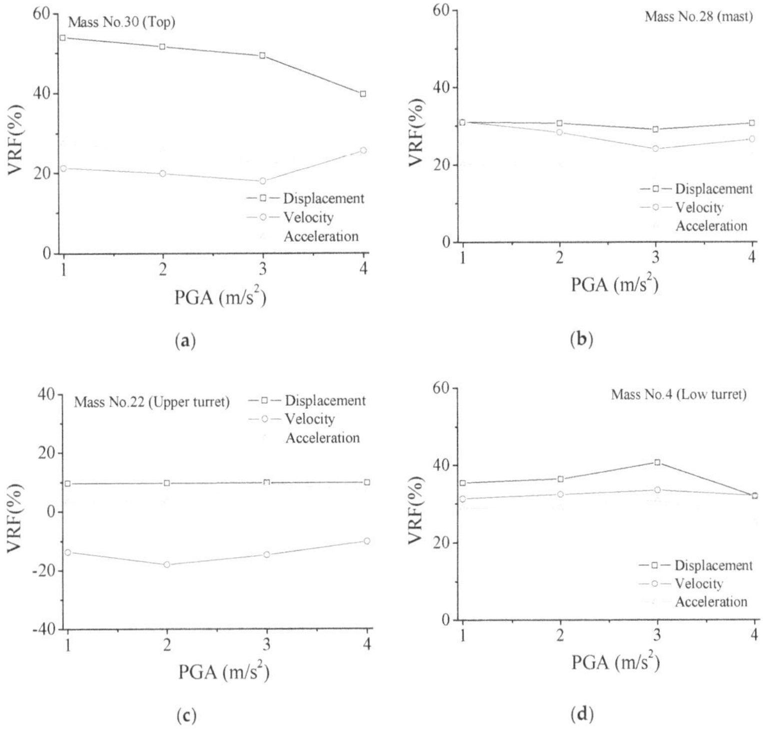

Figure 14. Variation of VRFs with seismic PGA. (**a**) VRF No. 30; (**b**) VRF No. 28; (**c**) VRF No. 22; (**d**) VRF No. 4.

Table 4. VRFs of the TV transmission tower under different earthquakes (%).

Location		Taft	El Centro	Kobe	Hachinohe	Northridge
Mass No. 30 (Tower top)	Displacement	39.64	31.95	22.38	23.24	29.03
	Velocity	25.57	13.87	18.70	33.89	5.59
	Acceleration	24.75	6.43	21.09	13.89	11.36
Mass No. 28 (Mast)	Displacement	30.70	19.46	23.64	22.56	10.70
	Velocity	26.59	−5.21	20.69	19.60	11.42
	Acceleration	22.48	−9.28	17.52	18.51	15.04
Mass No. 22 (Upper turret)	Displacement	9.72	9.24	8.31	8.91	10.77
	Velocity	−10.19	11.88	23.94	−1.05	−0.78
	Acceleration	6.95	8.86	10.48	−4.56	5.55
Mass No. 4 (Low turret)	Displacement	31.88	−2.81	35.34	2.69	4.04
	Velocity	32.14	−10.05	23.30	1.08	0.42
	Acceleration	25.30	−5.56	15.60	8.74	−1.64

6.6. Comparison between SMA Damper and Friction Damper

The dynamic responses of the tower with the commonly-used friction damper are computed and compared with those by SMA dampers. Friction dampers have been used to improve the seismic resistant performance of various structures during the past three decades. The Coulomb law is commonly adopted in the mathematical model of friction dampers. It presumes that the frictional force is independent of the velocity and the coefficient μ is the proportional constant considered in the mathematical relation between the frictional force and the clamping force N. The frictional force f_d equals the clamping force N multiplies the coefficient of friction μ. The clamping force of a friction damper is set in advance, and its frictional force is constant during the slip process. If the force of the friction damper f_d is no larger than the design frictional force, then the damper is sticking and the damper behaves as if it is a common brace. If the damper force f_d is larger than the design frictional force, the damper begins to slip and dissipate vibrant energy. During the structural vibration under an earthquake, if the relative displacement increment of the two ends of the friction element is large than zero, the control force is tensile. Otherwise, the control force is compressive. As displayed in Figures 10b and 11a, the peak force of an SMA damper is 150 kN. Thus, the peak control forces of friction dampers are set as the same for performance comparison. The number and location of friction dampers are also the same as those of SMA dampers.

The hysteresis loops of friction dampers, under different damper forces, are displayed in Figure 15. For a friction damper, its damper force cannot be changed during the structural vibration. If the damper slipping force is too large (250 kN), the friction damper is hard to slip and behaves similarly to a pure brace. If the damper slipping force is too small (50 kN), the damper is very easy to slip while the enclosed area is small, which means a limited energy-dissipating capacity of the friction damper. Thus, the optimal slipping force of a friction damper depends on the intensity of the external excitations.

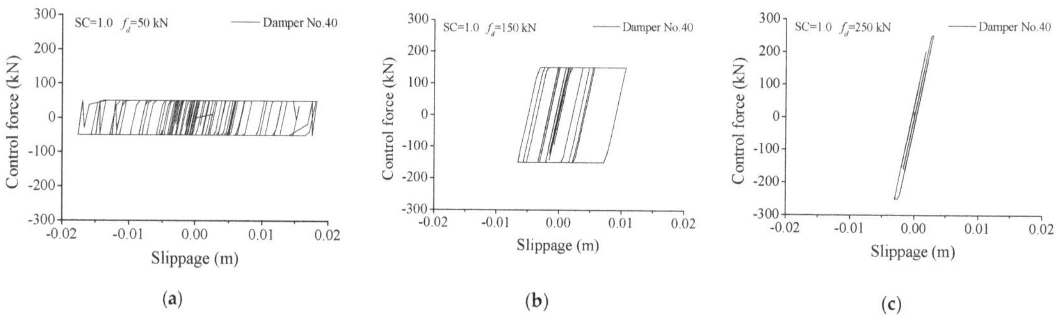

Figure 15. Hysteresis loops of friction dampers under different damper forces. (**a**) fd = 50 kN; (**b**) fd = 150 kN; (**c**) fd = 250 kN.

The hysteresis loops of friction dampers, under different earthquake intensities, are displayed in Figure 16. For a small seismic intensity (PGA = 1.0 or 2.0 m/s^2), the damper force of 150 kN is large enough, the damper is hard to slip, and it behaves as a brace. For an appropriate seismic intensity (PGA = 4.0 m/s^2), the hysteresis loop is plump, and the control performance is satisfactory. For a larger seismic intensity (PGA = 5.0 m/s^2), the damper can slip more easily, the enclosed area of the hysteresis loop is limited, and the energy-dissipating capacity and control performance is limited. The reason is that the slipping force of a friction damper is set in advance and cannot change with the varying excitation intensity. Therefore, the control robustness of friction dampers is unsatisfactory.

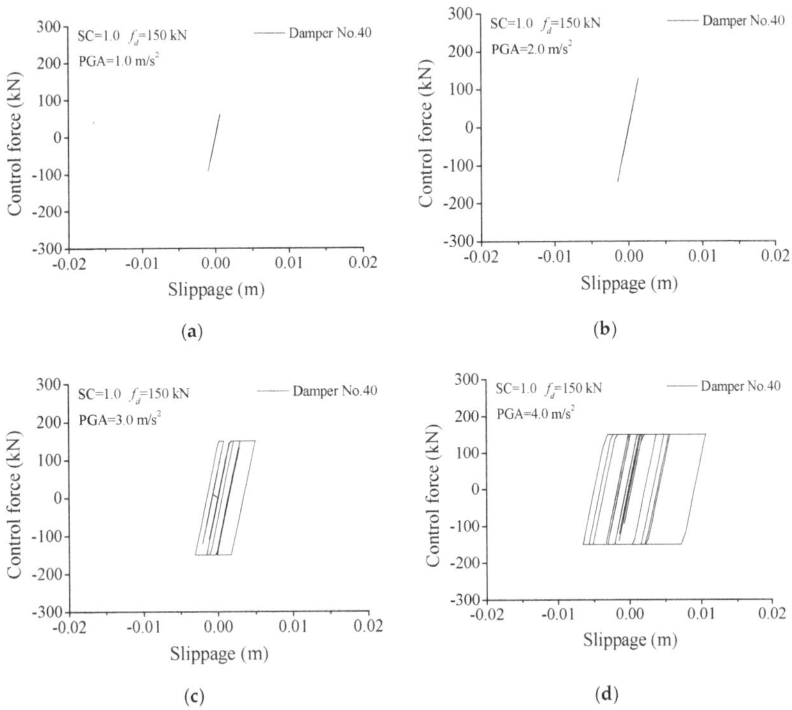

Figure 16. Hysteresis loops of friction dampers under different earthquake intensities. (**a**) PGA = 1.0 m/s^2; (**b**) PGA = 2.0 m/s^2; (**c**) PGA = 3.0 m/s^2; (**d**) PGA = 4.0 m/s^2.

The performance of SMA dampers is quite different in comparison with that of friction dampers. The variation of control force with seismic intensity is investigated and shown in Figure 17. It is seen that the control force of the SMA dampers can be modulated automatically with the earthquake-induced structural and damper responses. The control forces of SMA dampers quickly increase with the increasing seismic intensity. Thus, the SMA damper can move more easily under any seismic intensity. In addition, the shapes of hysteresis loops of SMA are similar, even if the seismic intensity increases. However, the performance of friction dampers is quite different in comparison with that of SMA dampers, as displayed in Figures 16 and 17. The control forces of the friction dampers are set at 150 kN in advance and cannot be changed during the whole vibration. For a small seismic intensity (PGA = 1.0 or 2.0 m/s^2), the friction dampers cannot slip, and they behave as a brace. Thus, the damper force is smaller than 150 kN. The energy-dissipating capacity of SMA dampers is stable and superior to that of friction dampers.

The structural VRFs, under different seismic intensities, are computed and compared, as displayed in Table 5. It is observed that the VRFs of SMA dampers are almost stable for different seismic intensities. The SMA damper-based control approach is robust and versatile for the TV transmission tower. However, the control efficacy of friction dampers with constant slipping force (150 kN) may dramatically reduce with the varying seismic intensity. If the damper force is a certain value and cannot be automatically changed, a friction damper is easy in the sticking stage and behaves much as a common brace does under a small seismic intensity. Thus, the friction damper is unable to fully slip, which causes unsatisfactory control performance. However, an SMA damper with proper parameters can automatically adjust its control force in line with the change of external loading and thereby ensure robust control performance.

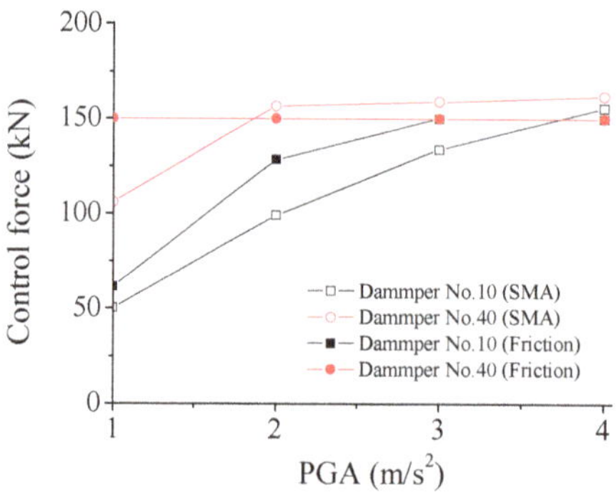

Figure 17. Variation of control force with seismic intensity.

Table 5. Comparison of VRFs with different control approaches under different earthquake intensities.

Location		SMA Damper PGA = 4.0	Friction Damper PGA = 4.0	SMA Damper PGA = 1.0	Friction Damper PGA = 1.0
Mass No. 30 (Tower top)	Displacement	39.64	39.87	53.92	40.69
	Velocity	25.57	38.84	21.37	33.72
	Acceleration	24.75	52.63	27.49	37.30
Mass No. 28 (Mast)	Displacement	30.70	47.07	30.99	8.61
	Velocity	26.59	49.51	31.12	31.33
	Acceleration	22.48	44.75	20.33	23.28
Mass No. 22 (Upper turret)	Displacement	9.72	8.61	9.68	8.57
	Velocity	−10.19	−3.93	−13.67	−5.77
	Acceleration	6.95	9.11	3.529	9.08
Mass No. 4 (Low turret)	Displacement	31.88	39.43	35.471	33.37
	Velocity	32.14	30.80	31.37	36.74
	Acceleration	25.30	29.61	29.02	23.16

7. Energy Responses of the Tower

7.1. Comparison of Energy Responses of the Tower

The vibrant intensity of the TV transmission tower can also be depicted from the viewpoint of energy. The energy properties of the example tower, subjected to various earthquakes, are displayed in Figure 18. For the uncontrolled tower under the Taft earthquake, the total inputted energy from the seismic excitations is absorbed by structural damping. Large kinetic and strain energy is seen because of the excessive vibration of the tower. Similar observations can be made from the other three ground motions. The magnitude of total inputted energy to the tower, under various earthquakes, is quite different even though their PGAs are the same. Thus, the frequency components of seismic excitations have substantial influences on structural dynamic responses.

Figure 19 displays the energy responses of the tower with SMA dampers in comparison with those without control. The application of SMA dampers can substantially reduce the vibrant energy of the TV transmission tower to a great extent. To compare the energy curves in Figures 18 and 19, one can find that the inputted energy of the controlled tower is smaller in comparison with the counterpart of the controlled tower. Owing to the aid of SMA dampers, the energy absorbed by structural damping is much less than that of the uncontrolled tower. The energy responses under Northridge earthquake have the similar properties.

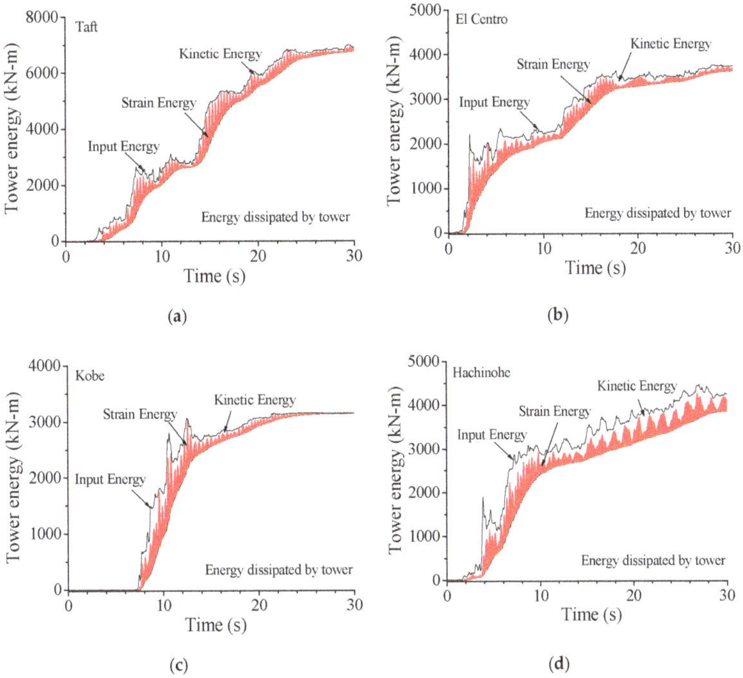

Figure 18. Energy responses of the tower without friction dampers. (**a**) Taft; (**b**) El Centro; (**c**) Kobe; (**d**) Hachinohe.

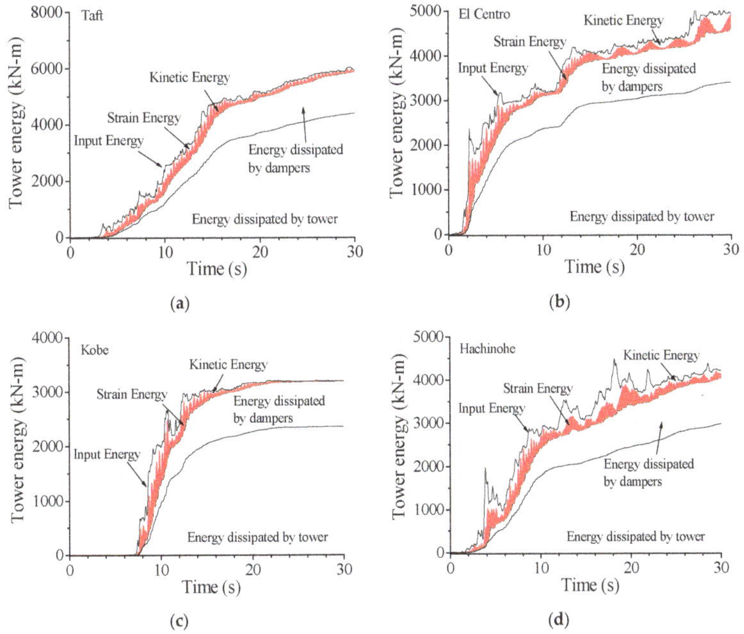

Figure 19. Energy responses of the tower with SMA dampers. (**a**) Taft; (**b**) El Centro; (**c**) Kobe; (**d**) Hachinohe.

7.2. Variation of Energy Response with Damper Stiffness

Curves in Figure 20 indicate that the total inputted energy from earthquakes E_E varies with the damper stiffness. For a small damper stiffness (SC = 0.6), the energy-dissipating capacity of the SMA dampers is poor, and the tower energy E_E is even slightly larger than that of the uncontrolled tower. The tower energy E_E gradually reduces with the increasing SC until approximately 1.0. A further increase in damper stiffness (SC = 2.0 or 3.0) may induce a large value of E_E. However, the energy dissipated by SMA dampers E_P is quite different. As displayed in Figure 20, E_P gradually increases with the increasing damper stiffness until SC reaches about 1.0. If stiffness coefficient SC reaches above 2.0, E_P reduces to some extent. The energy dissipated by SMA dampers E_P is limited if the damper stiffness is relatively small (SC = 0.6) or large (SC = 2.0 or 3.0). Besides, similar observations can be made from the energy curves of structural damping E_D. With the increase in damper stiffness, the energy E_D gradually reduces because more structural energy E_E is dissipated by SMA dampers. Thus, optimal damper stiffness can also be determined based on structural energy curves.

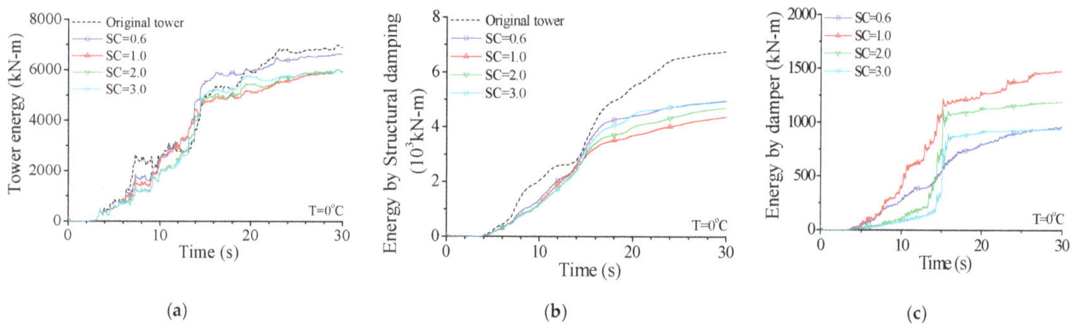

Figure 20. Variation of energy responses with damper stiffness. (**a**) Tower energy; (**b**) Energy by structural damping; (**c**) Energy by damper.

8. Concluding Remarks

The validity of the proposed control approach for TV transmission towers by SMA dampers is examined in detail. The PSD properties and energy responses of the tower are compared with those of the uncontrolled tower. The results indicate that SMA dampers, with optimal parameters, can be used in the control of TV transmission towers disturbed by earthquakes.

Dynamic responses of the TV tower are remarkably suppressed by SMA dampers particularly for the whiplash effects of the top mast. The overall control efficacy of scheme 1 is better than that of the other two schemes, which means that the optimal positions for SMA dampers are both two turrets. The hysteresis loops of SMA dampers are quite different because the damper stiffness can directly affect the damper's working status. The optimal damper stiffness can also be determined based on the configuration of hysteresis loops. The vibrant energy of the controlled tower is substantially reduced, and the inputted energy of the controlled tower is also mitigated to a great extent. The application of SMA dampers can remarkably improve the energy-dissipating mechanism of the TV transmission tower. Energy evaluation is an effective approach to determine the optimal damper parameters in vibration control.

It is not beneficial to set a very large SC value for SMA dampers to avoid unnecessary cost waste. Thus, an optimal SC value for all the responses does not exist, and optimal SC values for three kinds of peak responses at different places are different to some extent. Thus, in the real application of SMA dampers, an optimal damper stiffness should be determined based on the structural model of TV transmission tower through detailed parametric study, which may be time-consuming and complicated. Similarly, an optimal service temperature is required based on the structural model of the TV transmission tower through a detailed and complicated parametric study.

Author Contributions: Conceptualization, B.C.; methodology, B.C. and J.W.; software, J.W.; validation, B.C. and L.Z.; formal analysis, J.W. and X.S.; investigation, B.C. and J.W.; resources, J.W. and X.S.; data curation, J.W.; writing—original draft preparation, B.C. and J.W.; writing—review and editing, B.C. and L.Z.; visualization, B.C. and L.Z.; supervision, B.C. and L.Z.; project administration, B.C.; funding acquisition, B.C. All authors have read and agreed to the published version of the manuscript.

Funding: The authors are grateful for the financial support from the National Natural Science Foundation of China (51678463, 51978549).

Institutional Review Board Statement: Not applicable.

Informed Consent Statement: Not applicable.

Data Availability Statement: Data sharing is not applicable to this article.

Acknowledgments: The authors sincerely appreciate the selfless support provided by Wuhan University of Technology during the progress of this study.

Conflicts of Interest: The authors declare no conflict of interest.

References

1. Amiri, M.M.; Yahyai, M. Estimation of damping ratio of TV towers based on ambient vibration monitoring. *Struct. Des. Tall Spec. Build.* **2013**, *22*, 862–875. [CrossRef]
2. Chen, B.; Weng, S.; Zhi, L.H.; Li, D.M. Response control of a large transmission-tower line system under seismic excitations by using friction dampers. *Adv. Struct. Eng.* **2017**, *20*, 1155–1173. [CrossRef]
3. Li, T.; Fan, J.; Lv, Y.X. Collapse analysis of a steel TV tower under wind load. *Steel Constr.* **2017**, *32*, 80–83.
4. Liu, R.S.; Liu, J.L.; Yan, D.Q. Seismic damage investigation and analysis of electric power system in Lushan MS 7.0 earthquake. *J. Nat. Disasters* **2013**, *22*, 85–92.
5. Housner, G.W.; Bergman, L.A.; Caughey, T.K.; Chassiakos, A.G.; Claus, R.O.; Masri, S.F.; Skelton, R.E.; Soong, T.T.; Spencer, B.F., Jr.; Yao, T.P. Structural control: Past, present, and future. *J. Eng. Mech. ASCE* **1997**, *123*, 897–971. [CrossRef]
6. Xu, Y.L.; Qu, W.L.; Chen, B. Active/robust moment controllers for seismic response control of a large span building on top of ship lift towers. *J. Sound Vib.* **2003**, *261*, 277–296. [CrossRef]
7. Chen, B.; Xia, Y. Advanced technologies in disaster prevention and mitigation. *Adv. Struct. Eng.* **2017**, *20*, 1141–1142. [CrossRef]
8. Hitchcock, P.A.; Glanville, M.J.; Kwok, K.C.S.; Watkins, R.D. Damping properties and wind-induced response of a steel frame tower fitted with liquid column vibration absorbers. *J. Wind Eng. Ind. Aerodyn.* **1999**, *83*, 183–196. [CrossRef]
9. Balendra, T.; Wang, C.M.; Yan, N. Control of wind-excited towers by active tuned liquid column damper. *Eng. Struct.* **2001**, *23*, 1054–1067. [CrossRef]
10. Yang, N.; Wang, C.M.; Balendra, T. Composite mass dampers for vibration control of wind-excited towers. *J. Sound Vib.* **1998**, *213*, 301–316. [CrossRef]
11. Wu, J.C.; Yang, J.N. LQG control of lateral-torsional motion of Nanjing TV transmission tower. *Earthq. Eng. Struct. Dyn.* **2000**, *29*, 1111–1130. [CrossRef]
12. Ghorbani-Tanha, A.K.; Noorzad, A.; Rahimian, M. Mitigation of wind-induced motion of Milad Tower by tuned mass damper. *Struct. Des. Tall Spec. Build.* **2009**, *18*, 371–385. [CrossRef]
13. Lu, X.L.; Li, P.Z.; Guo, X.Q.; Shi, W.X.; Liu, J. Vibration control using ATMD and site measurements on the Shanghai World Financial Center Tower. *Struct. Des. Tall Spec. Build.* **2014**, *23*, 105–123. [CrossRef]
14. Pall, A.S.; Marsh, C. Response of friction damped braced frames. *J. Struct. Div. ASCE* **1982**, *108*, 1313–1323. [CrossRef]
15. Grigorian, C.E.; Yang, T.S.; Popov, E.P. Slotted bolted connection energy dissipators. *Earthq. Spectra* **1993**, *9*, 491–504. [CrossRef]
16. Chen, B.; Xiao, X.; Li, P.Y.; Zhong, W.L. Performance evaluation on transmission-tower line system with passive friction dampers subjected to wind excitations. *Shock. Vib.* **2015**, *2015*, 1–13. [CrossRef]
17. Quintana, H.C.; Petkovski, M. Optimum performance of structural control with friction dampers. *Eng. Struct.* **2018**, *172*, 154–162. [CrossRef]
18. Chen, B.; Yang, D.; Feng, K.; Ouyang, Y.Q. Response control of a high-rise television tower under seismic excitations by friction dampers. *Int. J. Struct. Stab. Dyn.* **2018**, *18*, 1850140. [CrossRef]
19. Zhang, Z.Q.; Li, A.Q. Seismic response vibration control analyses of Hefei TV Tower with fluid viscous damper. In Proceedings of the International Conference on Health Monitoring of Structure, Materials and Environment, Nanjing, China, 16 October 2007.
20. Xu, Y.L.; Qu, W.L.; Chen, Z.H. Control of wind-excited truss tower using semiactive friction damper. *J. Struct. Eng. ASCE* **2001**, *127*, 861–868. [CrossRef]

21. Chen, B.; Zheng, J.; Qu, W.L. Control of wind-induced response of transmission tower-line system by using magnetorheological dampers. *Int. J. Struct. Stab. Dyn.* **2009**, *9*, 661–685. [CrossRef]
22. Song, G.B.; Ma, N.; Li, H.N. Applications of shape memory alloys in civil structures. *Eng. Struct.* **2006**, *28*, 1266–1274. [CrossRef]
23. Song, G.B.; Ma, N. Robust control of a shape memory alloy wire actuated flap. *Smart Mater. Struct.* **2007**, *16*, 51–57. [CrossRef]
24. Ozbulut, O.; Hurlebaus, S.; Desroches, R. Seismic response control using shape memory alloys: A review. *J. Intell. Mater. Syst. Struct.* **2011**, *22*, 1531–1549. [CrossRef]
25. Fang, C.; Wang, W.; He, C.; Chen, Y. Self-centring behaviour of steel and steel-concrete composite connections equipped with NiTi SMA bolts. *Eng. Struct.* **2017**, *150*, 390–408. [CrossRef]
26. Fang, C.; Yam, M.C.; Lam, A.C.; Xie, L. Cyclic performance of extended end-plate connections equipped with shape memory alloy bolts. *J. Constr. Steel Res.* **2014**, *94*, 122–136. [CrossRef]
27. Zheng, Y.; Dong, Y.; Chen, B.; Ghazanfar, A.A. Seismic damage mitigation of bridges with self-adaptive SMA-cable-based bearings. *Smart Struct. Syst.* **2019**, *24*, 127–139.
28. Liu, Y.P.; Wang, H.Y.; Qiu, C.X.; Zhao, X.N. Seismic behavior of superelastic shape memory alloy spring in base isolation system of multi-story steel frame. *Materials* **2019**, *12*, 997. [CrossRef]
29. Brinson, L.C. One-dimensional constitutive behavior of shape memory alloys: Thermomechanical derivation with non-constant material functions and redefined martensite internal variable. *J. Intell. Mater. Syst. Struct.* **1993**, *4*, 229–242. [CrossRef]
30. Zheng, Y.; Dong, Y.; Li, Y. Resilience and life-cycle performance of smart bridges with Shape Memory Alloy (SMA)-cable-based bearings. *Constr. Build. Mater.* **2018**, *158*, 389–400. [CrossRef]
31. Qiu, C.; Fang, C.; Liang, D.; Du, X.; Yam, M.C.H. Behavior and application of self-centering dampers equipped with buckling-restrained SMA bars. *Smart Mater. Struct.* **2020**, *29*, 035009. [CrossRef]
32. Spencer, B.F., Jr.; Dyke, S.J.; Deoskar, H.S. Benchmark problems in structural control. I: Active mass driver system. *Earthq. Eng. Struct. Dyn.* **1998**, *27*, 1127–1139. [CrossRef]
33. Spencer, B.F., Jr.; Dyke, S.J.; Deoskar, H.S. Benchmark problems in structural control. II. Active tendon system. *Earthq. Eng. Struct. Dyn.* **1998**, *27*, 1141–1147. [CrossRef]
34. Ohtori, Y.; Christenson, R.E.; Spencer, B.F., Jr.; Dyke, S.J. Benchmark control problems for seismically excited nonlinear buildings. *J. Eng. Mech. ASCE* **2004**, *130*, 366–385. [CrossRef]
35. Chen, C.C.; Chen, G.D. Nonlinear control of a 20-story steel building with active piezoelectric friction dampers. *Struct. Eng. Mech.* **2002**, *14*, 21–38. [CrossRef]

materials

Article

Vibration Control of a Wind-Excited Transmission Tower-Line System by Shape Memory Alloy Dampers

Bo Chen [1], Xinxin Song [1,*], Wenbin Li [2] and Jingbo Wu [1]

[1] Key Laboratory of Roadway Bridge and Structural Engineering, Wuhan University of Technology, Wuhan 430070, China; cebchen@whut.edu.cn (B.C.); wujingbo0618@whut.edu.cn (J.W.)
[2] Guangdong Power Grid Energy Development, Co., Ltd., Guangzhou 510160, China; cemqdwhut@163.com
* Correspondence: sxxfx@whut.edu.cn

Abstract: To be typical electrical power infrastructures, high-rise tower-line systems are widely constructed for power transmission. These flexible tower structures commonly possess small damping and may suffer strong vibrations during external excitations. The control approaches based on various devices have been developed to protect transmission towers against strong vibrations, damages, and even failure. However, studies on the vibrant control of wind-excited tower-line systems equipped with SMA dampers have not yet been reported. To this end, the control approach for wind-excited tower-line systems using SMA dampers is conducted. The mechanical model of the tower-line system is established using Lagrange's equations by considering the dynamic interaction between transmission lines and towers. The vibration control method using SMA dampers for the tower-line coupled system is proposed. The control efficacy is verified in both the time domain and the frequency domain. Detailed parametric studies are conducted to examine the effects of physical parameters of SMA dampers on structural responses and hysteresis loops. In addition, the structural energy responses are computed to examine the control performance.

Keywords: transmission tower; wind excitation; SMA damper; energy response

Citation: Chen, B.; Song, X.; Li, W.; Wu, J. Vibration Control of a Wind-Excited Transmission Tower-Line System by Shape Memory Alloy Dampers. *Materials* **2022**, *15*, 1790. https://doi.org/10.3390/ma15051790

Academic Editor: Antonio Mattia Grande

Received: 20 January 2022
Accepted: 25 February 2022
Published: 27 February 2022

1. Introduction

High-rise truss towers, including television towers and transmission towers, are widely constructed for broadcast and electric energy supply. To be typical flexible structures, these truss towers commonly possess low damping and are prone to strong external excitations. If the load-induced strong vibration cannot be suppressed, possible damage and even failure are expected [1,2]. For example, a truss tower of more than 130 m in China collapsed when it was subjected to strong wind loadings [3]. The failure of transmission tower-line systems under strong earthquakes was also reported [4]. Therefore, many vibration control methods are developed to mitigate the excessive vibration of truss towers [5–7].

For television towers, vibration absorbers and dampers are firstly used for structural response control. The wind-excited television towers are protected by Yang et al. [8] and Wu et al. [9] by installing tuned mass dampers (TMD). In addition, the same devices are used in the vibration control of the 435 m Milad Tower in Tehran [10] and the 492 m World Financial Center Tower in Shanghai [11]. A similar wind-induced response control has been conducted using tuned liquid dampers [12,13]. The vibration control of television towers using dampers has also been performed in the past two decades [14–16]. Chen et al. [17] examined the seismic responses of a 340 m television tower reinforced using friction dampers. It was found that the implementation of friction dampers with optimal parameters can substantially reduce seismic responses. Zhang and Li [18] conducted seismic response control of a flexible truss tower using fluid dampers. They found that the vibration reduction effectiveness of fluid dampers was sensitive to the spectral characteristics of earthquake waves. For transmission towers, the response mitigation based on traditional control approaches was also performed. Chen et al. [19] examined the control performance of friction dampers on a wind-excited power transmission tower. The work on an example tower indicates that the application of friction dampers with optimal parameters could significantly reduce wind-induced responses of the transmission tower-line system for both the in-plane and out-of-plane vibration. In addition, they [20] also investigated the control efficacy of passive friction dampers on earthquake-disturbed transmission towers. They found that the best control performance of the transmission tower under a certain ground motion can be achieved only based on the optimal damper parameters and the control efficacy under different seismic excitations cannot keep optimal for all the time. The vibration control of a transmission tower under multi-component seismic excitations was conducted by Tian et al. [21] using TMD. Zhang et al. [22] applied a pounding TMD in a 55 m transmission tower for the seismic response control. Fluid dampers were accepted by Chen et al. [23] to reduce the impact responses of truss towers under cable rupture.

Various smart control devices, such as magnetorheological dampers, piezoelectric actuators, and shape memory alloy (SMA) devices, are recently gaining popularity in the vibration control of engineering structures [5,24,25]. The semi-active control of flexible truss towers under wind loading was carried out by Xu et al. [26] using piezoelectric friction dampers and by Chen et al. [27] using magnetorheological dampers. SMA is a typical smart material with many advantages including super-elasticity, fatigue resistance, and high strength [28,29]. Thus, SMA wires are widely used in vibration control to develop smart control devices, such as SMA dampers and SMA bracings [30–33]. Tian et al. [34] developed an SMA-based TMD for the seismic control of power transmission towers. Wu et al. [35] examined the seismic responses of a truss tower controlled by SMA dampers.

However, studies on the vibrant control of wind-excited tower-line systems equipped with SMA dampers have not yet been reported. To this end, the new control method of transmission tower-line systems under wind excitations is proposed using SMA dampers. The mechanical model of a real tower-line coupled system was proposed using Lagrange's equations by considering the dynamic interaction between transmission lines and towers. The vibration control method using SMA dampers for the tower-line coupled system was developed. The rational position of SMA dampers was determined by comparing three damper schemes. The feasibility of the proposed control method was verified through numerical analysis. Detailed parametric studies were conducted to investigate the effects of the physical parameters of SMA dampers on structural responses and hysteresis loops. Finally, the structural energy responses were computed to examine the control performance.

2. Model of a High-Rise Tower-Line Coupled System

2.1. Mathematical Model of Transmission Lines

The nonlinear equivalent method can be used to compute the dynamic responses of transmission lines using the Hamilton variational method and the Lagrange equation. As displayed in Figure 1, a transmission line is simulated by many masses and elements. Following the Hamilton variational principle, the line can be described as a series of generalized coordinates, namely the difference of the angle θ and element length l. If the line vibrates in the in-plane direction, the kinetic energy T_{line} is [27]:

$$T_{line} = \sum_{i=1}^{4} \frac{1}{2} m_i (\dot{x}_i^2 + \dot{y}_i^2) - T_{line}(\dot{\xi}_2, \dot{\xi}_3, \dot{\xi}_4, \delta\dot{l}_1, \delta\dot{l}_2, \delta\dot{l}_3, \delta\dot{l}_4, \delta\dot{l}_5) \quad (1)$$

where ξ_i $(i = 2,3,4)$ and δ_i $(i = 1,2,3,4,5)$ are the structural generalized coordinates related to the θ and l, respectively [27]; x_i and y_i are the horizontal and vertical displacement of the ith mass, respectively.

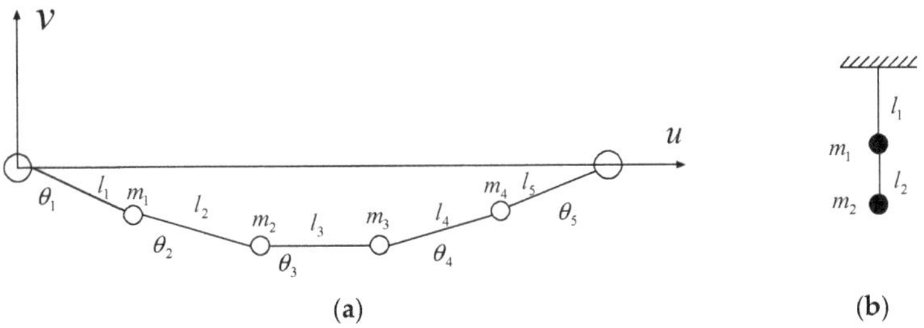

Figure 1. Analytical model of a transmission line. (**a**) In-plane vibration; (**b**) out-of-plane vibration.

Similarly, structural potential energy U_{line} is:

$$U_{line} = \sum_{i=1}^{4} m_i g y_i + \sum_{j=1}^{5} \frac{EA}{2}\left(\frac{(l_j^s + \delta l_j)^2}{l_j^0} - \frac{l_j^{s2}}{l_j^0}\right) \tag{2}$$

where E is Young's modulus of the line; A is the cross-sectional area of the line; l_j^0 and l_j^s are the initial and deformation length of the jth element.

The equation of motion is derived based on Hamilton's equation in terms of a set of generalized coordinates q_i, namely ξ and δ:

$$\int_{t_1}^{t_2} \delta[T_{line}(t) - U_{line}(t)]dt + \int_{t_1}^{t_2} \delta W_{line}(t)dt = 0 \tag{3}$$

In which $W_{line}(t)$ denotes the virtual work.

Then, computing the variation yields into Lagrange's equation:

$$\frac{d}{dt}\left(\frac{\partial T_{line}}{\partial \dot{q}_i}\right) - \frac{\partial T_{line}}{\partial q_i} + \frac{\partial U_{line}}{\partial q_i} = Q_i \tag{4}$$

where Q_i is the generalized forcing function of the line.

If the line vibrates in the in-plane direction, the stiffness matrix $\mathbf{K}_l^{in}$ is established by calculating the partial differential of the U_{line} to the generalized displacement $\partial U/\partial \xi_i$ and $\partial U/\partial \delta_i$. Similarly, the mass matrix $\mathbf{M}_l^{in}$ is established by calculating the partial differential of the T_{line} to the generalized velocity $\partial T/\partial \dot{\xi}_i$ and $\partial T/\partial \dot{\delta}_i$. If the line vibrates in the out-of-plane direction, the line can be simulated as a suspended pendulum, as shown in Figure 1b. The system matrices $\mathbf{M}_l{}^{out}$ and $\mathbf{K}_l{}^{out}$ of the line are deduced similarly, as follows:

$$\mathbf{M}_l^{out} = \begin{bmatrix} m_1 & \\ & m_2 \end{bmatrix} \tag{5}$$

$$\mathbf{K}_l^{out} = \begin{bmatrix} \frac{m_1 g}{l_1} & -\frac{m_1 g}{l_1} \\ -\frac{m_1 g}{l_1} & \frac{m_1 g}{l_1} + \frac{(m_1+m_2)g}{l_2} \end{bmatrix} \tag{6}$$

2.2. Mathematical Model of Tower-Line Coupled System

The three-dimensional (3D) model of a real transmission tower in China is constructed using ANSYS, as shown in Figure 2a. If a 3D finite element (FE) model is used for the large-scale tower and is incorporated with SMA dampers subjected to wind excitations, the step-by-step dynamic computation will be unbearably time-consuming. This may make the parametric study tedious and impractical. In addition, the numerical simulation of wind loading of the 3D FE model is commonly carried out using the spectral representation method, which requires enormous series calculus. In practice, a lumped mass model is commonly adopted for vibration control and parametric studies, as shown in Figure 2b.

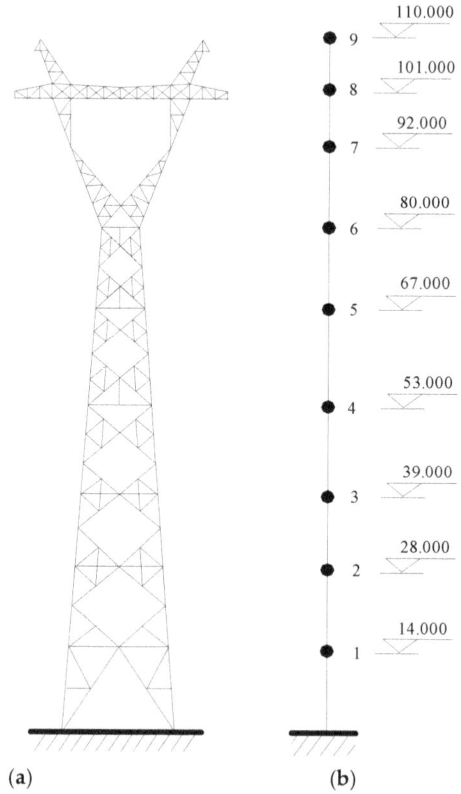

Figure 2. Analytical model of a large transmission tower. (**a**) 3D FE model; (**b**) 2D dynamic model.

A transmission tower-line coupled system is a complex continuous system consisting of many towers and lines. It is impossible and unnecessary to establish the system model considering all the lines and towers. Thus, the single tower associated with connected lines can be adopted in the dynamic analysis, as shown in Figure 3. The kinetic energy T and potential energy U of the tower-line system are expressed as follows:

$$T = T_t + \sum_{j=1}^{nl} T_l^{(j)} \tag{7}$$

$$U = U_t + \sum_{j=1}^{nl} U_l^{(j)} \tag{8}$$

where T_t and U_t are the kinetic and potential energy of the single tower; $T_l^{(j)}$ and $U_l^{(j)}$ are the kinetic and potential energy of the jth line; nl is the number of all the lines.

Then, Equations (7) and (8) can be substituted into the Lagrange equation. Similar to the computation of the transmission line expressed in Equation (4), the stiffness matrix of the entire coupled system $\mathbf{K}^{in}$ can be established in line with $\partial U/\partial \xi_i$ and $\partial U/\partial \delta_i$. The mass matrix $\mathbf{M}^{in}$ can be established in line with $\partial T/\partial \dot{\xi}_i$ and $\partial T/\partial \dot{\delta}_i$. For the out-of-plane vibration, the stiffness matrix $\mathbf{K}^{out}$ and mass matrix $\mathbf{M}^{out}$ of the entire coupled system are determined by combing the system matrices of the tower and lines.

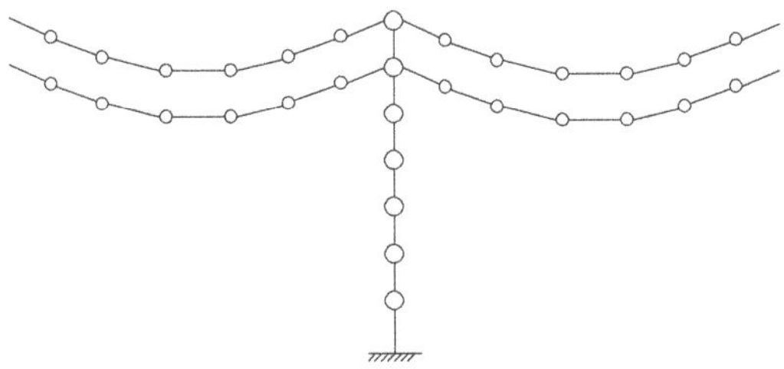

Figure 3. Analytical model of a transmission tower-line system.

3. Mathematical Model of SMA Dampers

SMA wires have excellent inherited properties and can be used to fabricate smart energy-dissipating dampers. The SMA material can be described by the widely-used constitutive model [36–38]. The 2D and 3D configuration of an SMA damper is displayed in Figure 4. The SMA damper consists of the outer tube, inner tube, and circular plates. The SMA wires are incorporated in tension to dissipate energy during its reciprocal movement in vibration.

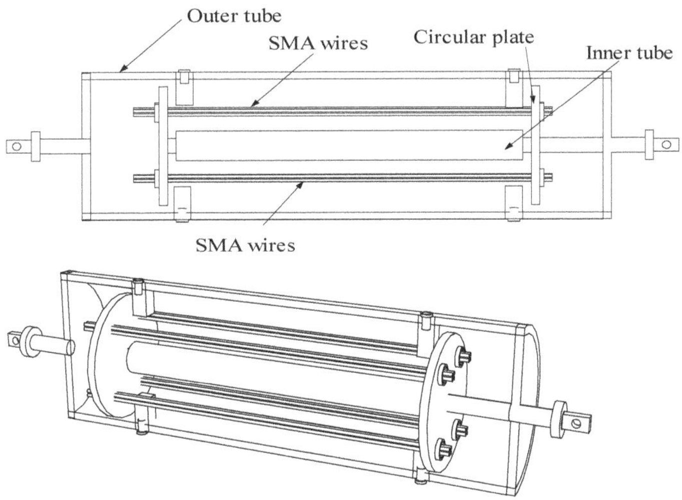

Figure 4. Configuration of an SMA damper.

Figure 5 shows the hysteretic model of the widely-used Ni-Ti SMA material. In the figure, M_f and M_s are the martensite finish and start temperature, respectively. A_f and A_s are the austenite finish and start temperature, respectively; σ_{M_s} and ε_{M_s} are the critical stress and strain at martensite start temperature, respectively; σ_{M_f} and ε_{M_f} are the critical stress and strain at martensite finish temperature, respectively; σ_{A_s} and ε_{A_s} are the critical stress and strain at austenite start temperature, respectively; σ_{A_f} and ε_{A_f} are the critical stress and strain at austenite finish temperature, respectively; ε_L is the maximum residual strain; E_A and E_M are Young's moduli at the austenite and martensite phases, respectively.

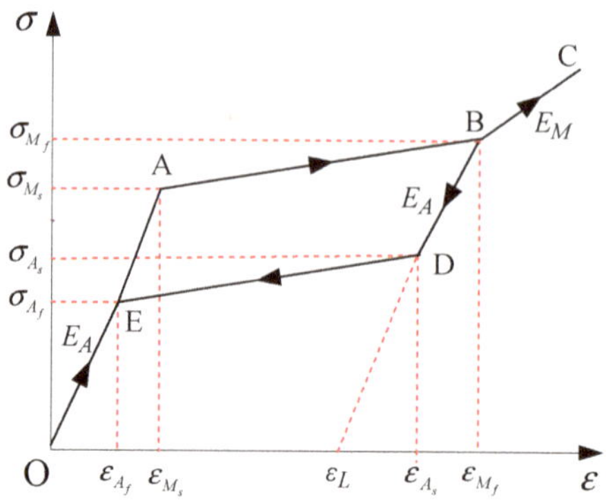

Figure 5. Hysteretic model of an SMA wire.

The relationships of strain and stress of the Ni-Ti SMA material shown in Figure 5 can be described based on different paths.

The elastic stages (Paths O-A and E-O) are the full austenite stage. The damper force is given by:

$$u(t) = \frac{E_A A}{l_w}(d(t) - l_w) \tag{9}$$

where $u(t)$ is damper force; $d(t)$ is the damper length with deformation; A is the cross-sectional area of a wire; l_w is the original length of a wire.

The forward transformation stage (Path A-B) is the loading stage and the damper force $u(t)$ is given by:

$$u(t) = \left[\sigma_{M_s} + \frac{\sigma_{M_f} - \sigma_{M_s}}{\varepsilon_{M_f} - \varepsilon_{M_s}}(\varepsilon(t) - \varepsilon_{M_s})\right]A \tag{10}$$

For the full martensite stage (Path B-C), the elastic deformation is observed and the damper force is:

$$u(t) = \frac{E_M A}{l_w}(d(t) - l_w) \tag{11}$$

For the full martensite stage (Path B-D), the unloading process is observed and the damper force is:

$$u(t) = \sigma_{M_f} A + E_M\left[(d(t) - l_w) - \varepsilon_{M_f}\right]A \tag{12}$$

The reverse transformation stage (Path D-E) is the unloading stage and the damper force $u(t)$ is given by:

$$u(t) = \left[\sigma_{M_s} + \frac{\sigma_{M_s} - \sigma_{A_f}}{\varepsilon_{A_s} - \varepsilon_{A_f}}(\varepsilon(t) - \varepsilon_{A_s})\right]A \tag{13}$$

in which $\varepsilon(t)$ is the stress of a wire.

4. Equation of Motion of the Controlled Structure

The equation of motion of the tower-line system equipped with SMA dampers is:

$$\mathbf{M}\ddot{\mathbf{x}}(t) + \mathbf{C}\dot{\mathbf{x}}(t) + \mathbf{K}\mathbf{x}(t) = \mathbf{W}(t) + \mathbf{H}\mathbf{u}(t) \tag{14}$$

$$\mathbf{M} = \begin{bmatrix} \mathbf{M}^{in} & 0 \\ 0 & \mathbf{M}^{out} \end{bmatrix} \tag{15}$$

$$\mathbf{C} = \begin{bmatrix} \mathbf{C}^{in} & 0 \\ 0 & \mathbf{C}^{out} \end{bmatrix} \tag{16}$$

$$\mathbf{K} = \begin{bmatrix} \mathbf{K}^{in} & 0 \\ 0 & \mathbf{K}^{out} \end{bmatrix} \quad (17)$$

$$\mathbf{W}(t) = \begin{bmatrix} \mathbf{W}^{in}(t); & \mathbf{W}^{out}(t) \end{bmatrix} \quad (18)$$

$$\mathbf{u}(t) = \begin{bmatrix} u_1 & u_2 & \cdots & u_n \end{bmatrix}^T \quad (19)$$

where $\mathbf{x}(t)$, $\dot{\mathbf{x}}(t)$, and $\ddot{\mathbf{x}}(t)$ are the displacement, velocity, and acceleration of the tower-line system, respectively; **M**, and **K** are the mass and stiffness matrices of the system, respectively; **C** is the Rayleigh damping matrix; **W**(t) is the wind-loading vector; $\mathbf{W}^{in}(t)$ and $\mathbf{W}^{out}(t)$ are the wind-loading vectors in the in-plane and out-of-plane direction, respectively; **u**(t) is the control force vector; **H** is the position matrix of **u**(t); n is the damper number.

The wind excitations acting on the structural system are simulated using the spectral representation method. The vibration of a tower-line coupled system can also be illustrated using energy responses. The energy equations of the entire coupled system without and with SMA dampers are formed by integrating Equation (15). The total input energy from wind loading to the structural system E_W is the sum of the kinetic energy E_K, the strain energy E_S, the energy dissipated by structural damping E_D, and the energy dissipated by SMA dampers E_C.

5. Case Study

5.1. Structural and Damper Parameters

A 110 m transmission tower is displayed in Figure 2 and the span of the transmission lines is 800 m. Six platforms are constructed in the tower body and a horizontal cross arm is placed on the top to connect transmission lines. The tower members are fabricated by Q235 steel, which is a typical type of ordinary carbon structural steel in China. Q represents the yield limit of this material. The following 235 refers to the yield value, which is about 235 MPa. The Q235 steel is widely used in civil engineering structures because of the moderate carbon content and good comprehensive properties, such as strength, plasticity, and welding. The chemical composition of Q235 steel includes C, Mn, Si, S, and P. According to the contents of the different chemical compositions, the Q235 steel can be divided into four categories, A, B, C, and D. The chemical composition of Q235 steel is listed in Table 1.

Table 1. Chemical composition of Q235 steel.

A	$C \leq 0.22\%$	$Mn \leq 1.4\%$	$Si \leq 0.35\%$	$S \leq 0.050\%$	$p \leq 0.045\%$
B	$C \leq 0.20\%$	$Mn \leq 1.4\%$	$Si \leq 0.35\%$	$S \leq 0.045\%$	$p \leq 0.045\%$
C	$C \leq 0.17\%$	$Mn \leq 1.4\%$	$Si \leq 0.35\%$	$S \leq 0.040\%$	$p \leq 0.040\%$
D	$C \leq 0.17\%$	$Mn \leq 1.4\%$	$Si \leq 0.35\%$	$S \leq 0.035\%$	$p \leq 0.035\%$

The axial stiffness EA of the transmission line is 4.88×10^4 kN. The weight per meter of the line is 1.43 kN/m. There are 1452 spatial beam elements and 353 nodes in the 3 D FE model of the example tower. The simplified dynamic model is established using a developed MATLAB program. The fundamental frequency of the single tower in the in-plane direction is 0.649 Hz and the counterpart in the out-of-plane direction is 0.643 Hz. The equation of motion is established using Rayleigh damping and solved using the Newmark-β method with a time interval of 0.02 s. The damping ratios of two fundamental frequencies are set as 0.01.

Eight SMA dampers are evenly distributed in the tower body, as shown in Figure 6. Four dampers are installed in the in-plane direction (No. 1–4) and the other four are equipped in the out-of-plane direction (No. 5–8). The Young's modulus of the SMA damper brace is 2.3×10^{11} N/m, and the cross-sectional area is 50 cm^2. Considering the configuration of the tower, an SMA damper with an axial brace can be connected to a structural member in parallel, as shown in Figure 6. The control forces provided by the

SMA damper directly act on the joint connection in the member's axial direction. The material parameters of SMA dampers are as follows: the M_f and M_s of SMA materials are −46 °C and −37.4 °C, respectively; the A_f and A_s of SMA materials are −6 °C and −18.5 °C, respectively; the C_M and C_A of SMA materials are 10 MPa/°C and 15.8 MPa/°C, respectively; the D_A and D_M of SMA materials are 75000 MPa and 29300 MPa, respectively; the maximum residual strain ε_L is 0.079.

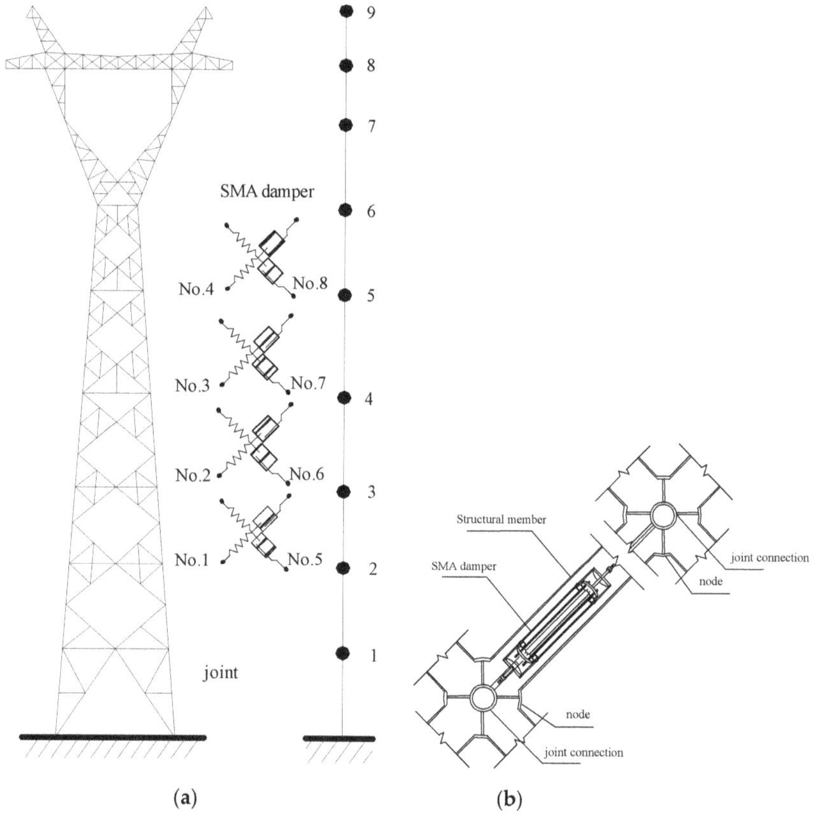

Figure 6. Installation scheme of SMA dampers. (**a**) Damper location; (**b**) Damper connection.

5.2. Peak Response Comparison

The vibration reduction factor (VRF) is adopted to assess the damper performance:

$$VRF = \frac{X_n - X_c}{X_n} \tag{20}$$

where X_n and X_c are the peak response without and with control, respectively.

Three damper location schemes are taken into consideration to compare the effects of damper position on control efficacy. For scheme 1, eight SMA dampers are installed on top of the tower body. Four dampers are installed in the in-plane direction with two dampers on the fifth floor and the other two on the sixth floor. Similarly, four dampers are installed in the out-of-plane direction with two dampers on the fifth floor and the other two on the sixth floor. For scheme 2, eight SMA dampers are placed at the bottom of the tower body. Four dampers are installed in the in-plane direction with two dampers on the first floor and the other two on the second floor. Similarly, four dampers are installed in the out-of-plane direction with two dampers on the first floor and the other two on the second floor. For

scheme 3, eight SMA dampers are evenly installed in the middle of the tower body from the third floor to the sixth floor, as shown in Figure 6.

The performance comparison of different control schemes is displayed in Figure 7. The structural peak responses are reduced substantially due to the installation of SMA dampers. The control performance of scheme 1 is slightly better than that of scheme 2. The control performance of scheme 3 is much worse than that of the other two schemes. Regarding scheme 2, eight SMA dampers are incorporated at the bottom of the tower body. The displacement responses of the tower bottom are less than those on top of the tower body. Relative small floor drifts at the tower bottom and small deformation of SMA can be observed. Thus, the energy dissipated by dampers is limited and the control efficacy is unsatisfactory. The overall control efficacy of scheme 1 is the best one, which is adopted in parametric studies and energy computation. The time histories of dynamic responses with and without SMA dampers are displayed in Figure 8. The controlled responses are much less than those of the original tower. The structural wind-induced responses are substantially suppressed for both two horizontal directions. The control efficacy of velocity is better than that of displacement and acceleration. The control performance of acceleration is slightly worse than that of displacement.

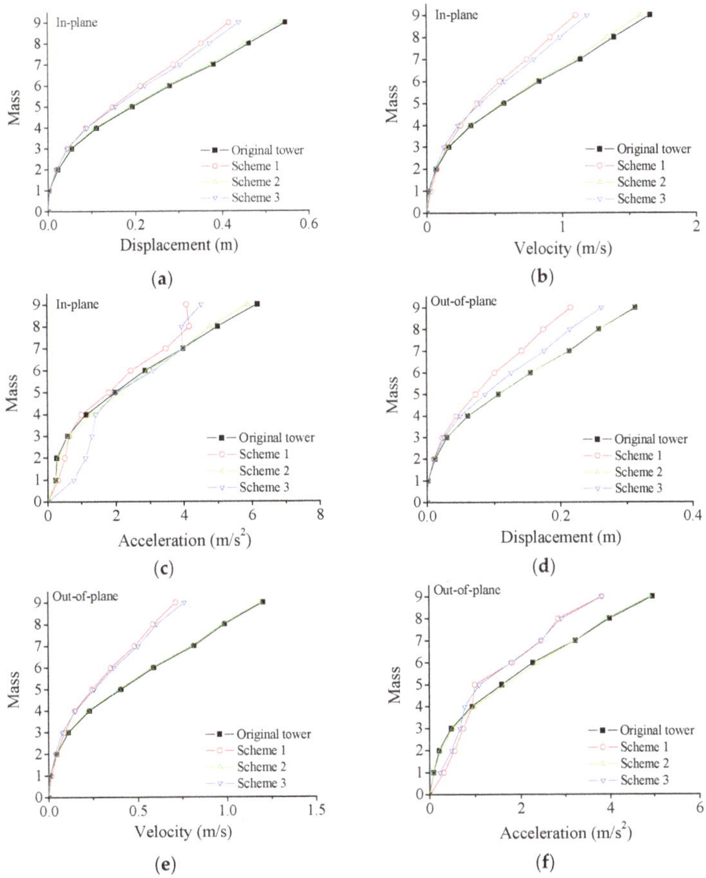

Figure 7. Comparison of different control schemes. (**a**) Peak displacement; (**b**) peak velocity; (**c**) peak acceleration; (**d**) peak displacement; (**e**) peak velocity; (**f**) peak acceleration.

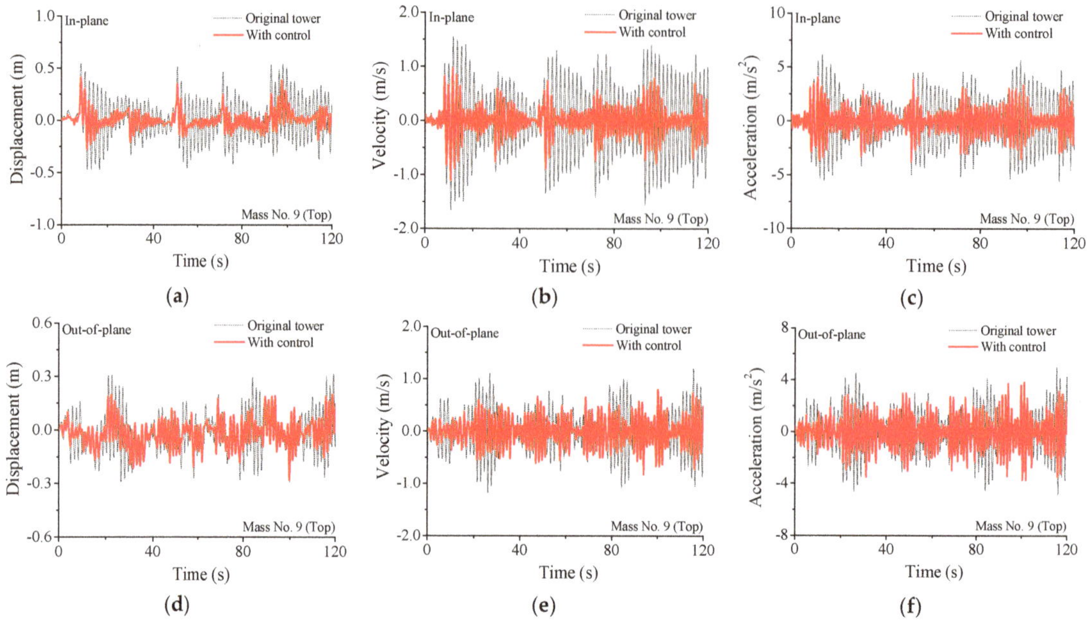

Figure 8. Time histories of dynamic responses of the transmission tower-line system. (**a**) Displacement response; (**b**) velocity response; (**c**) acceleration response; (**d**) displacement response; (**e**) velocity response; (**f**) acceleration response.

The power spectral density (PSD) curves of dynamic responses of the controlled transmission tower are also plotted in Figure 9. The PSD curves of the fundamental vibrant mode are much larger compared with the other modes. This means that the major contribution of dynamic responses of a flexible truss tower is the first vibration mode. The peak PSD values of the controlled tower are much smaller than those of the uncontrolled tower. In addition, it is observed that the properties of PSD curves for the out-of-plane vibration are quite similar to the counterpart in the in-plane direction, which means that the control efficacy of SMA dampers for two horizontal directions is close. Thus, from the viewpoint of the frequency domain, the wind-excited responses of the structural system can be substantially suppressed by SMA dampers.

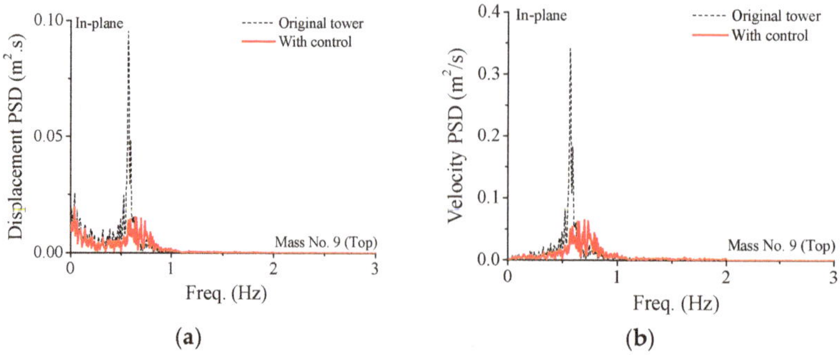

Figure 9. *Cont.*

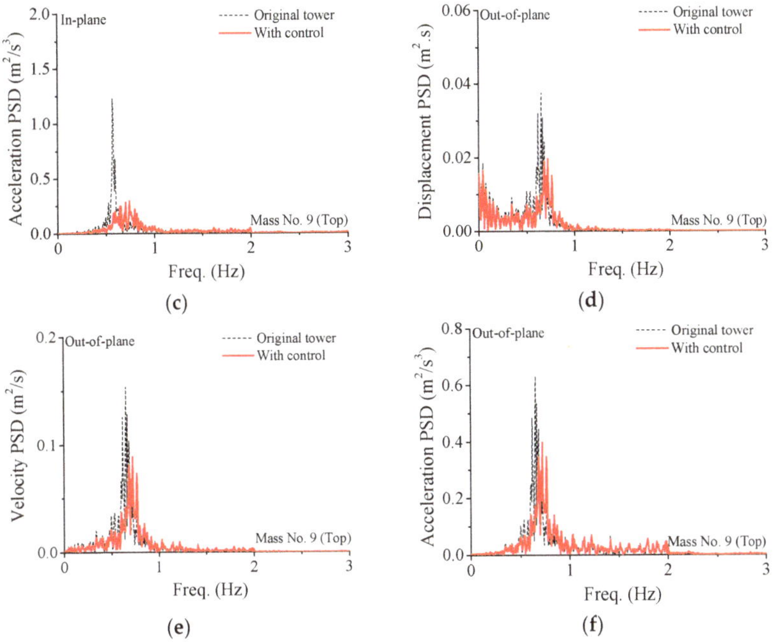

Figure 9. Comparison of PSD curves of the transmission tower-line system. (**a**) Displacement PSD curve; (**b**) velocity PSD curve; (**c**) acceleration PSD curve; (**d**) displacement PSD curve; (**e**) velocity PSD curve; (**f**) acceleration PSD curve.

6. Parametric Study on Control Efficacy

6.1. Effect of Damper Stiffness

The stiffness coefficient (SC) of an SMA damper is defined as:

$$SC = \frac{k_d^{SMA}}{k_0^{SMA}} \tag{21}$$

where k_0^{SMA} is the initial damper stiffness; k_d^{SMA} is the stiffness in the parametric study.

The influences of the damper stiffness on the structural peak responses are displayed in Figure 10. In the in-plane direction, the peak displacement gradually reduces with the increasing damper SC values until it increases to about 1.0. However, a further increment in SC cannot generate further significant displacement reduction. The varying trends of the peak velocity and acceleration are similar, as shown in Figure 10b,c. The optimal SC values for the peak velocity and acceleration are 2.0 and 1.0, respectively. Therefore, optimal SC values for various responses are different to some extent. Thus, it is not beneficial to accept a large stiffness coefficient to save fabrication costs. Similar observations are made in the out-of-plane direction, as shown in Figure 10d–f. In the out-of-plane direction of the structural system, the optimum SC values for the peak displacement, velocity, and acceleration are 1.0, 1.0, and 0.8, respectively. Therefore, the optimum SC value is selected as 1.0 considering the overall damper performance.

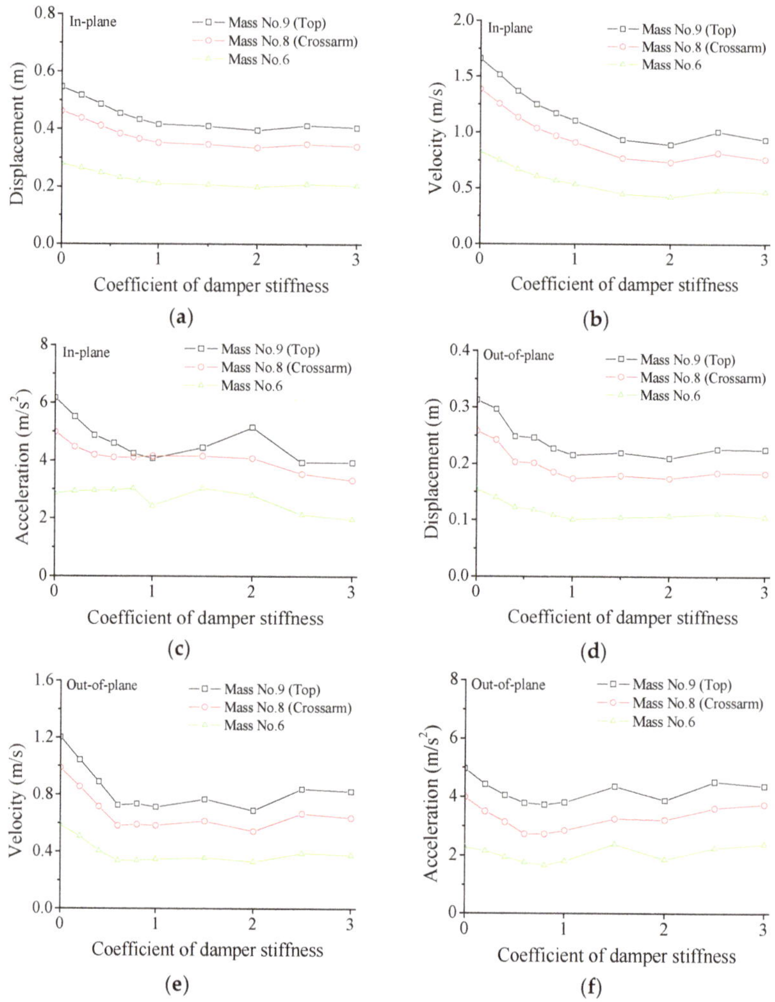

Figure 10. Stiffness effects of SMA damper on peak responses. (**a**) Peak displacement; (**b**) peak velocity; (**c**) peak acceleration; (**d**) peak displacement; (**e**) peak velocity; (**f**) peak acceleration.

The variations in damper force and deformation with damper stiffness are investigated and displayed in Figure 11 and Table 2. The peak forces of SMA dampers are proportional to the SC values for both two directions. The peak forces of SMA dampers in the in-plane direction are larger than those in the out-of-plane vibration. Similarly, the peak deformation of SMA dampers is also examined and shown in Figure 11c,d and Table 2 for the two horizontal directions. The peak damper deformation gradually reduces with the increasing SC values. With the increasing SC value, the relative variations in peak forces are much larger than those of peak deformation. A very large damper force is disadvantageous for the damper movement and energy dissipation, which makes the damper behave as a steel brace.

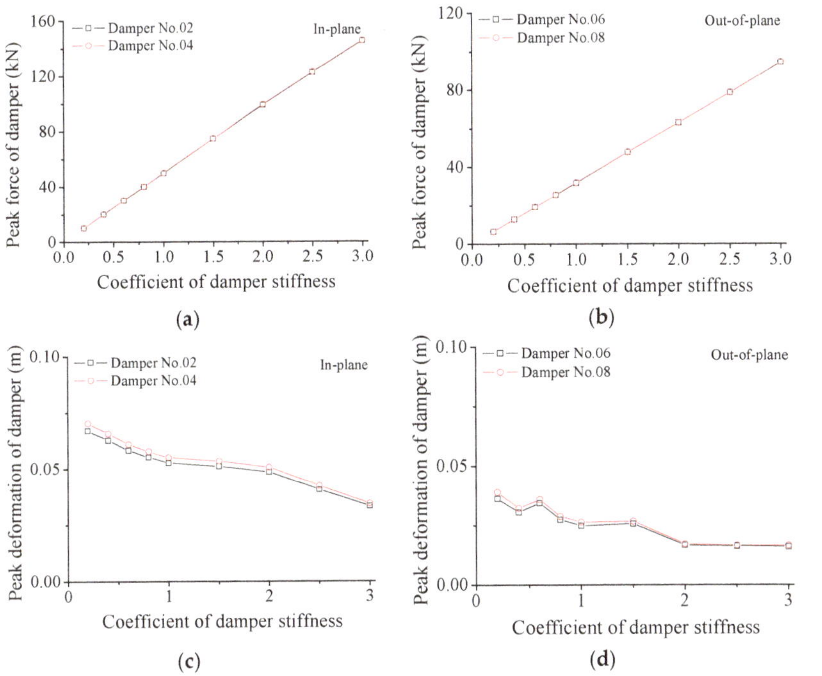

Figure 11. Variation in damper force and deformation with damper stiffness. (**a**) Damper force for the in-plane direction; (**b**) damper force for the out-of-plane direction; (**c**) damper deformation for the in-plane direction; (**d**) damper deformation for the out-of-plane direction.

Table 2. Variations in peak force and deformation of SMA dampers with damper stiffness.

Damper. No.		SC = 0.4	SC = 1.0	SC = 2.0	SC = 3.0
02 In-plane direction	Peak force	20.24 kN	49.86 kN	99.16 kN	145.44 kN
	Peak defomation	6.28 cm	5.27 cm	4.86 cm	3.33 cm
06 Out-of-plane direction	Peak force	12.87 kN	31.91 kN	63.04 kN	94.47 kN
	Peak defomation	3.05 cm	2.48 cm	1.66 cm	1.61 cm

6.2. *Influence of Damper Service Temperature*

The variations in structural peak responses with service temperature are displayed in Figure 12. It is observed that the influences on peak responses of the in-plane vibration are relatively slight in comparison with those of damper stiffness, as displayed in Figure 12a–c. An optimal service temperature for the peak displacement and velocity of the tower top and cross-arm may exist. However, the peak responses of the tower body keep stable with varying service temperature. A similar observation can be made from the peak responses of the structural system in the out-of-plane direction.

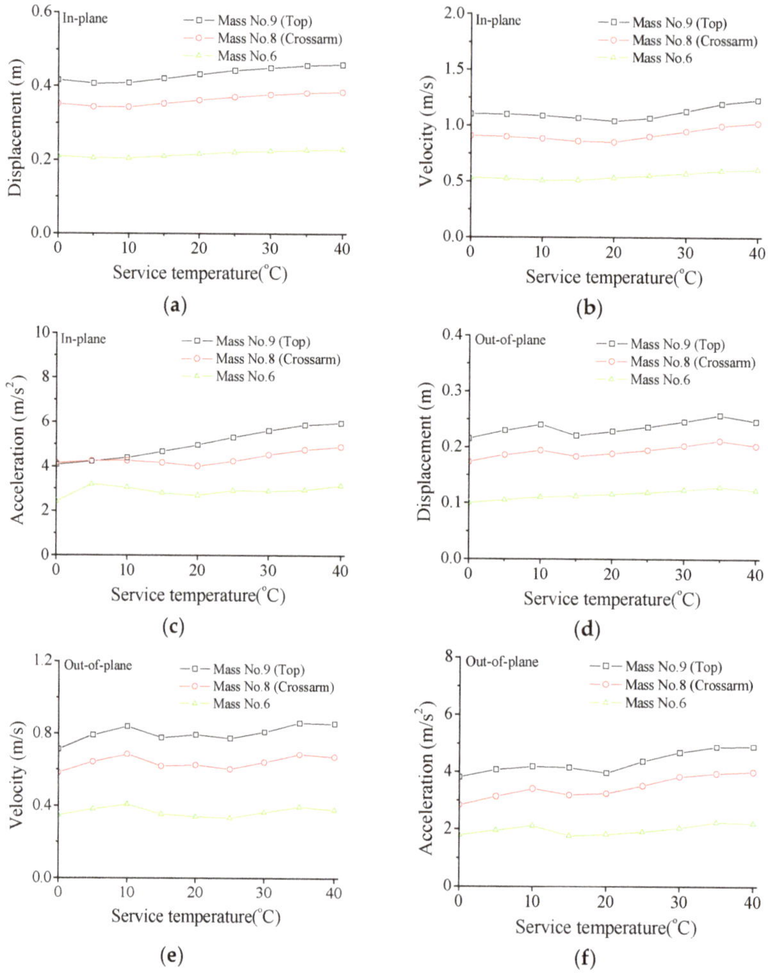

Figure 12. Effects of service temperature on maximum responses of the transmission tower. (**a**) Peak displacement; (**b**) peak velocity; (**c**) peak acceleration; (**d**) peak displacement; (**e**) peak velocity; (**f**) peak acceleration.

The variation trends in damper force and deformation are also examined and displayed in Figure 13 and Table 3. Similar to the effects of damper stiffness, the peak forces are proportional to the service temperature and the peak damper forces for the in-plane vibration are larger than those in the out-of-plane direction. However, the varying trend of the peak damper deformation is quite different from that of damper stiffness, as shown in Figure 13c,d and Table 3. With the increase in service temperature, the peak deformation of SMA dampers keeps stable for the in-plane movement. The peak damper deformation in the out-of-plane direction slightly increases with the increasing service temperature. Thus, the influence of service temperature on peak damper deformation is much smaller compared with that of damper stiffness.

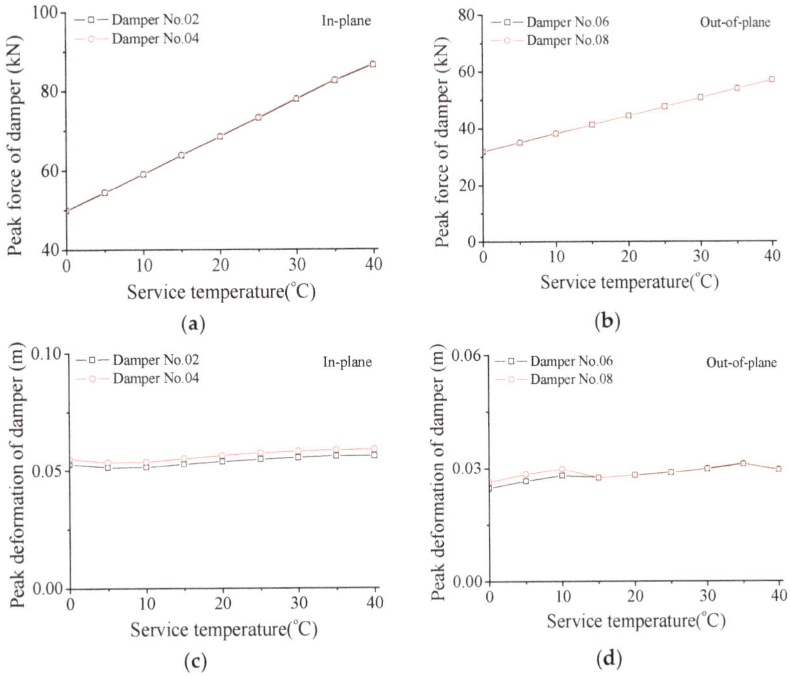

Figure 13. Variation in damper force and deformation with service temperature. (**a**) Damper force for the in-plane direction; (**b**) damper force for the out-of-plane direction; (**c**) damper deformation for the in-plane direction; (**d**) damper deformation for the out-of-plane direction.

Table 3. Variations in peak force and deformation of SMA dampers with service temperature.

Damper. No.		T = 0 °C	T = 10 °C	T = 20 °C	T = 40 °C
02 In-plane direction	Peak force	49.86 kN	59.04 kN	68.48 kN	86.39 kN
	Peak defomation	5.28 cm	5.14 cm	5.39 cm	5.62 cm
06 Out-of-plane direction	Peak force	31.91 kN	38.26 kN	44.47 kN	56.94 kN
	Peak defomation	2.49 cm	2.81 cm	2.82 cm	2.96 cm

6.3. Variation of Hysteresis Loop

Configuration of hysteresis loops can reflect the control performance of a damper under wind excitations. Displayed in Figure 14 are the variations in hysteresis loops with damper stiffness for the two orthogonal directions. If a small stiffness is adopted (SC = 0.2), the SMA damper is easy to move and a large deformation is expected, as shown in Figure 14a. The enclosed area of the hysteresis loop is very small, which reflects a poor energy-dissipating capacity. To increase the damper stiffness (SC = 1.0), the shape of the hysteresis loops can be changed to a great extent, as displayed in Figure 14b. The damper deformation is reduced and the damper force is remarkably improved. In addition, the enclosed area increases remarkably and the control performance is substantially improved. If the damper stiffness continues to increase (SC = 3.0), the damper force can increase accordingly while the deformation reduces, as shown in Figure 14c. In this circumstance, the enclosed area does not increase and the energy-dissipating ability cannot be improved. For the out-of-plane vibration, the same conclusion can be drawn, as shown in Figure 14d–f. Thus, an optimal SC value can also be selected in line with the shapes of hysteresis loops.

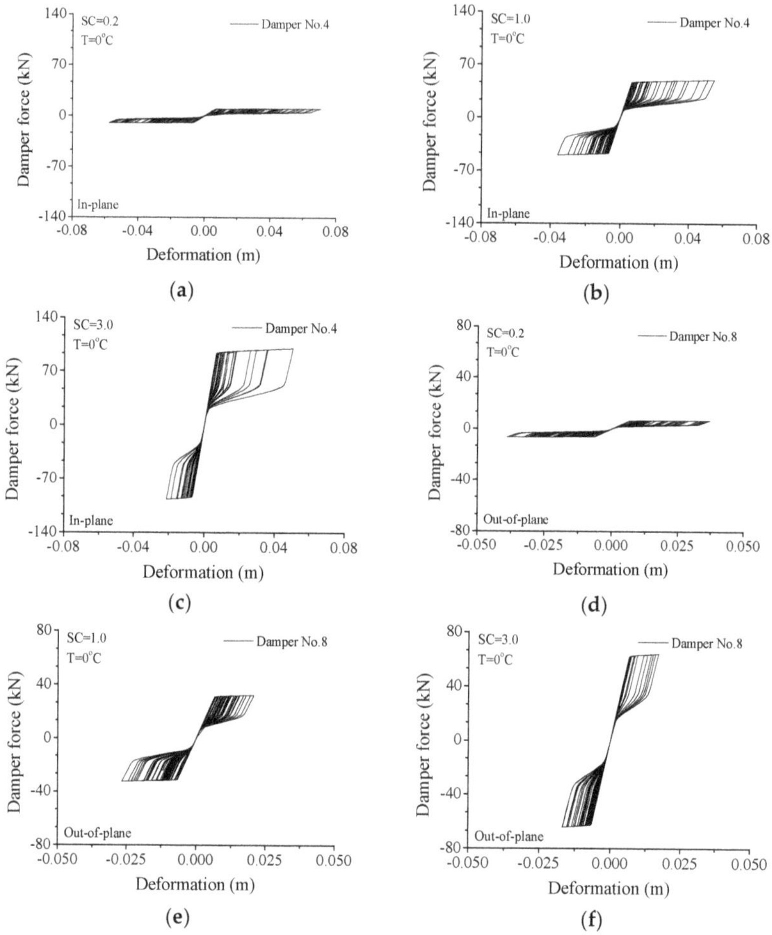

Figure 14. Variation in hysteresis loop with damper stiffness. (**a**) SC = 0.2; (**b**) SC = 1.0; (**c**) SC = 3.0; (**d**) SC = 0.2; (**e**) SC = 1.0; (**f**) SC = 3.0.

The variations in hysteresis loops with service temperature are also investigated through a detailed parametric study, as shown in Figure 15. Similar to the conclusions made from Figure 13, the peak damper forces quickly increase with the increasing service temperature for both two directions. If the service temperature is common (T = 0 °C), a relatively large damper deformation is observed and the hysteresis loop is plump which means satisfactory energy-dissipating capacity (See Figure 15a). If the service temperature gradually increases, the peak damper force also increases and the enclosed areas of SMA dampers reduce. If the service temperature reaches a relatively large value (T = 40 °C), the peak damper force increases quickly and, at the same time, the enclosed areas of SMA dampers dramatically reduce to a very small level, as shown in Figure 15c. In this circumstance, the SMA damper behaves like a steel brace. The peak displacement is reduced while the peak acceleration increases. This is due to the large peak damper force instead of a poor energy-dissipating capacity. Similar effects are observed from the hysteresis loops of SMA dampers for the out-of-plane vibration. Overall, the service temperature has a great influence on damper force instead of damper deformation. A very large service temperature is unnecessary for the improvement of the control performance of SMA dampers.

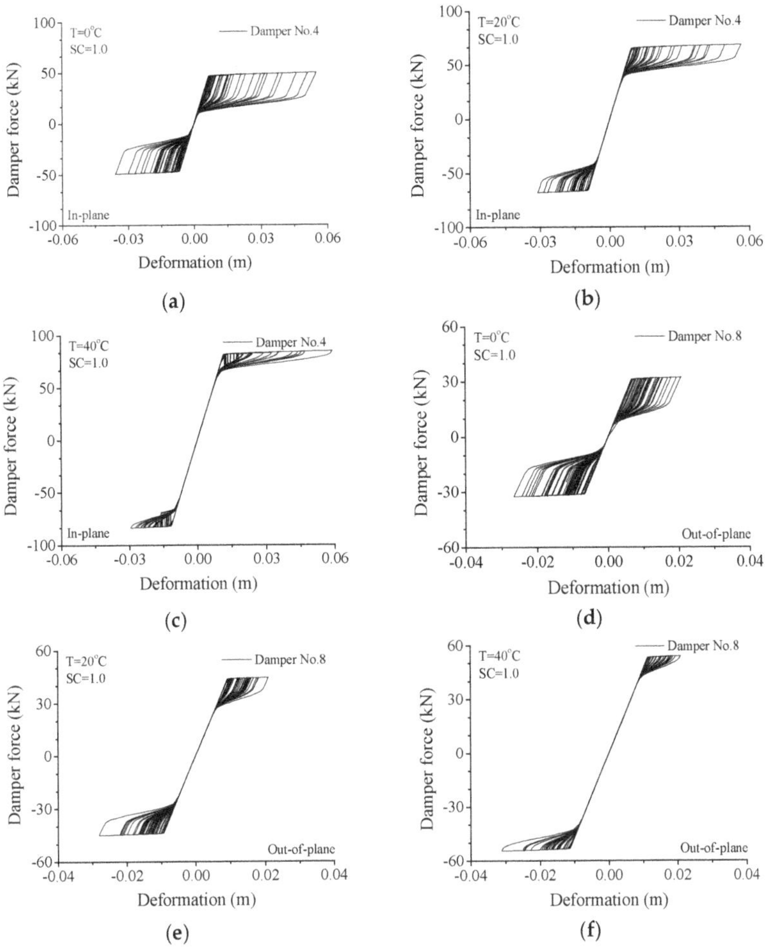

Figure 15. Variation in hysteresis loop with service temperature. (**a**) T = 0 °C; (**b**) T = 20 °C; (**c**) T = 40 °C; (**d**) T = 0 °C; (**e**) T = 20 °C; (**f**) T = 40 °C.

It is noted that the real application of various energy-dissipating dampers in civil engineering structures will depend on the damper configuration and cost. For the transmission tower with SMA dampers, the amount of alloy used and the cost of SMA dampers are crucial issues that should be taken into consideration. From the viewpoint of real application, satisfactory control efficacy with optimal damper stiffness is essential. Optimal damper stiffness can be determined through parametric studies. A very large damper force is disadvantageous for the damper movement and energy dissipation, which makes the damper behave as a steel brace. Thus, it is not beneficial to accept a large stiffness coefficient to save fabrication costs.

7. Properties of System Energy Responses

7.1. Energy Curves with Control

The control performance of SMA dampers on the structural system can also be illustrated by energy responses, as shown in Figure 16. For the uncontrolled transmission tower, the total input energy from wind loading E_W is the sum of the kinetic energy E_K, the strain energy E_S, and structural damping energy E_D, as shown in Figure 16a,b. Large E_K and

E_S are observed due to the strong vibration of the entire coupled system. The kinetic and strain energy can only be dissipated by structural damping.

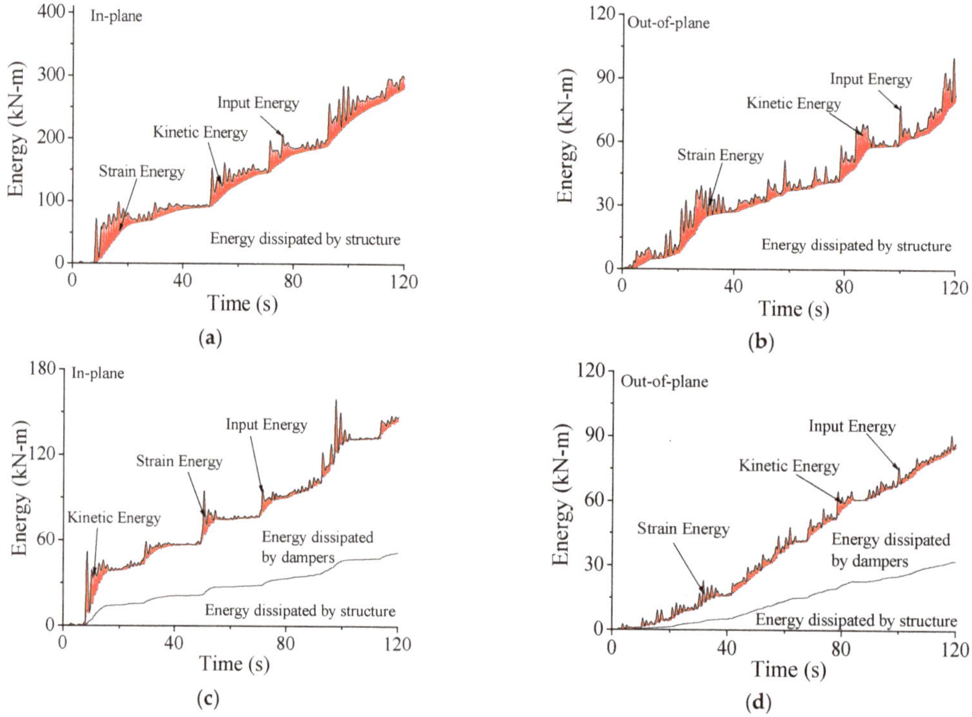

Figure 16. Energy responses of the transmission tower without/with SMA dampers. (**a**) Tower energy for in-plane vibration; (**b**) tower energy for out-of-plane vibration; (**c**) tower energy with control for in-plane vibration; (**d**) tower energy with control for out-of-plane vibration.

The case for the controlled transmission tower is quite different, as displayed in Figure 16c,d. The magnitude of the kinetic energy E_K and the strain energy E_S are remarkably mitigated because the total sum of dissipated energy is substantially increased. The vibrant energy can be absorbed simultaneously by both the structural damping and SMA dampers. Owing to the contribution of SMA dampers, the structural damping energy E_D is dramatically reduced. When comparing the energy curves without/with SMA dampers, the inputted energy E_W is smaller compared with that of the uncontrolled tower. This is because the inputted energy is directly related to structural vibrant intensity. The dynamic responses of the controlled system are much smaller than those of the uncontrolled system. Thus, the inputted energy from wind excitations to the controlled tower E_W is much smaller than that of the original tower.

7.2. Effect of Damper Stiffness on Energy Response

The effects of damper parameters on structural energy responses are investigated through a parametric study in detail. The variations in energy responses with damper stiffness are examined and displayed in Figure 17. If damper stiffness is too small (SC = 0.2), the damper capacity in energy-dissipating is limited. As displayed in Figure 17a, the energy dissipated by SMA dampers E_C quickly increases until the SC value reaches about 0.6. After that value, the E_C gradually reduces. The energy E_D is much smaller than that of the original tower, as shown in Figure 17b. The E_D quickly reduces with the increase in SC values. It is also seen that the E_D for SC = 1.0 is quite close to that for SC = 3.0, which means that a very large damper stiffness is unnecessary for the improvement of damper performance. The variations in total energy input from wind excitations E_W with damper stiffness are also investigated and plotted in Figure 17c. Similar to the observations made from structural damping, firstly, the E_W quickly reduces with the increasing SC values until SC reaches about 1.0. Then, a further increment in SC value cannot remarkably reduce the dynamic responses and the inputted energy E_W. Furthermore, optimal damper SC values can be selected using the energy curves.

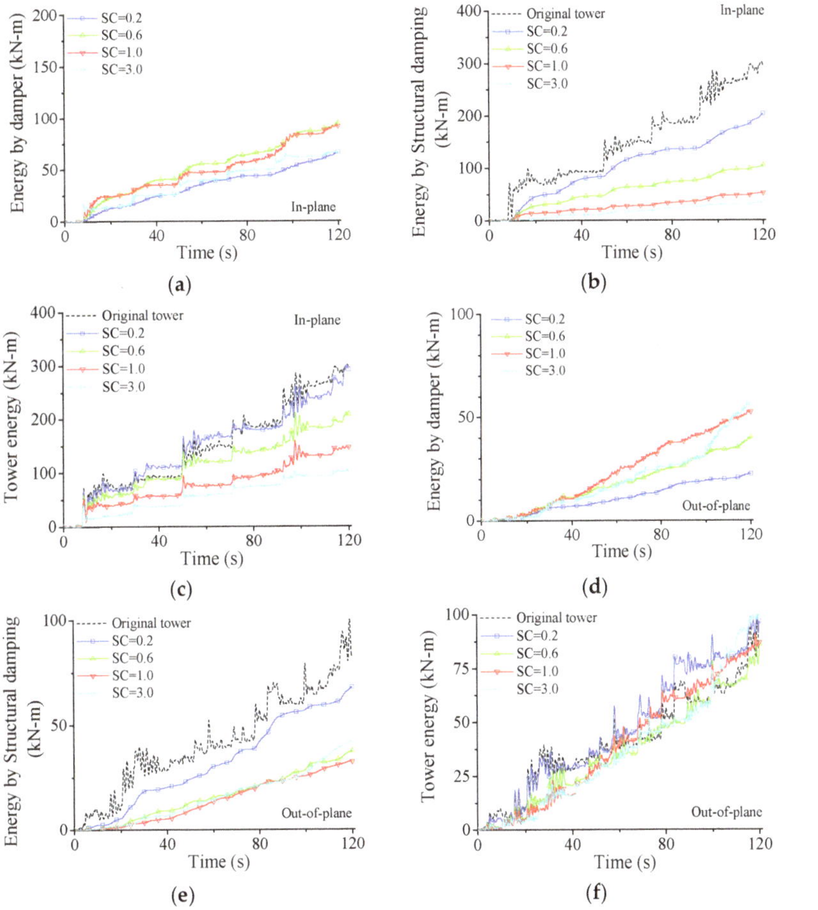

Figure 17. Variation in energy responses with damper stiffness. (**a**) Tower energy for in-plane vibration; (**b**) structural damping energy for in-plane vibration; (**c**) damper energy for in-plane vibration; (**d**) tower energy for out-of-plane vibration; (**e**) structural damping energy for out-of-plane vibration; (**f**) damper energy for out-of-plane vibration.

The variations in energy responses with damper stiffness for the out-of-plane vibration are also investigated, as displayed in Figure 17d–f and similar conclusions to the in-plane vibration can be drawn. It is noted that the optimal SC values for different energy responses may differ to some extent. Furthermore, optimal SC values for the two directions are slightly different due to the difference in structural dynamic responses. Overall, the optimal SC value of SMA dampers for the example system can be selected as 1.0 in line with the energy responses, which is the same as that based on the peak responses, as shown in Figure 10.

7.3. Variation in Energy Response with Service Temperature

Figure 18 displays the variation in energy responses with service temperature. As shown in Figure 18a,b, with the increase in the service temperature, the energy dissipated by SMA dampers E_P decreases while the energy dissipated by structural damping E_D increases gradually. If the service temperature is common (T = 0 °C), the hysteresis loop is plump which means satisfactory energy-dissipating capacity, as shown in Figure 15a. If the service temperature is relatively large (T = 40 °C), the peak damper force increases quickly and the SMA damper behaves like a steel brace, as shown in Figure 15c. The enclosed areas of the hysteresis loop dramatically decrease and the energy dissipated by SMA dampers E_P reduces, as shown in Figure 18a.

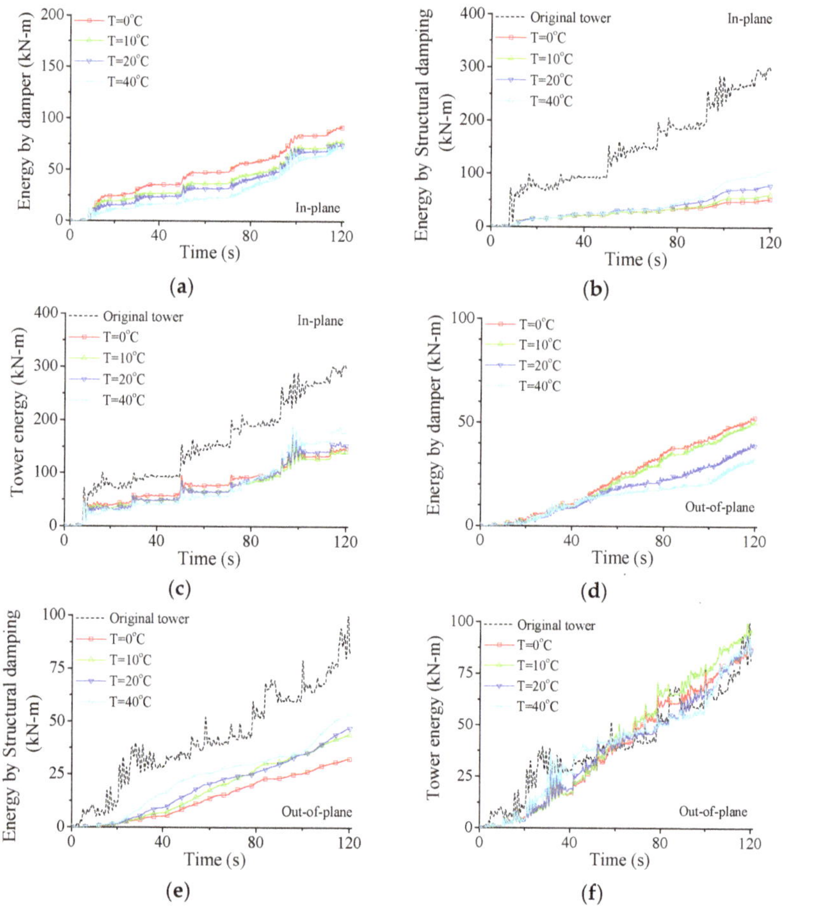

Figure 18. Variation in energy responses with service temperature. (**a**) Tower energy for in-plane vibration; (**b**) structural damping energy for in-plane vibration; (**c**) damper energy for in-plane vibration; (**d**) tower energy for out-of-plane vibration; (**e**) structural damping energy for out-of-plane vibration; (**f**) damper energy for out-of-plane vibration.

The variations in the inputted energy E_W with service temperature are shown in Figure 18c, which are different from those with damper stiffness, as displayed in Figure 17c. The increment in the service temperature cannot improve the E_C of SMA dampers but increase the damper force, as displayed in Figure 15. Thus, the effects of service temperature on structural peak responses are small (see Figure 12). As mentioned above, the inputted energy is directly related to structural vibrant intensity. The influence of service temperature on the inputted energy E_W is relatively small. The energy curves for the out-of-plane vibration present similar results, as shown in Figure 18d–f. On the whole, a very large service temperature is not beneficial for the performance of SMA dampers. An optimal service temperature of SMA dampers for the example structural system can be selected as T = 0 °C based on the energy curves, which is the same as that based on the peak responses, as shown in Figure 12.

8. Concluding Remarks

The vibration control of a tower-line coupled system disturbed by wind loading was conducted by SMA dampers. The analytical model of the entire system was established based on Lagrange's equations by considering the dynamic interaction between transmission lines and towers. The control efficacy was analyzed in both the time domain and the frequency domain. Detailed parametric studies were conducted to examine the influence of damper stiffness, service temperature on structural responses, and hysteresis loops. The following conclusions can be drawn:

(1) SMA dampers are beneficial in the vibration control of the tower-line coupled system disturbed by wind loading. The control efficacy on displacement and velocity is slightly better than acceleration. The peak PSD values of the controlled tower are much smaller than those of the uncontrolled tower.
(2) The peak responses gradually decrease with the increasing damper stiffness. For the in-plane vibration, the optimal SC values for the peak displacement, velocity, and acceleration are 1.0, 2.0, and 1.0, respectively. For the out-of-plane vibration, the optimal SC values for the peak displacement, velocity, and acceleration are 1.0, 1.0, and 0.8, respectively. Therefore, the optimum SC value is selected as 1.0 considering the overall damper performance. An optimum stiffness coefficient exists for the response control and it is unnecessary to set a very large stiffness coefficient to save fabrication cost.
(3) The influence of service temperature on peak damper deformation is much smaller compared with that of damper stiffness. If the service temperature reaches a relatively large value (T = 40 °C), the peak damper force increases quickly and, at the same time, the enclosed areas of SMA dampers dramatically reduce to a very small level. Overall, the service temperature has a great influence on damper force instead of damper deformation. A very large service temperature is unnecessary for the improvement of the control performance of SMA dampers.
(4) The control performance on wind-induced dynamic responses can also be depicted by energy responses. Furthermore, the optimal stiffness coefficient and service temperature of SMA dampers can also be determined in line with energy curves.

Author Contributions: Conceptualization, B.C.; methodology, B.C. and X.S.; software, W.L.; validation, B.C. and X.S.; formal analysis, X.S. and J.W.; investigation, B.C. and W.L.; resources, J.W. and X.S.; data curation, J.W.; writing—original draft preparation, B.C. and X.S.; writing—review and editing, B.C. and W.L.; visualization, B.C. and W.L.; supervision, B.C. and W.L.; project administration, B.C.; funding acquisition, B.C. All authors have read and agreed to the published version of the manuscript.

Funding: The authors are grateful for the financial support from the National Natural Science Foundation of China (51978549, 51678463).

Institutional Review Board Statement: Not applicable.

Informed Consent Statement: Not applicable.

Data Availability Statement: Date sharing is not applicable to this article.

Acknowledgments: The authors sincerely appreciate the selfless support provided by Wuhan University of Technology and Guangdong Power Grid Energy Development, Co., Ltd. during the progress of this study.

Conflicts of Interest: The authors declare no conflict of interest.

References

1. Amiri, M.M.; Yahyai, M. Estimation of damping ratio of TV towers based on ambient vibration monitoring. *Struct. Des. Tall Spec. Build.* **2013**, *22*, 862–875. [CrossRef]
2. Chen, B.; Guo, W.H.; Li, P.Y.; Xie, W.P. Dynamic responses and vibration control of the transmission tower-line system: A state-of-the-art-review. *Sci. World J.* **2014**, *538457*, 1–20. [CrossRef] [PubMed]
3. Li, T.; Fan, J.; Lv, Y.X. Collapse analysis of a steel TV tower under wind load. *Steel Constr.* **2017**, *32*, 80–83.
4. Liu, R.S.; Liu, J.L.; Yan, D.Q. Seismic damage investigation and analysis of electric power system in Lushan MS 7.0 earthquake. *J. Nat. Disasters* **2013**, *22*, 85–92.
5. Housner, G.W.; Bergman, L.A.; Caughey, T.K.; Chassiakos, A.G.; Claus, R.O.; Masri, S.F.; Skelton, R.E.; Soong, T.T.; Spencer, B.F., Jr.; Yao, T.P. Structural control: Past, present, and future. *J. Eng. Mech. ASCE* **1997**, *123*, 897–971. [CrossRef]
6. Xu, Y.L.; Qu, W.L.; Chen, B. Active/robust moment controllers for seismic response control of a large span building on top of ship lift towers. *J. Sound Vib.* **2003**, *261*, 277–296. [CrossRef]
7. Chen, B.; Xia, Y. Advanced technologies in disaster prevention and mitigation. *Adv. Struct. Eng.* **2017**, *20*, 1141–1142. [CrossRef]
8. Yang, N.; Wang, C.M.; Balendra, T. Composite mass dampers for vibration control of wind-excited towers. *J. Sound Vib.* **1998**, *213*, 301–316. [CrossRef]
9. Wu, J.C.; Yang, J.N. LQG control of lateral-torsional motion of Nanjing TV transmission tower. *Earthq. Eng. Struct. Dyn.* **2000**, *29*, 1111–1130. [CrossRef]
10. Ghorbani-Tanha, A.K.; Noorzad, A.; Rahimian, M. Mitigation of wind-induced motion of Milad Tower by tuned mass damper. *Struct. Des. Tall Spec. Build.* **2009**, *18*, 371–385. [CrossRef]
11. Lu, X.L.; Li, P.Z.; Guo, X.Q.; Shi, W.X.; Liu, J. Vibration control using ATMD and site measurements on the Shanghai World Financial Center Tower. *Struct. Des. Tall Spec. Build.* **2014**, *23*, 105–123. [CrossRef]
12. Hitchcock, P.A.; Glanville, M.J.; Kwok, K.C.S.; Watkins, R.D. Damping properties and wind-induced response of a steel frame tower fitted with liquid column vibration absorbers. *J. Wind Eng. Ind. Aerodyn.* **1999**, *83*, 183–196. [CrossRef]
13. Balendra, T.; Wang, C.M.; Yan, N. Control of wind-excited towers by active tuned liquid column damper. *Eng. Struct.* **2001**, *23*, 1054–1067. [CrossRef]
14. Pall, A.S.; Marsh, C. Response of friction damped braced frames. *J. Struct. Div. ASCE* **1982**, *108*, 1313–1323. [CrossRef]
15. Grigorian, C.E.; Yang, T.S.; Popov, E.P. Slotted bolted connection energy dissipators. *Earthq. Spectra* **1993**, *9*, 491–504. [CrossRef]
16. Quintana, H.C.; Petkovski, M. Optimum performance of structural control with friction dampers. *Eng. Struct.* **2018**, *172*, 154–162. [CrossRef]
17. Chen, B.; Yang, D.; Feng, K.; Ouyang, Y.Q. Response control of a high-rise television tower under seismic excitations by friction dampers. *Int. J. Struct. Stab. Dyn.* **2018**, *18*, 1850140. [CrossRef]
18. Zhang, Z.Q.; Li, A.Q. Seismic response vibration control analyses of Hefei TV Tower with fluid viscous damper. In Proceedings of the International Conference on Health Monitoring of Structure, Materials and Environment, Nanjing, China, 16 October 2007.
19. Chen, B.; Xiao, X.; Li, P.Y.; Zhong, W.L. Performance evaluation on transmission-tower line system with passive friction dampers subjected to wind excitations. *Shock Vib.* **2015**, *310458*, 1–13. [CrossRef]
20. Chen, B.; Weng, S.; Zhi, L.H.; Li, D.M. Response control of a large transmission-tower line system under seismic excitations by using friction dampers. *Adv. Struct. Eng.* **2017**, *20*, 1155–1173. [CrossRef]
21. Tian, L.; Yu, Q.Q.; Ma, R.S. Study on seismic control of power transmission tower-line coupled system under multicomponent excitations. *Math. Probl. Eng.* **2013**, *829415*, 1–12. [CrossRef]
22. Zhang, P.; Song, G.B.; Li, H.N.; Lin, Y.X. Seismic control of power transmission tower using pounding TMD. *J. Eng. Mech. ASCE* **2013**, *139*, 1395–1406. [CrossRef]
23. Chen, B.; Wu, J.B.; Ouyang, Y.Q.; Yang, D. Response evaluation and vibration control of a transmission tower-line system in mountain areas subjected to cable rupture. *Struct. Monit. Maint.* **2018**, *5*, 151–171.
24. Urreta, H.; Leicht, Z.; Sanchez, A.; Agirre, A.; Kuzhir, P.; Magnac, G. Hydrodynamic bearing lubricated with magnetic fluids. *J. Intell. Mater. Syst. Struct.* **2010**, *21*, 1491–1499. [CrossRef]
25. Urreta, H.; Aguirre, G.; Kuzhir, P.; de Lacalle, L.N.L. Seals based on magnetic fluids for high precision spindles of machine tools. *Int. J. Precis. Eng. Manuf.* **2018**, *19*, 495–503. [CrossRef]
26. Xu, Y.L.; Qu, W.L.; Chen, Z.H. Control of wind-excited truss tower using semiactive friction damper. *J. Struct. Eng. ASCE* **2001**, *127*, 861–868. [CrossRef]
27. Chen, B.; Zheng, J.; Qu, W.L. Control of wind-induced response of transmission tower-line system by using magnetorheological dampers. *Int. J. Struct. Stab. Dyn.* **2009**, *9*, 661–685. [CrossRef]
28. Song, G.B.; Ma, N.; Li, H.N. Applications of shape memory alloys in civil structures. *Eng. Struct.* **2006**, *28*, 1266–1274. [CrossRef]

29. Fang, C.; Wang, W.; He, C.; Chen, Y. Self-centring behaviour of steel and steel-concrete composite connections equipped with NiTi SMA bolts. *Eng. Struct.* **2017**, *150*, 390–408. [CrossRef]
30. Han, Y.L.; Li, Q.S.; Li, A.Q.; Leung, A.Y.T.; Lin, P.H. Structural vibration control by shape memory alloy damper. *Earthq. Eng. Struct. Dyn.* **2003**, *32*, 483–494. [CrossRef]
31. Zheng, Y.; Dong, Y.; Chen, B.; Ghazanfar, A.A. Seismic damage mitigation of bridges with self-adaptive SMA-cable-based bearings. *Smart Struct. Syst.* **2019**, *24*, 127–139.
32. Tabrizikahou, A.; Kuczma, M.; Lasecka-Plura, M.; Noroozinejad Farsangi, E. Cyclic behavior of masonry shear walls retrofitted with engineered cementitious composite and pseudoelastic shape memory alloy. *Sensors* **2022**, *22*, 511. [CrossRef] [PubMed]
33. Li, X.H.; Zhu, Z.W. Nonlinear dynamic characteristics and stability analysis of energy storage flywheel rotor with shape memory alloy damper. *J. Energy Storage* **2022**, *45*, 103392. [CrossRef]
34. Tian, L.; Zhou, M.Y.; Qiu, C.X.; Pan, H.Y.; Rong, K.J. Seismic response control of transmission tower-line system using SMA-based TMD. *Struct. Eng. Mech.* **2020**, *74*, 129–143.
35. Wu, J.B.; Chen, B.; Zhi, L.H.; Song, X.X. Seismic response mitigation of a television transmission tower by shape memory alloy dampers. *Materials* **2021**, *14*, 6987. [CrossRef]
36. Brinson, L.C. One-dimensional constitutive behavior of shape memory alloys: Thermomechanical derivation with non-constant material functions and redefined martensite internal variable. *J. Intell. Mater. Syst. Struct.* **1993**, *4*, 229–242. [CrossRef]
37. Zheng, Y.; Dong, Y.; Li, Y. Resilience and life-cycle performance of smart bridges with shape memory alloy (SMA)-cable-based bearings. *Constr. Build. Mater.* **2018**, *158*, 389–400. [CrossRef]
38. Qiu, C.; Fang, C.; Liang, D.; Du, X.; Yam, M.C.H. Behavior and application of self-centering dampers equipped with buckling-restrained SMA bars. *Smart Mater. Struct.* **2020**, *29*, 035009. [CrossRef]

materials

MDPI

Article

Performance-Based Assessment of Bridges with Novel SMA-Washer-Based Self-Centering Rocking Piers

Jiawei Chen [1], Dong Liang [1], Xin You [2,*] and Hao Liang [3,*]

1 College of Civil Engineering, Tongji University, Shanghai 200092, China
2 College of Civil Engineering, Qinghai University, Xining 810016, China
3 College of Civil Engineering, Southeast Jiaotong University, Chengdu 611756, China
* Correspondence: 2022740001@qhu.edu.cn (X.Y.); liangboju@my.swjtu.edu.cn (H.L.); Tel.: +86-138-9741-0751 (X.Y.); +86-152-0922-0506 (H.L.)

Abstract: This study discussed a novel self-centering rocking (SCR) bridge system equipped with shape memory alloy (SMA)-based piers, with a particular focus on the benefit of the SCR bridge system in a life-cycle context. The study commences with an introduction of the SCR bridge system; subsequently, a life-cycle loss and resilience assessment framework for the SCR bridge system is presented. Specifically, the seismic fragility, resilience, and life-cycle loss associated with the SCR and conventional bridge systems were addressed. The proposed life-cycle assessment framework was finally applied to two highway bridges with and without SMA washer-based rocking piers, considering the representative hazard scenarios that could happen within the investigated regions. The results revealed that the novel SCR pier bridge system slightly increased the bearing displacement but extensively reduced the pier curvature ductility due to the rocking mechanism. The SCR bridge system kept a lower life-cycle loss level and exhibited more resilient performance than the conventional bridge, especially in the region with higher seismic intensities. Indirect loss can be significantly larger than the direct loss, specifically for the earthquakes with a relatively low probability of occurrence. The SCR bridge system outperformed the conventional system in terms of recovery time, where a quick recovery after an earthquake and drastically decreased the social and economic losses.

Keywords: self-centering rocking (SCR) piers; shape memory alloy (SMA); seismic fragility; resilience; life-cycle loss

Citation: Chen, J.; Liang, D.; You, X.; Liang, H. Performance-Based Assessment of Bridges with Novel SMA-Washer-Based Self-Centering Rocking Piers. *Materials* **2022**, *15*, 6589. https://doi.org/10.3390/ma15196589

Academic Editors: Cheng Fang, Canxing Qiu and Yue Zheng

Received: 8 August 2022
Accepted: 12 September 2022
Published: 22 September 2022

1. Introduction

Although ductility-based seismic design philosophy has been employed for a long time, severe damage to bridges (e.g., unseating of girders) in recent earthquakes (e.g., Maduo earthquake, China, 2021) indicates its limitation. Overly large residual deformation may compromise the normal function of the bridges after earthquakes, and this issue is gaining increasing attention. To ensure normal operation of the lifeline systems, new design principles associated with residual deformation in seismic codes have been appended in many countries (e.g., US, Japan, and New Zealand) [1]. For instance, rocking bridge piers have been gaining attraction because of their small residual deformation property [2] and fast precast construction characteristic [3–6]. The objective of rocking is to remarkably decrease the input energy due to earthquakes by elongating the fundamental period of vibration. Some experimental studies [7] have been successfully carried out to verify the concept. The design allows the piers to rock around the foundation/footing, resulting in the alleviation of seismic damage. In order to avoid overturning of rocking bridges, typical post-tensioned (PT) rocking piers have been proposed together with test verification [8–11]. To further improve the performance of the rocking piers subjected to severe earthquakes, a series of novel supplementary self-centering and/or energy dissipation devices have been developed and examined [12–16].

Although combining PT tendons with energy dissipaters can remarkably decrease structural damage and residual deformation, repair or replacement of the energy dissipaters after earthquakes is time consuming and costly. Corrosion may also be an issue for metal energy dissipaters. In this regard, shape memory alloy (SMA), a novel class of metal, has been recently employed in bridge structures [17–22] as well as other engineering structural systems [23–34] to enhance their seismic resilience. At the austenite phase, SMA can exhibit superelasticity at room temperature, and is capable of recovering large strains (up to 8~10% strain) after experiencing earthquake excitation. A representative investigation was conducted by Varela and Saiidi [20], who examined the feasibility of using SMA bars at the plastic hinge zone of RC piers. The results indicated that except for the bucking of the SMA bars, the RC piers experienced almost no damage during severe earthquakes. Two encouraging examples of using SMA-based components for real construction projects have been reported, of which the SMA bars were used in the plastic hinge zones of the RC piers in the State Route 99 Off Ramp Bridge in USA [35], and the SMA-cable-based bearings were installed in the Datianba #2 Bridge in China [36,37]. However, lifecycle assessment of these novel bridge systems, especially their direct, indirect, and long-term economic loss performances, is still insufficient.

With initial confidence gained from the successful practical applications of SMA in bridges, this paper further discussed a novel type of bridge system employing SCR bridge piers, where superelastic shape memory alloy (SMA) washer springs serve as kernel functional components providing self-centering capability and energy dissipation. This new system significantly extends the scope of the practical application of SMA elements in infrastructure. The present study also offers a comprehensive life-cycle assessment framework that evaluates the performance of the new system from both structural and economic perspectives. In the following discussions, the working mechanism of the SCR bridge with the SMA washer-based pier is introduced first. Quasi-static tests on a 1/4 scaled SMA-washer-based RC pier specimen carried out previously by the authors and coworkers are briefly introduced. Subsequently, a performance-based life-cycle assessment flowchart for the SCR bridge system subjected to earthquakes is proposed. Fragility curve, life-cycle loss assessment, and resilience assessment of the SCR bridge are introduced in detail. Finally, a prototype SCR bridge and a conventional bridge are designed and taken as two examples to illustrate the assessment framework. The assessment results, including the fragility curves, life-cycle loss, and resilience performances, are comprehensively discussed.

2. Brief Description of Bridge Systems with SMA-Washer-Based Rocking Piers

2.1. Working Mechanism and Design Objectives

According to post-earthquake field investigation reports, many bridge systems that followed the displacement-based seismic design philosophy in earthquake-prone areas suffered catastrophic damage or even collapsed. To enhance the resilience of newly designed bridges in earthquake regions, an SMA-washer-based SCR bridge pier was proposed in the previous work by the authors and co-workers [38], as shown in Figure 1. The rocking control capability of the novel bridge system is enabled by its SMA-washer-based SCR piers. The SCR pier is mainly composed of three parts: the upper pile cap, the lower pile cap, and the SMA washer springs (also known as disc springs), which are the kernel components providing self-centering capability and energy dissipation for the bridge system. These washer springs can be stacked either in parallel or in series (or in combination), which makes them flexible in terms of load resistance and available deformability. More technical details of the SMA washer springs can be found elsewhere [39].

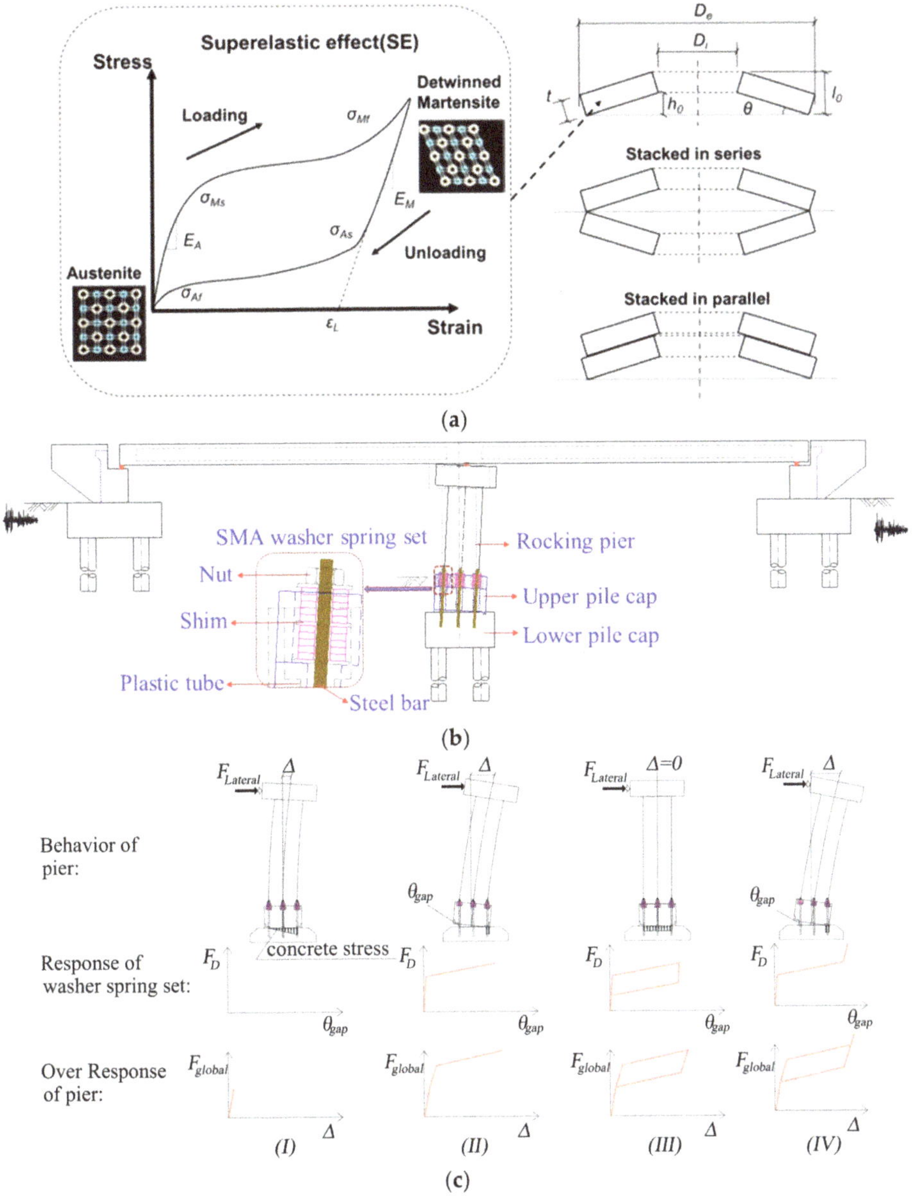

Figure 1. SMA-washer-based SCR bridge pier: (**a**) illustration of the SMA washer springs, (**b**) illustration of the bridge system, (**c**) working principle of the SCR bridge pier.

In the new pier, each SMA washer spring set consists of several SMA washers, steel bars embedded in the lower pile cap, plastic tubes cast in the concrete of the upper pile cap, and several nuts and shims. As illustrated in Figure 1c, the seismic behavior of such a pier with increasing lateral load could be divided into three stages: (1) decompression stage (I) (where the gap is just about to open), (2) post-decompression stage (II) (III), and (3) locking stage (IV). In the first stage, the behavior of the pier behaves similar to a fixed pier, i.e., the lateral deformation of the pier relies on the elastic deformation of the pier due to the application of appropriate preload of the SMA washer springs. In the subsequent stage, the lifting force exceeds the decompression force provided by SMA washer spring sets, and the pier starts rocking. When the allowable deformation of washer sets is consumed, the pier is "locked". Further lateral displacement may rely on the nonlinear deformation of the pier. In light of the above, three basic design goals could be set: (1) the pier does not uplift during small earthquakes; (2) a maximum drift ratio of the bridge is less than the "lock rotation" at moderate (E1) earthquake level; and (3) collapse is prevented at large (E2) earthquake level.

2.2. Experimental Verification of SMA-Washer-Based SCR Pier

Figure 2 schematically depicts the test arrangement for the pier specimen. The specimen was held down via four anchor bars passing through the designated slots in the pier base. In order to consider the dead weight of the bridge's superstructure, a PT tendon was used to apply the axial force. A double-action electro-hydraulic servo actuator was used to provide the lateral load to the loading head, and the lever arm, or the distance between the loading head's centroid and the rocking interface, was 1625 mm. The RC pier's diameter and height were 0.3 and 1.05 m, respectively, and other relevant dimensions are shown in Figure 2, with more details given in Fang et al. [38]. The typical test result (hysteretic response) is illustrated in Figure 3. The specimen displayed stable flag-shaped hysteretic curves under cyclic load, with no noticeable decrease in strength and stiffness responses and negligible residual drift. Numerical simulation and system-level analysis of a novel bridge system incorporating the new bridge pier are discussed in detail in Section 4.

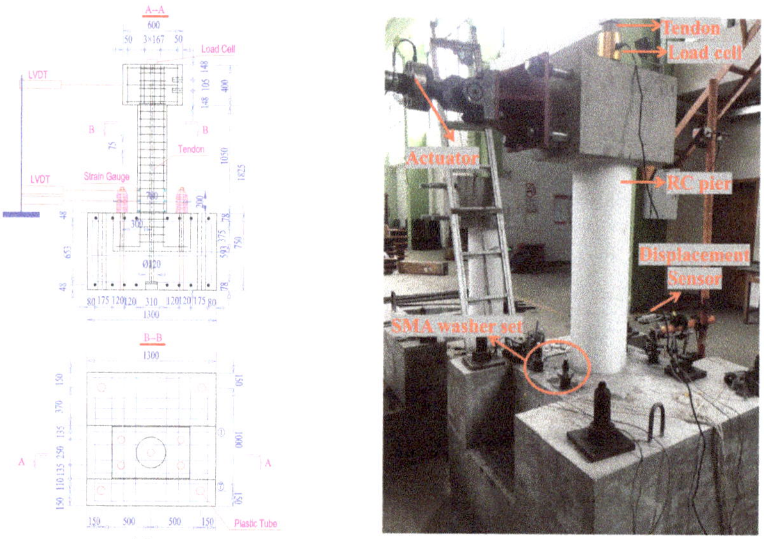

(a) (b)

Figure 2. Test of the SCR pier: (**a**) drawing of the SCR pier specimen, (**b**) photo of the test setup.

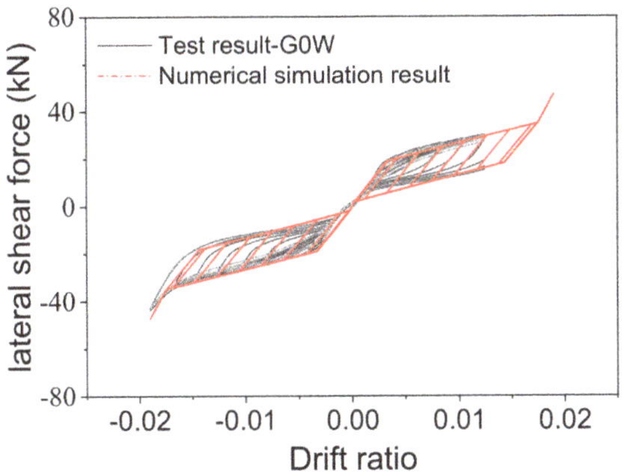

Figure 3. Shear force–drift ratio hysteretic curves of the test and numerical simulation results.

3. Methodology of Performance-Based Assessment

The main steps of the analysis framework are shown in Figure 4. Seismic fragility analysis and loss assessment are two key steps: the former gives the probabilities of exceeding certain component or system damage levels and the latter enables translation from the damage level to economic loss quantities.

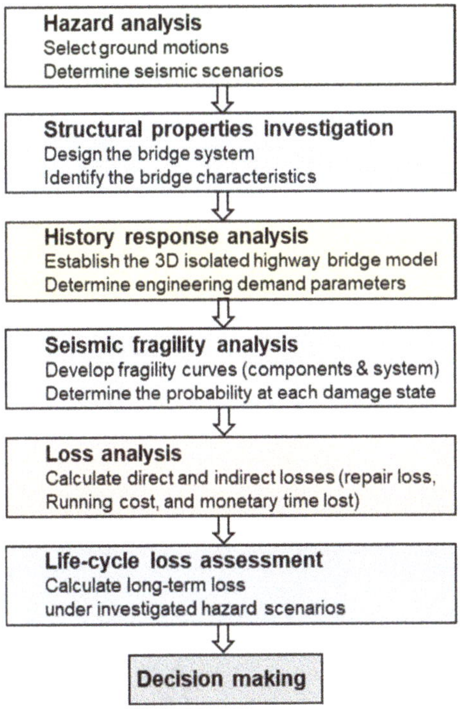

Figure 4. Flowchart of the performance-based life-cycle assessment of bridge systems under seismic hazard.

3.1. Seismic Fragility Analysis

Structural seismic fragility assessment was carried out first according to the flowchart shown in Figure 4. Fragility analysis is a frequently used technique in the seismic risk assessment in order to calculate the conditional probability of a structure's or component's demand reaching or beyond its corresponding capacity [40]. Analytical fragility curves were derived using the probabilistic seismic demand model (PSDM) based on nonlinear time history analysis series. A PSDM is typically developed using two methods: incremental dynamic analysis (IDA) [41] or a cloud technique [42]. The former method requires scaling all the ground motions to specific intensity measurement (*IM*) and conducting a nonlinear time history analysis at each level. In the later procedure, a collection of un-scaled ground motion data is used in the nonlinear time history analysis. Both methods are dependent on *IM*s, and extensive research has been carried out on the selection of suitable *IM*s, considering, for example, Peak ground acceleration (PGA), Peak ground velocity (PGV), and response spectrum type at a specific period. The optimal selection of *IM* may vary with different characteristics of structures [43].

The probabilistic seismic demand model (PSDM) is the probability distribution of structural demand conditioned on specified *IM* and based on the cloud technique. The probability that a structure's seismic demand (D) exceeds its capacity (C) may be written as follows:

$$P[D \geq \mathrm{C}|IM] = P[\frac{\mathrm{D}}{\mathrm{C}} \geq 1] \tag{1}$$

Equation (1) might be rewritten as a lognormal cumulative probability density function provided that C and D have a two-parameter lognormal distribution:

$$P[D \geq \mathrm{C}|IM] = \Phi(\frac{\ln(S_d) - \ln(S_c)}{\sqrt{\beta_{d|IM}^2 + \beta_c^2}}) \tag{2}$$

where S_c signifies the median structural capacity estimate and β_c denotes the standard deviation. Lognormal median estimate and standard deviation of structural demand in terms of an *IM* are represented by S_d and $\beta_{d\,|\,IM}$, respectively. Regression analysis was used to determine the relationship between *IM* and S_d. The median value of seismic demand, according to Cornell's power exponent model, may be stated as:

$$S_d = aIM^b \text{ or } \ln(S_d) = \ln(a) + b\ln(IM) \tag{3}$$

where a and b represent the regression parameters obtained from the response analysis. $\beta_{d\,|\,IM}$ can be characterized as:

$$\beta_{d|IM} \cong \sqrt{\frac{\sum\limits_{i=1}^{n} (In(d_i) - In(S_d))^2}{N-2}} \tag{4}$$

where d_i represents the structural demand, also known as the seismic response of components, and *i*th represents the earthquake-model sample that corresponds to it. The following steps need to be taken to obtain $\beta_{d\,|\,IM}$:

Given the fragility curves of the components, the fragility curve of a bridge system can be developed according to the first-order reliability theory (explained in Equations (5)–(7)). Such a theory assesses structural performance as an overall system by accounting for the relationship between the vulnerable components. Equation (5) provides the upper and lower bounds of the system fragility functions. The lower bound assumes complete correlation among components, while the upper bound refers to the components with no correlation.

$$\max_{i=1}^{n}[P(F_k)] \leq P(F_{sys}) \leq 1 - \prod_{i=1}^{n}(1 - P[F_{component,i}]) \tag{5}$$

where n is the total number of components that might fail, $P(F_k)$ is the probability that the component in concern will fail, $P[F_{component,\,i}]$ and $P[F_{sys}]$ are the failure probabilities of the i_{th} component and system, respectively, and Π is the product operator.

If a bridge is supposed to operate as a serial system, with each component executing an essential function separately, any component failure will result in system failure at the same level. As a result, the most significant damage state at the component level is as follows:

$$DS_{sys} = \max(DS_{Pier}, DS_{Bearing}) \tag{6}$$

When a bridge is supposed to be a parallel system, however, it will attain a specific damage state once all of its components have reached that condition. As a result, the system damage state DS_{sys} is determined by the component with the minimum damage state:

$$DS_{sys} = \min(DS_{Pier}, DS_{Bearing}) \tag{7}$$

The intersection of component probability and its lower and upper limits, as shown in Equation (8), yields the failure probability of a parallel system:

$$\prod_{i=1}^{n} P(F_i) \leq P(F_{sys}) \leq \min[P(F_i)] \tag{8}$$

Independent components are represented by the lower bound, whereas the upper limit represents entirely correlated components. These boundaries are often quite broad, showing the importance of component correlation. In fact, a bridge is neither a parallel nor a serial system, and component responses are often coupled to some degree. The first-order constraints in Equations (5)–(7), which assume total correlation or perfect independence between components, cannot accurately estimate the bridge system's failure probability. According to the work of Kim et al. [44], the bearing damage due to the load moving to other components of a multi-span simply supported bridge has a substantial impact on the bridge's overall seismic behavior. The global damage state is hence located in-between the limits set by Equations (5)–(7).

A composite *DS* based on component *DSs*, proposed by Zhang and Huo [45], was employed in this work. Piers and isolation devices were given a weighted ratio of 0.75 and 0.25, respectively, based on their proportional value for load carrying and maintenance cost. This ratio highlights that piers are more important than isolation devices and, as a result, should be given greater weight. However, since either excessive bearing displacement or pier collapse damage ($DS = 4$) might cause a single span or the whole bridge to collapse, a serial mechanism for the collapse damage was used. The following equation summarizes the resulting composite DS_{sys} for system behavior:

$$\begin{array}{ll} DS_{sys} = int(0.75 \cdot DS_{Pier} + 0.25 \cdot DS_{Bearing}) & DS_{Pier}, DS_{Bearing} < 4 \\ DS_{sys} = 4 & DS_{Pier\ or\ DS_{Bearing}} = 4 \end{array} \tag{9}$$

3.2. Life-Cycle Loss Assessment

The fragility analysis could be followed by a life-cycle loss assessment, a framework that was initially proposed by the Pacific Earthquake Engineering Research (PEER) Center. Life-cycle loss assessment is an effective tool to evaluate the long-term benefit of the newly proposed bridge system. Direct loss (mostly repair loss) and indirect losses (e.g., running cost and property loss) are important qualities in the life-cycle loss assessment.

The selected seismic events should cover both frequent low-magnitude events with a high probability of occurrence and the high-magnitude earthquakes with a low probability of occurrence. Six hazard events with return periods of 225 years (E1), 475 years (E2), 975 years (E3), 1500 years (E4), 2475 years (E5), and 5000 years (E6) were considered [46]. The relationships between earthquake intensity measurement (*IM*) and the frequency of occurrence for the location of the bridge can be obtained from the USGS national seismic hazard map [47].

The obtained fragility curves were then used to quantify the damage probability of the bridge system. Under a given hazard event, the seismic loss can be calculated by summing up the consequences weighted with the damage probability. Equation (10) gives the expression of the expected annual loss under a specific hazard [48].

$$R = \sum_{LS} C_{LS} P_{LS|IM} \tag{10}$$

where C_{LS} and $P_{LS\,|\,IM}$ are the consequences at a specific limit state of the bridge and the conditional probability of the bridge at a limit state for a given *IM*, respectively. Direct and indirect losses are the two types of consequences considered in the present study, and the consequences were evaluated in terms of monetary values.

3.2.1. Direct Loss

It is assumed that the necessary repair cost at a certain limit state, *i*, is proportional to the cost needed to rebuild the bridge, as expressed as [49]:

$$C_{REP,i} = R_{rcr} \cdot c_{reb} \cdot W \cdot L \tag{11}$$

The total repair cost of a bridge, $C_{REP,i}$, at damage state *i* can be obtained by multiplying the rebuilding cost per square meter (unit: $/m^2$) c_{reb} by the width *W* and length *L* of the bridge (unit: m), with an extra consideration of the repair cost ratio R_{rcr} at damage state *i*. As suggested by Mander [50], the repair cost ratios at the slight, moderate, extensive, and collapse levels can be taken as 0.1, 0.3, 0.75, and 1.0, respectively.

3.2.2. Indirect Loss

Societal and economic issues often occur following a seismic hazard, and these consequences result in indirect loss which can be even higher than the direct loss (i.e., repair cost) for highway bridges [51]. The indirect loss after an earthquake is somehow related to structural damage which, for example, affects the traffic flow in the route as the drivers are forced to detour during the closure of the bridge. In this study, the running cost, C_{RUN}, and the monetary value converted from the time loss for users (i.e., vehicle drivers) through the detour, C_{TL}, were considered as the indirect loss. C_{RUN} under a given limit state *i* can be expressed as [49]:

$$C_{RUN,i} = [c_{Run,car}(1 - \frac{T_0}{100}) + c_{Run,truck}\frac{T_0}{100}] \cdot D_l \cdot ADT \cdot d_i \tag{12}$$

where $c_{Run,car}$ and $c_{Run,truck}$ are the average running costs for cars and trucks per kilometer ($/km), respectively; T_0 is the average daily truck traffic, defined as the total volume of vehicle traffic of a highway or road for a year divided by 365 days; D_l is the detour length (km); *ADT* is the average daily traffic to detour, which is the average detour distance of all vehicles influenced by the bridge damage. ADT is generally determined by the bridge damage level; d_i is the duration of the downtime associated with the damage levels, where 7, 30, 120, and 400 days are typically adopted corresponding to the slight, moderate, extensive, and complete damage states, respectively [52]. The monetary value of detour-induced time loss, C_{TL}, can be calculated from:

$$C_{TL,i} = [c_{AW} o_{car}(1 - \frac{T_0}{100}) + (c_{ATC} o_{truck} + c_{goods})\frac{T_0}{100}] \cdot [\frac{D_l \cdot ADT}{S} + ADTE \cdot (\frac{l}{S_D} - \frac{l}{S_0})] d_i \tag{13}$$

where c_{AW} and c_{ATC} are the average wage plus compensation per hour ($/h) for car and truck drivers, respectively; o_{car} and o_{truck} are the average vehicle occupancies for cars and trucks, respectively; c_{goods} is the time value of the goods transported in a cargo ($/h); *S* is the average detour speed (km/h); *l* is the route segment containing the bridge (km); S_0 and S_D represent the average speed on the intact link and damaged link (km/h), respectively.

3.2.3. Long-Term Loss

By substituting Equations (11)–(13) into Equation (10), the annual loss of the bridge under a specific hazard event can be calculated. By assuming a Poisson distribution for the occurrence of an earthquake during an investigated time interval (0, t_{int}), the total life-cycle loss of the bridge can be expressed as [53]:

$$LCL_i(t_{\text{int}}) = \sum_{i=1}^{N(t_{\text{int}})} L_i(t_k) \cdot e^{-\tau t_k} \tag{14}$$

where $L_i(t_k)$ is the expected annual loss at time t_k, and τ is the monetary discount rate. The total expected lifetime failure loss of the bridge during the time interval, t_{int}, can be expressed as:

$$E[LCL_i(t_{\text{int}})] = \frac{\lambda_f \cdot E(L_i)}{\tau} \cdot (1 - e^{-\tau t_{\text{int}}}) \tag{15}$$

where λ_f denotes the mean rate of the Poisson model. The values of all necessary parameters mentioned above are summarized in Table 1.

Table 1. Parameters associated with consequences.

Parameters	Notation	Value	References
Average daily truck traffic	$ADTT$	19,750	[51]
Average daily traffic on the damage link to average daily traffic	$ADTE/ADT$	0.12	[51]
Daily truck traffic ratio	T_0	13%	[53]
Link length (km)	l	6	[53]
Detour additional distance (km)	D_1	2	[53]
Vehicle occupancies for cars	o_{car}	1.5	[49,53]
Vehicle occupancies for trucks	o_{truck}	1.05	[49,53]
Wage for car drivers ($/h)	c_{AW}	11.91	[49,53]
Compensation for truck drivers ($/h)	c_{ATC}	29.87	[49,53]
Operating costs for cars ($/km)	$c_{Run,car}$	0.4	[49,53]
Operating costs for trucks ($/km)	$c_{Run,truck}$	0.57	[49,53]
Rebuilding costs ($/m^2)	c_{reb}	2306	[50]
Detour speed (km/h)	S	50	[53]
Link speed (km/h)	S_0	80	[53]
Time value of a cargo ($/h)	c_{goods}	4	[53]
Monetary discount rate	τ	2%	[53]

3.3. *Resilience Assessment*

Resilience is another important structural performance indicator defining the capability of a civil infrastructure system of maintaining its post-hazard functionality. Generally, resilience includes four gradients, namely, rapidity, robustness, redundancy, and resourcefulness. In this paper, the resilience of the bridges under the considered seismic hazards was assessed, where the functionality is deemed to be resumed when the traffic becomes normal after earthquake.

It is worth mentioning that various functionality levels may be considered for different periods, e.g., emergency response and post-earthquake recovery stages. In the former stage, the main focus should be on the capability of transferring the resources to the disaster area. In the latter phase, the functionality of the bridge can be defined with different service statuses, e.g., "closed", "limited use", and "open". The resilience of the bridge can be evaluated through its recovery pattern. As illustrated graphically in Figure 5, one of the most widely adopted approaches to quantify resilience is [54,55] expressed as follows:

$$R_{Resi} = \frac{1}{\Delta t_r} \int_{t_0}^{t_0 + \Delta t_r} Q(t) dt \tag{16}$$

where $Q(t)$ is the functionality of a bridge at time t (e.g., days); t_0 is the initial time at the investigated point; and Δt_r is the investigated time interval (e.g., days or months).

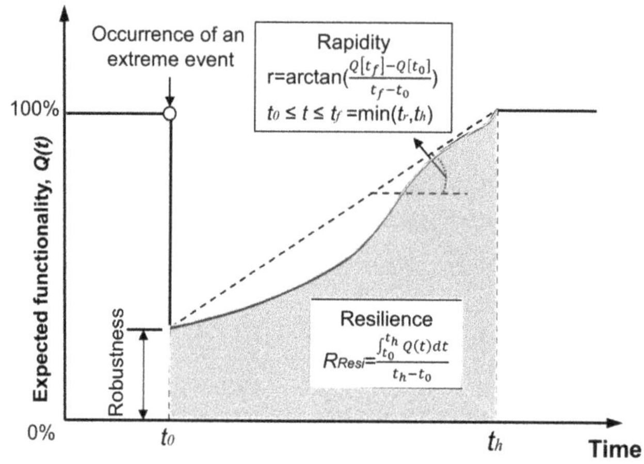

Figure 5. Schematic representation of resilience.

The shape of the recovery pattern curve is related to the repair and recovery efforts. Generally, bridge functionality can be assessed by defining the damage state to a value between 0 and 1.0, where a value of 0 means collapse of the bridge. Considering various levels of damage state, the bridge functionality can be expressed as:

$$Q = \sum_{i=1}^{5} FR_i \cdot P_{DS_i|IM} \tag{17}$$

where FR_i is the functionality ratio (Func) associated with damage state i. For a typical bridge, possible scenarios include immediate access (Func $\geq$ 0.9), weight restriction (0.6 $\leq$ Func < 0.9), one lane open only (0.4 $\leq$ Func < 0.6), emergency access only (0.1 $\leq$ Func < 0.4), and bridge closure (Func < 0.1). Similar concepts have also been adopted by Padgett and DesRoches [56] and Decò et al. [57]. Once repair actions are initiated, the functionality of the bridge starts to recover and the performance restoration curve starts to rise. The Applied Technology Council (ATC-13) report proposed an approach to quantify the change of functionality $Q_j(t)$ during the recovery phase [58], expressed as follows:

$$Q_j(t) = \frac{1}{\sigma_j\sqrt{2\pi}}\int_{-\infty}^{t} \exp[-\frac{(\tau - \mu_j \cdot A_t)^2}{2\sigma_j^2}]d\tau \tag{18}$$

where μ_j and σ_j are the mean and standard deviation of the recovery time for the jth damage state, as listed in Table 2; and A_t is an amplification factor considering the increase of mean recovery time due to underwater repair work. The functionality of the considered bridge on a daily basis during the recovery phase can be calculated with the above-mentioned parameters and methodology. The probabilities of the bridge staying in different damage states can be used to obtain the expected recovery functionality $Q(t)$, and the resilience can be assessed by Equation (18).

Table 2. Parameters associated with bridge restoration functionality [58].

Damage State	Mean (Days)				Coefficient of Variation
	Lower Limit	Upper Limit	Mode	Distribution Type	
Slight	0.2	1	0.6	Triangular	1
Moderate	1	5	2.5	Triangular	1
Major	30	120	75	Triangular	0.56
Complete	120	360	230	Triangular	0.48

4. Case Study

4.1. Description of Prototype Bridges

A continuous RC bridge with two equal spans (20 + 20 m) supported by a middle pier was developed to investigate the life-cycle loss and resilience performances of the novel SMA-washer-based SCR bridge system. Figure 6a presents the bridge's geometric layout. Sliding bearings were placed on each abutment, and fixed bearings were placed on the bent cap above the middle pier. A sufficient separation space between the bridge deck and the abutment was assumed [59–61]. The concrete used for the box girder has a compressive strength of 50 MPa, and that for the abutment, pier, and rocking pile caps have a compressive strength of 40 MPa. The longitudinal reinforcement and stirrup utilized in the RC pier have diameters of 32 mm and 16 mm, respectively, with yield strengths of 440 MPa and 300 MPa, respectively.

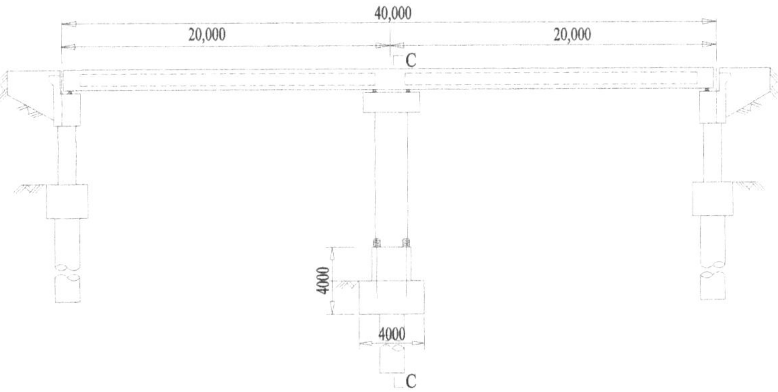

(**a**)

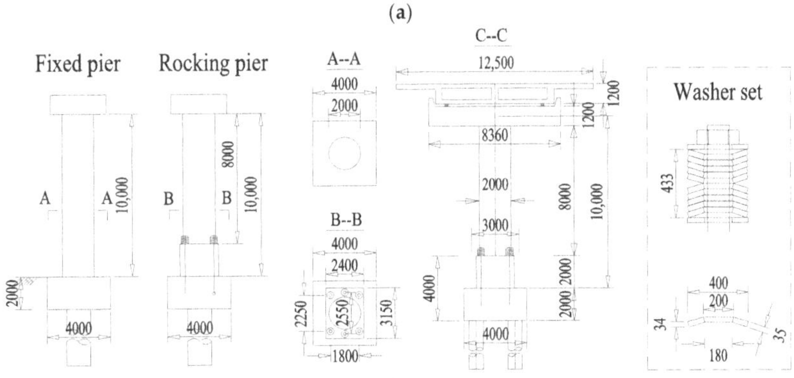

(**b**)

Figure 6. Geometric layout of two-span continuous concrete-girder bridge (Unit: mm): (**a**) general layout, (**b**) detailed dimensions.

A total of 55 longitudinal steel bars are equally distributed around the pier perimeter, corresponding to a 1.74% reinforcement ratio. Spiral stirrups are 100 mm apart. The RC pier has a 65-mm-thick concrete cover layer. The rocking pier employed six SMA washer sets, each of which has eight washers, 4 in parallel × 2 in series. Figure 6b shows the key characteristics of each unique SMA washer, where a maximum deformation of 32 mm and a maximum compressive resistance of roughly 250 kN are provided by every single washer. Each SMA washer spring set was preloaded with 960 kN, corresponding to a precompression deformation of 22 mm, to guarantee that the rocking pier does not uplift under normal service loads or small earthquakes. In other words, each SMA washer spring set has remaining deformability of 42 mm prior to the fully compressed status. When the SMA washer sets were totally flattened, a "locking" mechanism was produced, and no further rocking was allowed beyond this allowable angle. After locking, damage to the RC pier is expected.

For comparison, an extra conventional bridge was evaluated with a fixed-base RC pier that is 10 m tall, i.e., equal to the height of the rocking pier from the bottom surface of the bent cap to the rocking interface. The other design of the bridge remains the same.

4.2. Numerical Models

Figure 7 shows the behavior of a nonlinear FE model created in OpenSees [62]. Fiber Beam-Column components were used to simulate the pier's main body, taking nonlinear material features into account. A uniaxial Menegotto–Pinto constitutive model was used to model the behavior of the reinforcement [63,64]. The uniaxial Kent–Scott–Park concrete model was utilized to simulate both the unconfined and confined concrete [65]. The circular portion was separated into eight layers along the radius direction, and each layer was consistently split into 24 fiber components. In addition to the material property, co-rotational geometric transformations were used to account for the geometric nonlinearity [66]. For the pier, the yield curvature was approximately 0.0031 based on the pier's moment–curvature relationship. By providing an initial strain to the material model of the zero-length element, the preload given to each SMA washer set was considered. As seen in Figure 3, the numerical simulation result of the pier closely matches the actual test results. Rigid beam components were used to simulate the bent cap and the top pile cap. Six pairs of contact point sets were positioned at the interface, and each pair was given a zero-length element to record the change in pressure over time across the rocking interface. Each SMA washer group's behavior was "lumped" into the zero-length element to account for its overall force-deformation hysteretic behavior. To account for soil–structure interaction (SSI) between the abutment/pile and the soil, a series of zero-length spring components were incorporated [67].

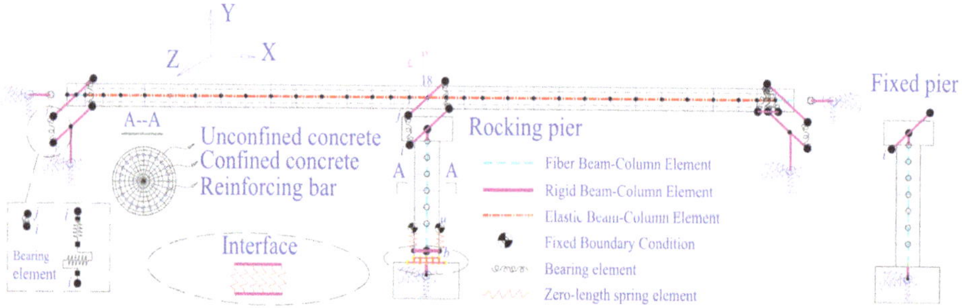

Figure 7. Schematic illustration of the FE bridge model and modeling details.

4.3. Selection of Ground Motions

A sufficient number of nonlinear time history analyses should be performed to determine the fragility functions. Using the cloud approach, a set of 60 original earthquake data was chosen from the PEER Next Generation Attenuation (NGA) Project ground motion collection. To match the target design spectrum and to offer a relatively wide range of IMs for time history analysis, 60 additional ground motions with a scale factor of 2.0 were included. In other words, this research used a total of 120 ground motions for nonlinear time history analysis. The spectral acceleration at one second (Sa1.0) of all the records, as shown in Figure 8a, has a broad range of values ranging from 0 to 1.6. Figure 8b shows the response spectra of the chosen recordings with the average spectra.

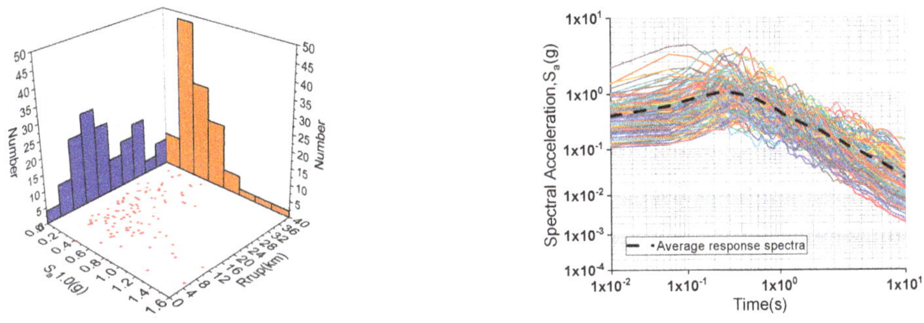

(a) (b)

Figure 8. Information of selected earthquake records: (**a**) Sa1.0 and fault distance; (**b**) spectral acceleration.

4.4. Capacity Models of Bridge Components

The RC pier and bearing are two key components determining the system fragility of the bridges under consideration [68]. In the present research, pier curvature ductility (μ_f) and bearing displacement (δ_b) were regarded as the engineering demand parameters (EDPs), and their peak responses were regarded as the damage indicator. For each EDP, four degrees of damage state were used: slight, moderate, extensive, and complete damage [69], with the damage state treated as a random variable to allow for uncertainty. Based on available test findings, curvature ductility is defined as a range of 0.8 to 7.0 for damage state ranging from slight damage to complete damage [70]. The damage status of the bearings is divided into four levels: 50, 100, 150, and 255 mm bearing displacement. HAZUS proposes a dispersion measure to account for the damage state's fluctuation, as seen in Table 3.

Table 3. Engineering demand parameters and damage states.

Component	DS_1 (Slight)		DS_2 (Moderate)		DS_3 (Extensive)		DS_4 (Complete)	
	S_c	β_c	S_c	β_c	S_c	β_c	S_c	β_c
Column curvature ductility (μ_f)	0.8	0.005	2	0.005	4	0.005	7	0.005
Bearing displacement(mm)	50	0.25	100	0.25	150	0.46	255	0.46

5. Analysis Results and Discussions

5.1. Typical Seismic Response of Bridge Components

To take a close look at the pier response, the typical time-history responses of bearing displacement and curvature ductility of the pier are shown in Figure 9a,b, respectively. It could be observed that the introduction of the novel SCR pier bridge system slightly increased the bearing displacement but extensively reduced the pier curvature ductility due to the rocking mechanism. The typical time history responses of the pier's lateral seismic force versus drift ratio in the conventional and novel bridges are further illustrated in Figure 9c. As anticipated, a flag-shaped hysteretic response with negligible residual deformation was observed for the novel bridge. The SMA washer sets provide moderate energy dissipation. On the other hand, the conventional pier exhibited a fuller hysteresis but was accompanied by damage accumulation. The degree of damage can be further understood by the typical bending moment versus curvature responses at the plastic hinge region of the pier, as shown in Figure 9d. The maximum curvature of the fixed pier was 0.0226, which is significantly larger than that of the rocking pier (i.e., 0.0079). This reaffirms that the fixed pier undergoes more extensive damage than the rocking pier.

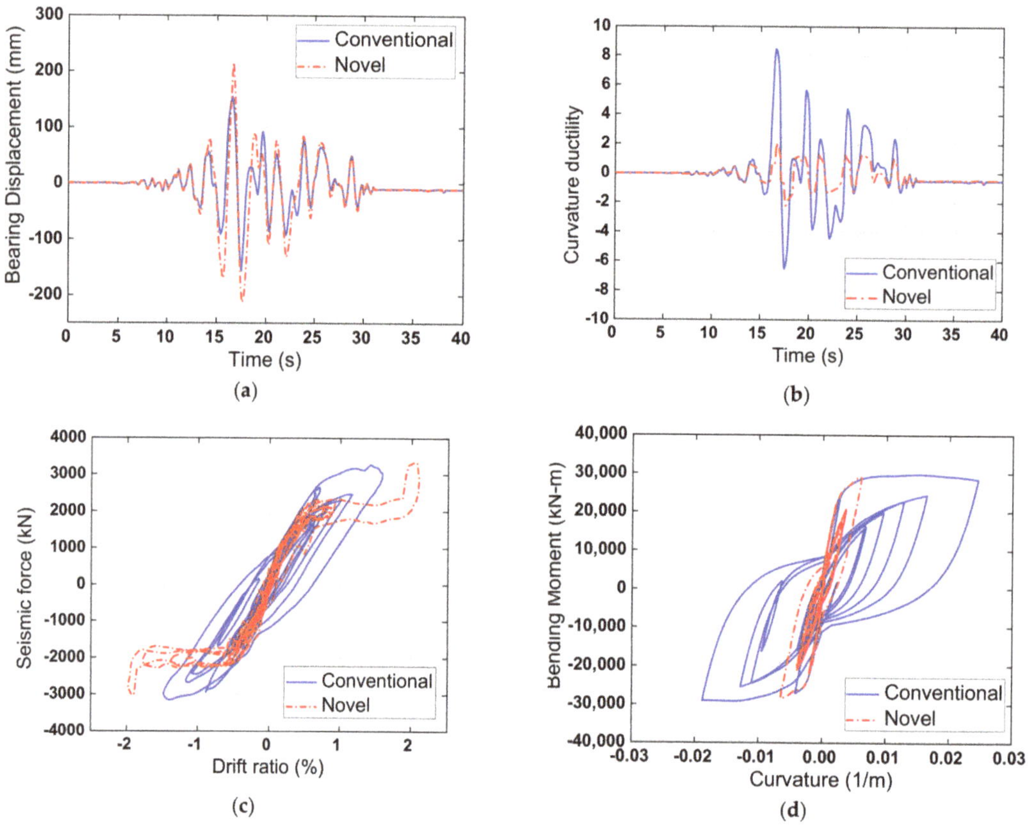

Figure 9. Seismic response of (**a**) bearing displacement, (**b**) curvature ductility, (**c**) seismic force versus drift ratio behavior, and (**d**) curvature versus bending moment behavior.

5.2. Regression Analysis and Optimum IM

When generating the fragility curves for bridges, proper *IM* selection is critical, and much prior research has looked into the optimal *IM* selection for probabilistic seismic risk assessment. The PGA and the spectral acceleration at one second (Sa1.0) are two *IMs* that have been frequently employed in recent research [71,72], although their performance varies depending on the scenario. Both PGA and Sa1.0 were evaluated here. As previously indicated, pier curvature ductility (μ_f) and bearing deformation (δ_b) are the main damage indicators. In a log-transformed space, a linear regression of demand-IM pairs for both μ_f and δ_b was performed (as expressed by $\ln(EDP) = \ln a + b\ln(IM)$, and the results are displayed in Figures 10 and 11.

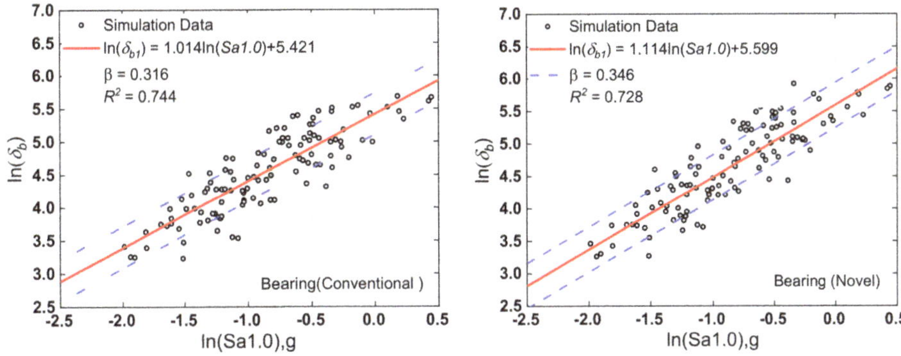

(**a**)

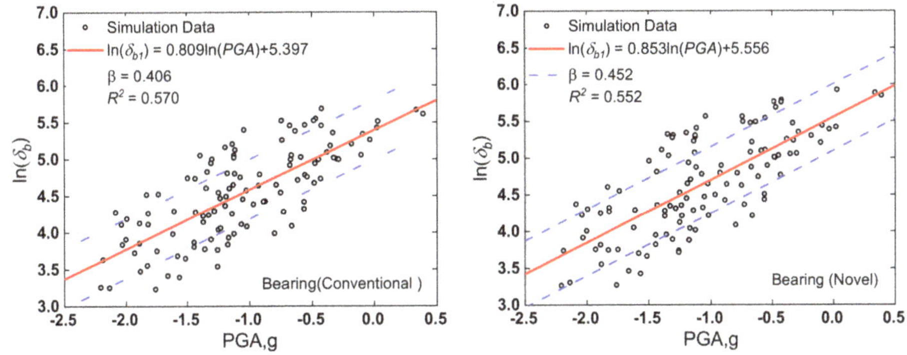

(**b**)

Figure 10. Probabilistic seismic demand models of bearing: (**a**) Sa1.0 versus bearing displacement responses, (**b**) PGA versus bearing displacement responses.

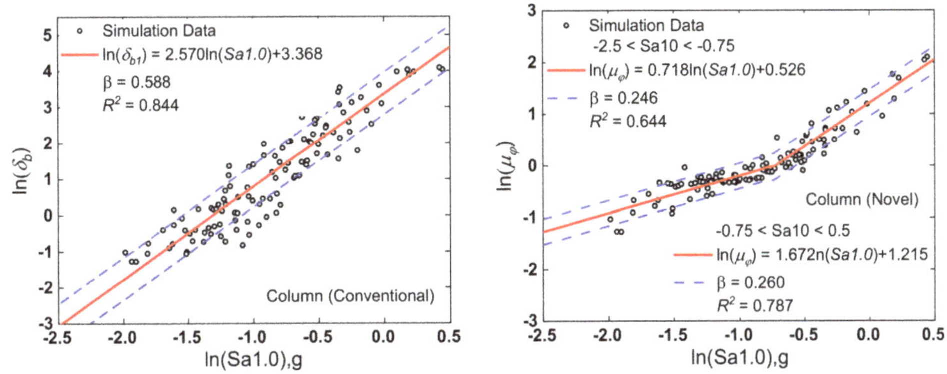

(**a**)

Figure 11. *Cont.*

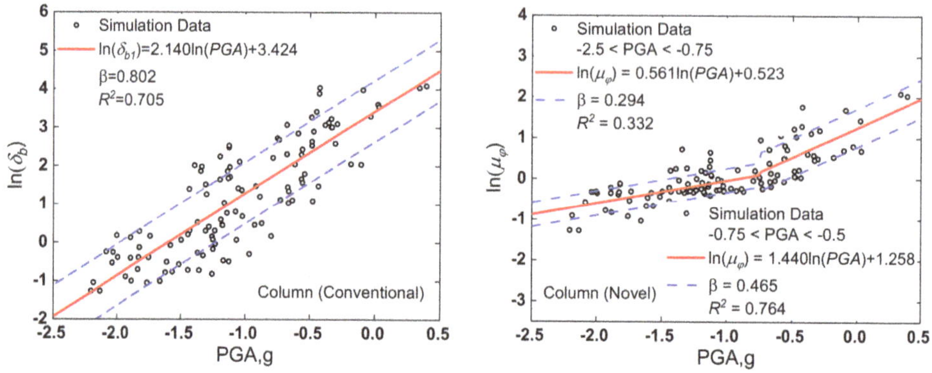

(**b**)

Figure 11. Probabilistic seismic demand models of column (pier): (**a**) Sa1.0 versus column curvature ductility responses, (**b**) PGA versus column curvature ductility responses.

For optimum *IM* selection, Padgett et al. [71] offered four criteria: efficiency, practicality, proficiency, sufficiency, and hazard computability. Efficiency means less dispersion about the estimated median in the nonlinear time history analysis results and was represented by a lower β. In addition, the coefficient of determination (R^2) was added, which reflects the degree of regression equation fitting. The greater the value of this coefficient, the better the regression. The results showed an evident linear correlation between *EDP* and *IM*, with the result linked with PGA having a more significant degree of dispersion than the result related to Sa1.0. Figure 10 shows, for example, that PSDM's R^2 values linked with Sa1.0 were more than 0.7, whereas the values related to PGA were less than 0.6. According to the previously established assessment criteria for optimum *IM*, it may be determined that Sa1.0 outperformed PGA in the present investigation.

5.3. Fragility Curves

The component fragility curves were directly constructed using Equation (2), which utilizes the capacity models of each limit state, as tabulated in Table 3, and the PSDM parameters determined from the regression analysis. Figure 12 shows the column (pier) and bearing fragility curves for slight, moderate, severe, and total damage states. For slight damage states, the damage probability of bearings associated with conventional and new systems is virtually the same, as shown in Figure 12a. This may be explained by the fact that both types of bridge systems exhibited identical behavior prior to washer spring set decompression. However, since the rocking behavior increases the superstructure displacement, the conditional exceeding probability of the other three bearing damage states for the novel system was somewhat higher than the conventional system. Furthermore, the locking mechanism of the washer spring set helps prevent excessive bearing displacement owing to the controlled rocking behavior. Due to the incorporation of the rocking mechanism, the damage probability of the pier with the novel system was lower than that of the conventional system. The positive impact was more evident considering more severe damage states. For example, assuming Sa1.0 = 0.7 g, the damage probability of a pier with the novel bridge system is 40% lower than that of a conventional bridge for the moderate damage state, and the chances of exceeding extensive and complete pier damage are practically avoided in the novel system.

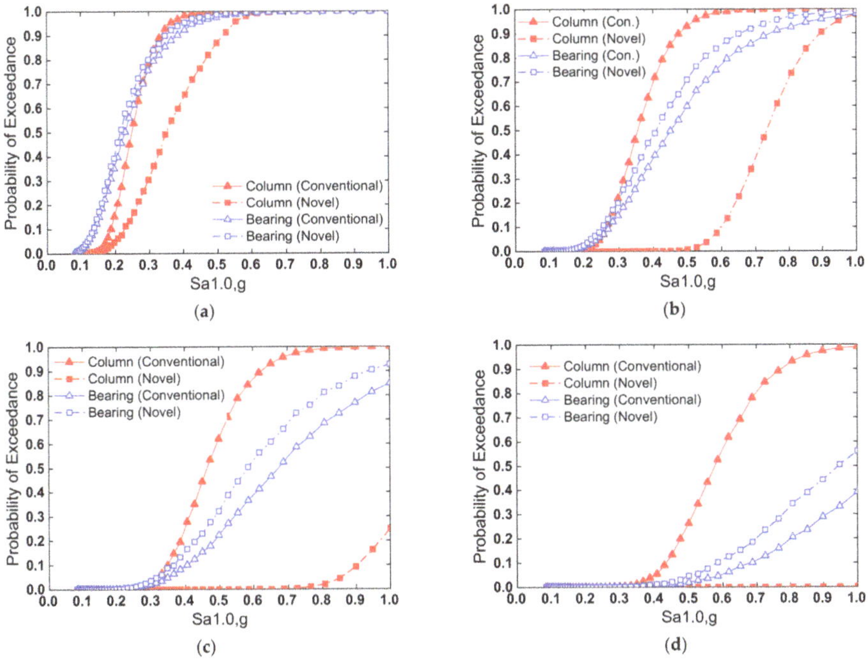

Figure 12. Component fragility for (**a**) slight damage, (**b**) moderate damage, (**c**) extensive damage, and (**d**) complete damage.

Figure 13 depicts the system fragility curves associated with the four damage states for both the conventional and novel systems. For slight damage state, the failure probability of the novel bridge system was similar to that of the conventional system, which could be explained by the identical behavior of the two types of systems before decompression of the SMA-washer-based SCR pier. For moderate, severe, and complete damage states, however, it is clear that the seismic performance of the novel system was better than the conventional system, which is due to the mitigated damage to the SMA-washer-based SCR pier and concurrently minor damage to the bearings, according to component fragility curves.

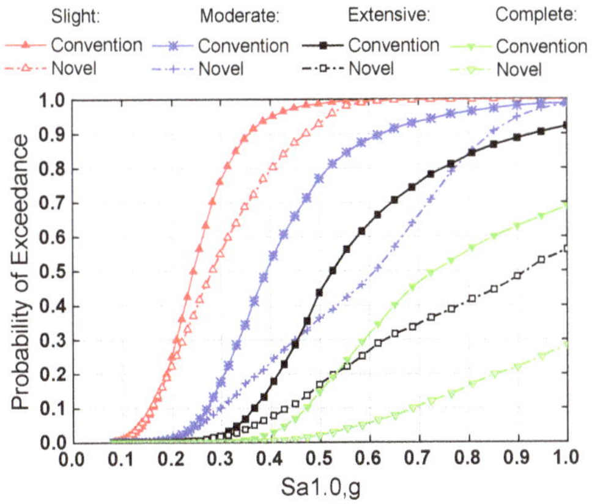

Figure 13. System fragility curve of conventional and novel bridges.

5.4. Performance-Based Long-Term Loss Assessment

Following the fragility analysis, an economic loss assessment was carried out, considering the consequences of various damage states. As previously indicated, six hazard scenarios were explored, and the loss associated with these seismic events was calculated. Salt Lake City and Los Angeles are supposed to be the sites of the investigated bridges for illustration purposes. The seismic level of Los Angeles is greater than that of Salt Lake City according to the hazard curve parameters provided by the United States Geological Survey [47]. The Sa1.0 values for Los Angeles is 0.181, 0.279, 0.402, 0.493, 0.604, and 0.791 g, when the return period is 225 years, 475 years, 975 years, 1500 years, and 2475 years, respectively. The Sa1.0 values for Salt Lake City considering the six return periods are 0.070, 0.148, 0.271, 0.362, 0.485, and 0.672 g, respectively. The total direct and indirect cost from all consequences associated with the probability of the bridge being in various damage states were estimated using different hazard scenarios.

Equations (10)–(18) calculate the direct and indirect losses associated with each damage state. The conditional probability of the bridge being in various damage states was calculated directly from the system fragility curve. Figure 14a,b show the total predicted direct and indirect loss from all the six earthquake events examined. Employing the novel bridge system lowers the overall direct and indirect costs, particularly for events 3 through 6. The average decrease rate of the direct loss for these four events was 41.1% in Salt Lake City and 45.7% in Los Angeles. The average decrease rate of indirect loss in Salt Lake City and Los Angeles was 51.2% and 63.0%, respectively. The effectiveness of the SCR bridge system for decreasing the indirect loss was greater than that for decreasing the direct repair loss, as shown in Figure 14b. This is probably because the indirect loss from extensive and complete damage states accounts for a more significant share of the total loss than the direct loss, and the utilization of the SCR bridge system reduces the conditional probability of a bridge system being in a more severe damage state. This also explains why the decrease rate is more significant in locations with higher seismic levels since the investigated bridge has a higher failure chance in such areas. It is also worth noting that when the hazard intensity rises, the indirect loss from social and economic aspects is considerably more significant than the direct loss from repair work, which emphasizes the social and economic consequences of traffic restrictions induced by bridge damage.

In Figure 14c,d, the direct and indirect loss of the conventional and SCR bridge systems under the four damage states are illustrated considering event 5 (i.e., 2475-year return period) and event 6 (i.e., 5000-year return period). It can be found that the loss associated with significant and complete damage states accounted for most of the overall direct or indirect loss for the conventional bridge system. As shown in Figure 14c, the degree of direct loss related to extensive and complete damage remained constant across the conventional and SCR bridge systems. However, the result did not reveal a substantial difference between the two locations under either event 5 or 6. The SCR bridge system reduced the direct loss corresponding to destruction at both sites under events 5 and 6. The primary reason is that using the SCR bridge system minimized the conditional probability of bridge collapse. The SCR bridge system successfully reduces the indirect loss corresponding to all the four damage states to a reasonably low level at the two sites, as illustrated in Figure 14d.

The estimated long-term loss of both the conventional and novel bridge systems is shown in Figure 15a. A life span of 75 years and a monetary discount rate of 2.0% are considered to represent the investigated time period and monetary discount rate respectively. At both sites, the novel bridge system reduced the predicted long-term loss compared with the conventional bridge system. From event 1 through event 6, the long-term loss of the conventional and SCR bridge systems at Salt Lake City kept rising. On the other hand, the long-term loss for Los Angeles peaked in event 5 (i.e., 2475-year return time) and subsequently fell in event 6. From event 3 (i.e., 975-year return period) through event 5 (i.e., 2475-year return period), the peak value of the expected long-term loss of the SCR bridge system remained stable at a relatively low level.

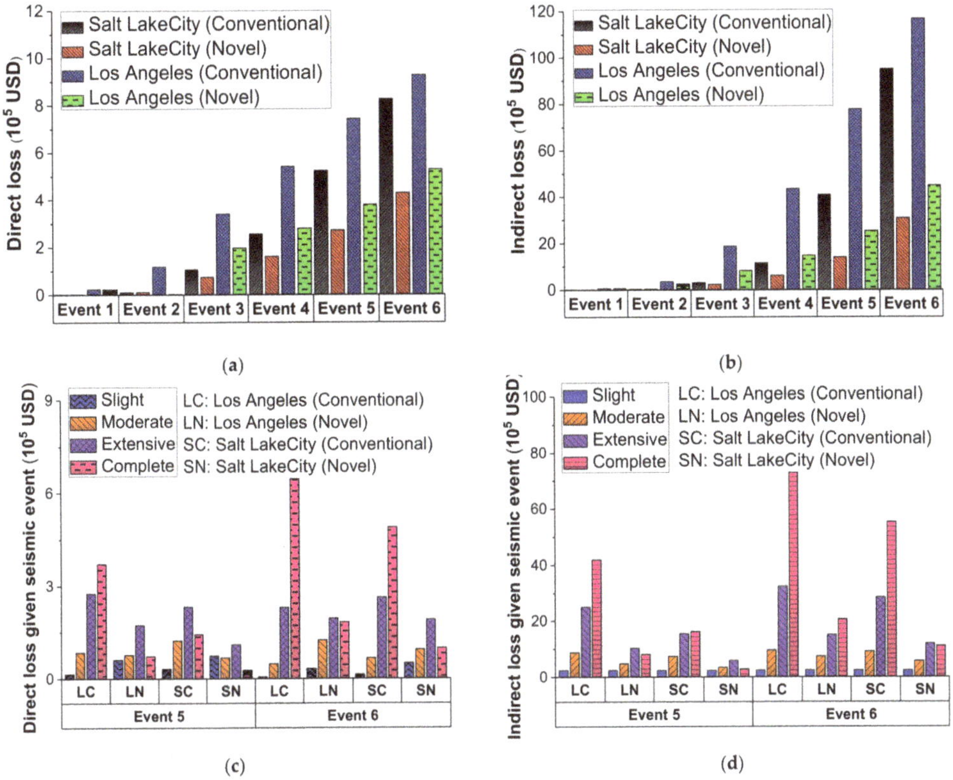

Figure 14. Predicted losses of bridges under different hazard events: (**a**) total direct loss, (**b**) total indirect loss, (**c**) direct loss under four damage states, and (**d**) indirect loss under four damage states.

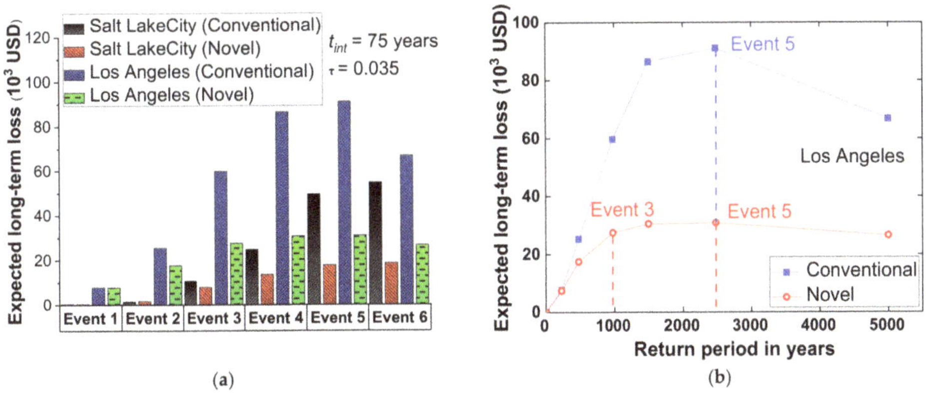

Figure 15. Expected long-term loss of bridges (**a**) under six events (**b**) with 50, 475, 2475, and 5000-year return periods.

5.5. Resilience Assessment

This section evaluates the resilience of the conventional and novel bridges. Resilience is now one of the most important structural performance indicators [73–78]. For the conventional bridge system in Los Angeles, the residual functionality for the six analyzed events was 0.95, 0.76, 0.46, 0.30, 0.18, and 0.09, respectively. In contrast, the residual functionality for the SCR bridge system was 0.95, 0.84, 0.66, 0.56, 0.46, and 0.31, respectively. As anticipated, an increasing seismic intensity leads to decreased residual functionality, and the SCR bridge successfully increases the residual functionality more than the conventional system. The residual functionality of the conventional bridge system at Salt Lake City was 1.00, 0.98, 0.78, 0.56, 0.31, and 0.14, respectively, whereas the novel system's values were 1.00, 0.98, 0.86, 0.71, 0.57, and 0.41, which, again, were much increased.

The resilience of the bridge was further assessed using Equation (16) and through the functionality restoration function, i.e., Equation (18). Figure 16a,b depict the functionality of the damaged bridge systems after recovery procedures, considering events 5 (i.e., 2475-year return period) and 6 (i.e., 5000-year return period), respectively. It can be seen that when the recovery effort was initiated, the predicted functionality of the damaged bridge systems grew with time. Importantly, the SCR bridge system outperformed the conventional system in terms of recovery time. As shown in Figure 16a, the predicted functionality of the SCR bridge system in Los Angeles returned to 0.9, i.e., rapid access, after 110 days following the occurrence of an earthquake under event 5. In contrast, the conventional bridge system took 290 days. A similar pattern is seen in Figure 16b. This phenomenon may be explained by the fact that the indirect loss of the SCR bridge system is substantially lower than that of the conventional system. In particular, quick recovery after an earthquake may drastically decrease the social and economic losses. As previously stated, the SCR bridge system considerably improved the resilience, and the improvement is more significant with increased hazard intensity. In addition, the resilience enhancement in a more seismically active location, such as Los Angeles, is more substantial.

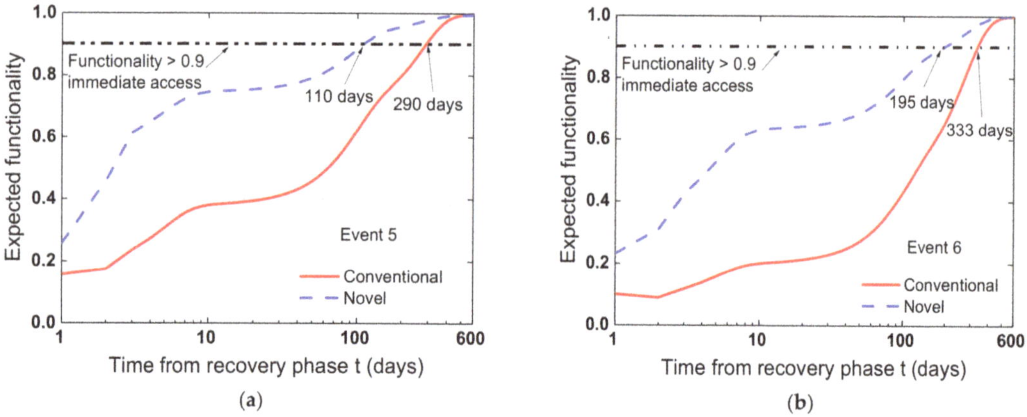

Figure 16. Expected functionality of the bridge from the recovery phase under (**a**) event 5 and (**b**) event 6.

6. Conclusions

This paper presented a performance-based framework for evaluating the life-cycle loss and resilience of a conventional bridge and a novel SCR bridge with SMA washer-based piers, and successfully demonstrated the benefit associated with the adoption of this novel pier. Fragility curves of the conventional and SCR bridge systems were obtained by conducting nonlinear time-history analysis. Life-cycle loss and resilience under the investigated hazard scenarios were assessed, considering the direct and indirect loss and hazard recovery pattern. The proposed assessment framework was applied to continuous RC bridges with and without incorporating the SCR pier. This work provided an efficient decision-making tool for the application of new bridge systems in the structural seismic design process. The following conclusions can be obtained.

- The introduction of the novel SCR pier bridge system slightly increased the bearing displacement but extensively reduced the pier curvature ductility due to the rocking mechanism; the damage probability of the pier with the novel system was lower than that of the conventional system. The positive impact was more evident considering more severe damage states.
- The long-term loss of the conventional and SCR bridge systems within the investigated time was assessed and compared. The indirect loss increased with a higher hazard intensity. Indirect loss is much larger than the direct loss, specifically under the earthquakes with a relatively low probability of occurrence. As can be concluded from the results, the life-cycle loss of the bridge using the SCR bridge piers can be reduced significantly. The investigated time interval can affect the life-cycle loss remarkably.
- The performance benefit of resilience associated with the SCR bridge system increase with a larger investigated hazard intensity. The SCR bridge system is more beneficial for the bridges located in seismic zones with higher hazard intensity.
- The investigated seismic intensity can affect the resilience significantly. The SCR bridge system outperformed the conventional system in terms of recovery time. Quick recovery after an earthquake may drastically decrease the social and economic losses. The difference of the resilience performance between the conventional and SCR bridge systems increased with the increase of the investigated hazard intensity.

Author Contributions: Conceptualization, X.Y. and H.L.; Data curation, J.C. and H.L.; Formal analysis, D.L. and H.L. All authors have read and agreed to the published version of the manuscript.

Funding: This research received no external funding.

Institutional Review Board Statement: Not applicable.

Informed Consent Statement: Not applicable.

Data Availability Statement: Data presented in this study are available in this article.

Conflicts of Interest: The authors declare no conflict of interest.

References

1. Buckle, I.G. Overview of seismic design methods for bridges in different countries and future directions. In Proceedings of the Eleventh World Conference on Earthquake Engineering, Acapulco, Mexico, 23–28 June 1996.
2. Mander, J.B.; Cheng, C.T. *Seismic Resistance of Bridge Piers Based on Damage Avoidance Design*; Technical report NCEER-97-0014; State University of New York: Buffalo, NY, USA, 1997; p. 109.
3. Hieber, D.G.; Wacker, J.M.; Eberhard, M.O.; Stanton, J.F. *Precast Concrete Pier Systems for Rapid Construction of Bridges in Seismic Regions*; Report No. WA-RD 611.1; University of Washington: Seattle, WA, USA, 2005.
4. Pang, J.B.; Steuck, K.P.; Cohagen, L.; Eberhard, M.O.; Stanton, J.F. *Rapidly Constructible Large-Bar Precast Bridge-Bent Connection*; Report No. WA-RD 684.2; University of Washington: Seattle, WA, USA, 2008.
5. Nguyen, W.; Trono, W.; Panagiotou, M.; Ostertag, C.P. *Seismic Response of a Hybrid Fiber-Reinforced Concrete Bridge Column Detailed for Accelerated Bridge Construction*; Report No. PEER 2014/19; University of California: Berkeley, CA, USA, 2014.
6. Schaefer, J.A.; Kennedy, B.; Eberhard, M.O.; Stanton, J.F. *Unbonded Pretensioned Bridge Columns with Rocking Detail*; Report No. PEER 2014/08; University of Washington: Seattle, WA, USA, 2014.
7. Saad, A.S.; Sanders, D.H.; Buckle, I.G. Experimental evaluation of bridge column foundation rocking behavior. *J. Bridge Eng.* **2018**, *23*, 04018088. [CrossRef]
8. Palermo, A.; Pampanin, S.; Calvi, G.M. Concept and development of hybrid solutions for seismic resistant bridge systems. *J. Earthq. Eng.* **2005**, *9*, 899–921. [CrossRef]
9. Cheng, C.T. Shaking table tests of a self-centering designed bridge substructure. *Eng. Struct.* **2008**, *30*, 3426–3433. [CrossRef]
10. Trono, W.; Jen, G.; Panagiotou, M.; Schoettler, M.; Ostertag, C.P. Seismic response of a damage-resistant recentering posttensioned-HYFRP bridge column. *J. Bridge Eng.* **2015**, *20*, 04014096. [CrossRef]
11. Sakai, J.; Mahin, S.A. Mitigation of residual displacements of circular reinforced concrete bridge columns. In Proceedings of the 13th World Conference on Earthquake Engineering, Vancouver, BC, Canada, 1–6 August 2004.
12. Palermo, A.; Pampanin, S.; Marriott, D. Design, modeling, and experimental response of seismic resistant bridge piers with posttensioned dissipating connections. *J. Struct. Eng.* **2007**, *133*, 1648–1661. [CrossRef]
13. Pollino, M.; Bruneau, M. Seismic retrofit of bridge steel truss piers using a controlled rocking approach. *J. Bridge Eng.* **2007**, *12*, 600–610. [CrossRef]

14. Kam, W.Y.; Pampanin, S.; Palermo, A.; Carr, A.J. Self-centering structural systems with combination of hysteretic and viscous energy dissipations. *Earthq. Eng. Struct. Dyn.* **2010**, *39*, 1083–1108. [CrossRef]
15. Marriott, D.; Pampanin, S.; Palermo, A. Quasi-static and pseudo-dynamic testing of unbonded post-tensioned rocking bridge piers with external replaceable dissipators. *Earthq. Eng. Struct. Dyn.* **2009**, *38*, 331–354. [CrossRef]
16. White, S.; Palermo, A. Quasi-static testing of posttensioned nonemulative column-footing connections for bridge piers. *J. Bridge Eng.* **2016**, *21*, 04016025. [CrossRef]
17. DesRoches, R.; Delemont, M. Seismic retrofit of simply supported bridges using shape memory alloys. *Eng. Struct.* **2002**, *24*, 325–332. [CrossRef]
18. Dezfuli, F.H.; Alam, M.S. Seismic vulnerability assessment of a steel-girder highway bridge equipped with different SMA wire-based smart elastomeric isolators. *Smart Mater. Struct.* **2016**, *25*, 075039. [CrossRef]
19. Roh, H.; Reinhorn, A.M. Hysteretic behavior of precast segmental bridge piers with superelastic shape memory alloy bars. *Eng. Struct.* **2010**, *32*, 3394–3403. [CrossRef]
20. Varela, S.; Saiidi, M.S. A bridge column with superelastic NiTi SMA and replaceable rubber hinge for earthquake damage mitigation. *Smart Mater. Struct.* **2016**, *25*, 075012. [CrossRef]
21. Zheng, Y.; Dong, Y.; Li, Y. Resilience and life-cycle performance of smart bridges with shape memory alloy (SMA)-cable-based bearings. *Const. Build. Mater.* **2018**, *158*, 389–400. [CrossRef]
22. Fang, C.; Zheng, Y.; Chen, J.; Yam, M.C.H.; Wang, W. Superelastic NiTi SMA cables: Thermal-mechanical behavior, hysteretic modelling and seismic application. *Eng. Struct.* **2019**, *183*, 533–549. [CrossRef]
23. Fang, C.; Yam, M.C.H.; Lam, A.C.C.; Xie, L.K. Cyclic performance of extended end-plate connections equipped with shape memory alloy bolts. *J. Constr. Steel Res.* **2014**, *94*, 122–136. [CrossRef]
24. Fang, C.; Yam, M.C.H.; Lam, A.C.C.; Zhang, Y.Y. Feasibility study of shape memory alloy ring spring systems for self-centring seismic resisting devices. *Smart Mater. Struct.* **2015**, *24*, 075024. [CrossRef]
25. Fang, C.; Yam, M.C.H.; Ma, H.W.; Chung, K.F. Tests on superelastic Ni–Ti SMA bars under cyclic tension and direct-shear: Towards practical recentring connections. *Mater. Struct.* **2015**, *48*, 1013–1030. [CrossRef]
26. Fang, C.; Ping, Y.; Chen, Y.; Yam, M.C.H.; Chen, J.; Wang, W. Seismic performance of self-centering steel frames with SMA-viscoelastic hybrid braces. *J. Earthq. Eng.* **2022**, *26*, 5004–5031. [CrossRef]
27. Fang, C.; Ping, Y.; Zheng, Y.; Chen, Y. Probabilistic economic seismic loss estimation of steel braced frames incorporating emerging self-centering technologies. *Eng. Struct.* **2021**, *241*, 112486. [CrossRef]
28. Yam, M.C.H.; Fang, C.; Lam, A.C.C.; Zhang, Y.Y. Numerical study and practical design of beam-to-column connections with shape memory alloys. *J. Constr. Steel Res.* **2015**, *104*, 177–192. [CrossRef]
29. Wang, W.; Fang, C.; Liu, J. Large size superelastic SMA bars: Heat treatment strategy, mechanical property and seismic application. *Smart Mater. Struct.* **2016**, *25*, 075001. [CrossRef]
30. Wang, W.; Fang, C.; Liu, J. Self-Centering Beam-to-Column Connections with Combined Superelastic SMA Bolts and Steel Angles. *J. Struct. Eng.* **2017**, *143*, 04016175. [CrossRef]
31. Chen, J.; Fang, C.; Wang, W.; Liu, Y. Variable-friction self-centering energy-dissipation braces (VF-SCEDBs) with NiTi SMA cables for seismic resilience. *J. Constr. Steel Res.* **2020**, *175*, 106318. [CrossRef]
32. Chen, J.; Wang, W.; Fang, C. Manufacturing, testing and simulation of novel SMA-based variable friction dampers with enhanced deformability. *J. Build. Eng.* **2022**, *45*, 103513. [CrossRef]
33. Qiu, C.; Zhu, S. Shake table test and numerical study of self-centering steel frame with SMA braces. *Earthq. Eng. Struct. Dyn.* **2017**, *46*, 117–137. [CrossRef]
34. Wang, B.; Zhu, S. Superelastic SMA U-shaped dampers with self-centering functions. *Smart Mater. Struct.* **2018**, *27*, 055003. [CrossRef]
35. Ge, J.; Saiidi, M.S.; Varela, S. Computational studies on the seismic response of the State Route 99 bridge in Seattle with SMA/ECC plastic hinges. *F. Struct. Civ. Eng.* **2019**, *13*, 149–164. [CrossRef]
36. Fang, C.; Liang, D.; Zheng, Y.; Lu, S. Seismic performance of bridges with novel SMA cable-restrained high damping rubber bearings against near-fault ground motions. *Earthq. Eng. Struct. Dyn.* **2022**, *51*, 44–65. [CrossRef]
37. Liang, D.; Zheng, Y.; Fang, C.; Yam, M.C.; Zhang, C. Shape memory alloy (SMA)-cable-controlled sliding bearings: Development, testing, and system behavior. *Smart Mater. Struct.* **2020**, *29*, 085006. [CrossRef]
38. Fang, C.; Liang, D.; Zheng, Y.; Yam, M.C.; Sun, R. Rocking bridge piers equipped with shape memory alloy (SMA) washer springs. *Eng. Struct.* **2020**, *214*, 110651. [CrossRef]
39. Fang, C.; Zhou, X.Y.; Osofero, A.I.; Shu, Z.; Corradi, M. Superelastic SMA Belleville washers for seismic resisting applications: Experimental study and modelling strategy. *Smart Mater. Struct.* **2016**, *25*, 105013. [CrossRef]
40. Mangalathu, S.; Jeon, J.S.; Padgett, J.E.; Des Roches, R. ANCOVA-based grouping of bridge classes for seismic fragility assessment. *Eng. Struct.* **2016**, *123*, 379–394. [CrossRef]
41. Alam, M.S.; Bhuiyan, M.R.; Billah, A.M. Seismic fragility assessment of SMA-bar restrained multi-span continuous highway bridge isolated by different laminated rubber bearings in medium to strong seismic risk zones. *Bull. Earthq. Eng.* **2012**, *10*, 1885–1909. [CrossRef]
42. Miano, A.; Jalayer, F.; Ebrahimian, H.; Prota, A. Cloud to IDA: Efficient fragility assessment with limited scaling. *Earthq. Eng. Struct. Dyn.* **2018**, *47*, 1124–1147. [CrossRef]

43. Wei, B.; Hu, Z.; He, X.; Jiang, L. Evaluation of optimal ground motion intensity measures and seismic fragility analysis of a multi-pylon cable-stayed bridge with super-high piers in Mountainous Areas. *Soil Dyn. Earthq. Eng.* **2020**, *129*, 105945. [CrossRef]
44. Kim, S.H.; Mha, H.S.; Lee, S.W. Effects of bearing damage upon seismic behaviors of a multi-span girder bridge. *Eng. Struct.* **2006**, *28*, 1071–1080. [CrossRef]
45. Zhang, J.; Huo, Y. Evaluating effectiveness and optimum design of isolation devices for highway bridges using the fragility function method. *Eng. Struct.* **2009**, *31*, 1648–1660. [CrossRef]
46. Zheng, Y.; Dong, Y. Performance-based assessment of bridges with steel-SMA reinforced piers in a life-cycle context by numerical approach. *Bull. Earthq. Eng.* **2018**, *17*, 1667–1688. [CrossRef]
47. USGS. Unified Hazard Tool. 2017. Available online: https://earthquake.usgs.gov/hazards/interactive/ (accessed on 11 September 2022).
48. Dong, Y.; Frangopol, D.M.; Saydam, D. Time-variant sustainability assessment of seismically vulnerable bridges subjected to multiple hazards. *Earthq. Eng. Struct. Dyn.* **2013**, *42*, 1451–1467. [CrossRef]
49. Stein, S.M.; Young, G.K.; Trent, R.E.; Pearson, D.R. Prioritizing scour vulnerable bridges using risk. *J. Infrastruct. Syst.* **1999**, *5*, 95–101. [CrossRef]
50. Mander, J.B. *Fragility Curve Development for Assessing the Seismic Vulnerability of Highway Bridges*; Technical report; State University of New York: New York, NY, USA, 1999; p. 89.
51. Dong, Y.; Frangopol, D.M. Risk and resilience assessment of bridges under mainshock and aftershocks incorporating uncertainties. *Eng. Struct.* **2015**, *83*, 198–208. [CrossRef]
52. Padgett, J.E.; Ghosh, J.; Dennemann, K. Sustainable infrastructure subjected to multiple threats. In Proceedings of the ASCE TCLEE 2009: Lifeline Earthquake Engineering in a Multi-Hazard Environment, Oakland, CA, USA, 28 June–1 July 2009.
53. Dong, Y.; Frangopol, D.M. Probabilistic time-dependent multihazard life-cycle assessment and resilience of bridges considering climate change. *J. Perform. Constr. Fac.* **2016**, *30*, 1–12. [CrossRef]
54. Cimellaro, G.P.; Reinhorn, A.M.; Bruneau, M. Seismic resilience of a hospital system. *Struct. Infrastruct. Eng.* **2010**, *6*, 127–144. [CrossRef]
55. Frangopol, D.M.; Bocchini, P. Resilience as optimization criterion for the rehabilitation of bridges belonging to a transportation network subjected to earthquake. In Proceedings of the SEI-ASCE 2011 Structures Congress, Las Vegas, NV, USA, 14–16 April 2011.
56. Padgett, J.E.; DesRoches, R. Bridge functionality relationships for improved seismic risk assessment of transportation networks. *Earthq. Spectra* **2007**, *23*, 115–130. [CrossRef]
57. Decò, A.; Bocchini, P.; Frangopol, D.M. A probabilistic approach for the prediction of seismic resilience of bridges. *Earthq. Eng. Struct. Dyn.* **2013**, *42*, 1469–1487. [CrossRef]
58. ATC. *Earthquake Damage Evaluation Data for California*; Technical Report ATC-13; Applied Technology Council: Redwood City, CA, USA, 1999.
59. Ozbulut, O.E.; Hurlebaus, S. Optimal design of superelastic-friction base isolators for seismic protection of highway bridges against near-field earthquakes. *Earthq. Eng. Struct. Dyn.* **2011**, *40*, 273–291. [CrossRef]
60. Ozbulut, O.E.; Hurlebaus, S. A comparative study on the seismic performance of superelastic-friction base isolators against near-field earthquakes. *Earthq. Spectra* **2022**, *28*, 1147–1163. [CrossRef]
61. Wang, J.Q.; Li, S.; Dezfuli, F.H.; Alam, M.S. Sensitivity analysis and multi-criteria optimization of SMA cable restrainers for longitudinal seismic protection of isolated simply supported highway bridges. *Eng. Struct.* **2019**, *189*, 509–522. [CrossRef]
62. Zheng, Y.; Fang, C.; Liang, D.; Sun, R. An innovative seismic-resilient bridge with shape memory alloy-washer-based footing rocking RC piers. *J. Intell. Mat. Syst. Str.* **2021**, *32*, 549–567. [CrossRef]
63. Menegotto, M.; Pinto, P.E. Method for analysis of cyclically loaded reinforced concrete plane frames including changes in geometry and non-elastic behavior of elements under combined normal force and bending. In Proceedings of the IABSE Symposium on Resistance and Ultimate Deformability of Structures Acted on by Well-Defined Repeated Loads, Lisbon, Portugal, 1973.
64. Barbato, M.; Conte, J.P. Finite element structural response sensitivity and reliability analysis using smooth versus non-smooth material constitutive models. *Int. J. Reliab. Saf.* **2006**, *1*, 3–39. [CrossRef]
65. Scott, B.D.; Park, P.; Priestley, M.J.N. Stress-strain behavior of concrete confined by overlapping hoops at low and high strain rates. *J. Am. Concrete Inst.* **1982**, *79*, 13–27.
66. Crisfield, M.A. A consistent co-rotational formulation for non-linear three-dimensional beam elements. *Comput. Methods Appl. Mech. Eng.* **1990**, *81*, 131–150. [CrossRef]
67. Caltrans Seismic Design Criteria of California. *Caltrans SD Seismic Design Criteria, Structural Design, Version 1.7*; Caltrans: Sacramento, CA, USA, 2004.
68. Ramanathan, K.N. *Next Generation Seismic Fragility Curves for California Bridges Incorporating the Evolution in Seismic Design Philosophy*; Georgia Institute of Technology: Atlanta, GA, USA, 2012.
69. Ramanathan, K.; Padgett, J.E.; DesRoches, R. Temporal evolution of seismic fragility curves for concrete box-girder bridges in California. *Eng. Struct.* **2015**, *97*, 29–46. [CrossRef]
70. Choi, E.; DesRoches, R.; Nielson, B. Seismic fragility of typical bridges in moderate seismic zones. *Eng. Struct.* **2004**, *26*, 187–199. [CrossRef]
71. Padgett, J.E.; Nielson, B.G.; DesRoches, R. Selection of optimal intensity measures in probabilistic seismic demand models of highway bridge portfolios. *Earthq. Eng. Struct. Dyn.* **2008**, *37*, 711–725. [CrossRef]

72. Wang, X.; Shafieezadeh, A.; Ye, A. Optimal intensity measures for probabilistic seismic demand modeling of extended pile-shaft-supported bridges in liquefied and laterally spreading ground. *Bull. Earthq. Eng.* **2018**, *16*, 229–257. [CrossRef]
73. Fang, C. SMAs for infrastructures in seismic zones: A critical review of latest trends and future needs. *J. Build Eng.* **2022**, *57*, 104918. [CrossRef]
74. Fang, C.; Wang, W.; Qiu, C.; Hu, S.; MacRae, G.A.; Eatherton, M.R. Seismic resilient steel structures: A review of research, practice, challenges and opportunities. *J. Constr. Steel Res.* **2022**, *191*, 107172. [CrossRef]
75. Shu, Z.; Gan, Z.; Fang, C.; MacRae, G.; Dong, H.; Xie, Y. Replaceable rotational viscoelastic dampers for improving structural damping and resilience of steel frames. *J. Earthq. Eng.* **2021**, *2021*, 1–23. [CrossRef]
76. Fang, C.; Wang, W.; Shen, D. Development and experimental study of disc spring–based self-centering devices for seismic resilience. *J. Struct. Eng.* **2021**, *147*, 04021094. [CrossRef]
77. Wang, W.; Fang, C.; Zhao, Y.; Sause, R.; Hu, S.; Ricles, J. Self-centering friction spring dampers for seismic resilience. *Earthq. Eng. Struct. Dyn.* **2019**, *48*, 1045–1065. [CrossRef]
78. Wang, W.; Fang, C.; Zhang, A.; Liu, X. Manufacturing and performance of a novel self-centring damper with shape memory alloy ring springs for seismic resilience. *Struct. Control Health Monitor.* **2019**, *26*, e2337. [CrossRef]

materials

Article

Seismic Response of Resilient Bridges with SMA-Based Rocking ECC-Reinforced Piers

Xiaogang Li [1], Ke Chen [2,*], Ji Chen [1], Yi Li [1] and Dong Yang [1]

[1] PowerChina Huadong Engineering Corporation Limited, Hangzhou 311122, China; li_xg2@hdec.com (X.L.); chen_j7@hdec.com (J.C.); li_y11@hdec.com (Y.L.); yang_d2@hdec.com (D.Y.)
[2] Research Institute of Highway, Ministry of Transport, Beijing 100088, China
* Correspondence: k.chen@rioh.cn; Tel.: +86-138-1099-7548

Abstract: Post-earthquake investigation shows that numerous reinforced concrete (RC) bridges were demolished because of large residual displacements. Improving the self-centering capability and hence resilience of these bridges located in earthquake-prone regions is essential. In this regard, a resilient bridge system incorporating engineered cementitious composites (ECC) reinforced piers and shape memory alloy (SMA) energy dissipation components, i.e., SMA washers, is proposed to enhance its resilience when subjected to strong earthquakes. This study commences with a detailed introduction of the resilient SMA-washer-based rocking bridge system with ECC-reinforced piers. Subsequently, a constitutive model of the ECC material is implemented into OpenSees and the constitutive model is validated by test data. The working principle and constitutive model of the SMA washers are also introduced. A series of dynamic analysis on the conventional and resilient rocking bridge systems with ECC-reinforced piers under a suite of ground motions at E1 and E2 earthquake levels are conducted. The analysis results indicate that the resilient rocking bridge system with ECC-reinforced piers has superior resilience and damage control capacities over the conventional one.

Keywords: seismic analysis; rocking pier; shape memory alloy; ECC material; bridge engineering

Citation: Li, X.; Chen, K.; Chen, J.; Li, Y.; Yang, D. Seismic Response of Resilient Bridges with SMA-Based Rocking ECC-Reinforced Piers. *Materials* **2021**, *14*, 6500. https://doi.org/10.3390/ma14216500

Academic Editors: Marco Corradi and Raffaele Landolfo

Received: 30 August 2021
Accepted: 23 October 2021
Published: 29 October 2021

1. Introduction

Post-earthquake field investigations on the damaged bridges revealed that many reinforced concrete (RC) bridges, although designed conforming to the ductility design philosophy, commonly experienced overly large residual displacement which is difficult to recover. For example, more than 100 RC piers were demolished because they suffered from large permanent drift ratio (i.e., 1.5%) after the 1995 Kobe earthquake [1]. Lessons drawn from these events enlighten us that only satisfying the seismic ductility demand is not enough for engineering structures because their residual deformation after earthquake still significantly jeopardizes their normal functionality [2–4]. To guarantee service operation of the structures after earthquakes, resilient capacity is being paid more attention in the seismic codes of many countries (e.g., US, Japan, and New Zealand) [5]. Rocking component, as a resilient structural member, has been attracting extensive experimental and numerical studies [6–10]. For instance, the conventional post-tensioned (PT) rocking bridges have been studied by shake table tests recently [11–13]. These studies revealed that these self-centering bridge systems were capable of sustaining a large drift ratio of up to 10% but only experienced small residual drift ratio (i.e., 0.5%) with non-critical damages [14]. Subsequently, a series of innovative devices had been presented to further improve the self-centering and energy dissipation capacities of the rocking piers under extreme earthquake events [15–18]. Although the PT tendons together with various energy dissipaters can provide excellent recoverability and energy dissipation capacity to the rocking pier [19,20], the energy dissipater may be damaged and thus should be replaced after earthquakes, leading to compromised rescue efficiency. Additionally, some damage patterns such as relaxation and environmental corrosion of the PT tendons are difficult to fix. In this regard, shape memory alloy (SMA) that is characterized by super-elasticity has been recently considered for various devices (i.e., SMA tendons, bars, and springs) used in resilient bridge structures [21–31] as well as other types of structural systems [32–39]. In particular, a bridge system with SMA-washer based rocking pier was recently proposed to achieve self-centering functionality during earthquakes [40]. The SMA washers provided restoring force for the RC pier, which can eliminate some inherent shortcomings, such as corrosion and relaxation, induced by the PT tendon. However, the reinforcing steel embedded in the plastic hinge of the pier was still vulnerable to yield due to large bending moment during severe earthquakes. Varela and Saiidi [41] integrated SMA bars with elastomeric rubber bearing to replace the conventional plastic hinge of the RC pier. The test results indicated that, except for the bucking of the SMA bars, the RC pier experienced almost no damage even under 2.5 times the design's seismic loading. One drawback of the SMA bars is their sensitivity to manufacture imperfections and vulnerable to brittle fracture when sustaining large strains [42,43]. Apart from the SMA material, an innovative engineered cementitious composites (ECC) material was used to replace the conventional concrete in the plastic hinge of the pier to enhance its strength and ductility in both tension and compression [44]. The metal-like strain strengthening property of the ECC material makes it an attractive alternative to the conventional concrete for earthquake resilient structures.

Inspired by the existing studies on rocking piers and the application of emerging materials such as ECC and SMA, a resilient self-centering structural bridge system with SMA-washer-based ECC-reinforced rocking piers is proposed. The SMA washers allows the pier to rock in a large rotation angle under reliable control. The ECC material in the pier can significantly enhance the resilience of the rocking piers and remarkably reduces structural cracks during earthquakes. In this paper, the configuration and self-centering rocking mechanism of the resilient bridge system with ECC reinforced pier is firstly proposed. Subsequently, the constitutive model of the ECC material is introduced and validated by an experimental study on a 1/5 scaled ECC-reinforced pier. An FE model of the prototype rocking bridge with the SMA-washer-based ECC-reinforced rocking pier is established in OpenSees [45] to study the superiority of the proposed solution in alleviating seismic damage over the conventional bridges.

2. Resilient Self-Centering Rocking Bridge System with ECC-Reinforced Pier

2.1. Configuration of the Rocking Bridge System

The configuration of the resilient bridge system with the ECC-reinforced rocking pier is presented in Figure 1. The main body of the pier and two separated pile caps (i.e., the upper pile cap and the lower pile cap) are the most critical components in the resilient bridge system. Each SMA washer set consists of a group of SMA washer springs and associated nuts, anchor rebar and shim plates. The anchor rebar was casted into the concrete of the lower pile cap and passed through the upper one via a plastic tube which was embedded in the upper pile cap. The SMA washers which were assembled in appropriate styles (e.g., in parallel, in series or in combination) passed through the anchor rebar and were tightly precompressed by the nuts. This SMA-washer based rocking solution is capable of providing stable recoverability and moderate energy dissipation capacity for the bridge system when subjected to severe earthquake events. The fixed bearings were installed on the top of the bent cap but sliding bearings are installed on the top of two abutments for the girder deformation induced by temperature variance. A comprehensive experimental study on reduced scale SMA-washer based rocking piers (with normal concrete) has been recently conducted, where the details are reported elsewhere [40]. When ECC is used for the pier, extra benefits are enabled such as enhanced tension and compression strength and ductility to sustain flexural and shear deformation during seismic excitations. In this way, the residual deformation of the resilient bridge system tends to decrease and the damage to the pier will be alleviated. For the prototype bridge, six sets of the SMA washer devices are equipped in the pile. The configuration of a typical SMA washer set is shown in Figure 1b.

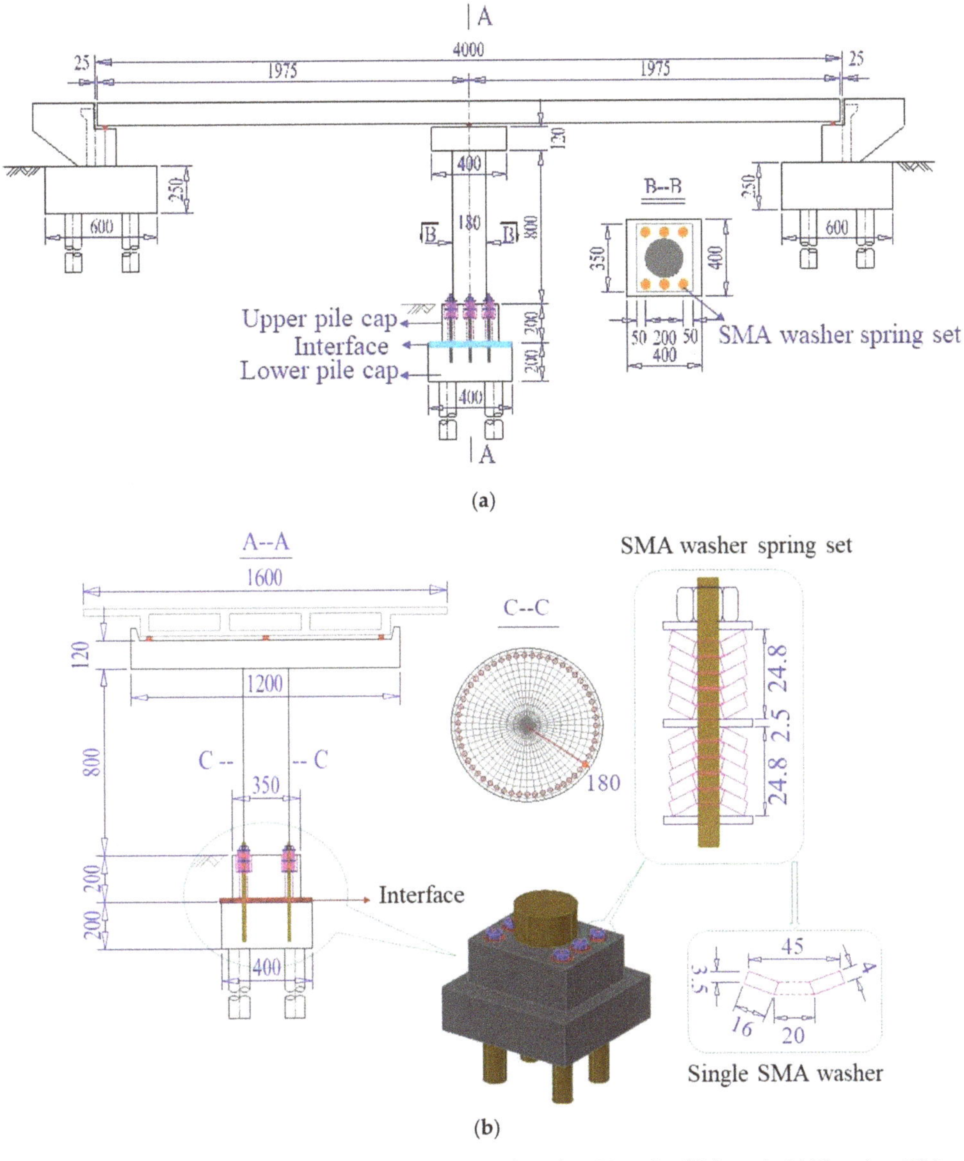

Figure 1. Configuration of the resilient bridge with ECC-reinforced rocking pier (Unit: cm). (**a**) Elevation; (**b**) Layout diagram of the SMA washer spring devices.

2.2. Rocking Mechanism and Seismic Design Objectives of Resilient Bridge System

There are three seismic design objectives under different earthquake intensities (i.e., small earthquake, moderate earthquake and large earthquake). The first objective is that the interface between two pile caps is close under small earthquakes, which ensures that the rocking bridge system exhibits similar function to the conventional bridge with fixed base pier. The second one is that the maximum drift ratio of the girder is not more than 1.0% under moderate earthquakes (i.e., E1 level). Such a small drift angle would induce limited damage to the structure. The last one is that the maximum drift ratio of the girder is not more than 2.0% under large earthquakes (i.e., E2 level). When the bridge subjects to severe earthquakes, the resulting inertial force will cause the bridge to rock around the two base corners of the upper pile cap and meanwhile the compressed SMA washers together with the gravity of the bridge will provide restoring force for the bridge to return to its original state. The maximum rocking angle of the pier can be controlled by designing appropriate assembles (i.e., series, parallel or both of them) of the SMA washer set. The natural period of the resilient bridge is much larger than that of the conventional bridge, which makes it far away from the dominant periods of the earthquake ground motions. Thus, the resilient bridge system is expected to significantly reduce seismic damage.

The rocking mechanism of the pier and the seismic design objectives of the bridge can be interpreted by Figure 2. The total height of the pier and the pier cap is H. To satisfy the first seismic design objective, an appropriate prestressed force should be imposed on the SMA washer sets, as shown in Figure 2a. The uniform distribution of the resulting reacting force at the bottom of the upper pile cap is also shown in Figure 2a. The upper pile cap and the bottom pile cap always touch tightly under dead load of the superstructure (i.e., G) and servicing loads such as vehicle and temperature. When the horizontal force F_h continuously increases, the reacting force at one base side gradually reduces to zero, as shown in Figure 2b. If F_h further increases, the pier will uplift with a drift ratio of $\alpha = \Delta/H$, where Δ is the horizontal displacement of the pier, as shown in Figure 2c. The horizontal displacement is commonly composed of two parts, of which one part is the deformation of the pier and another part is the contribution of rocking. If the horizontal deformation of the RC pier is too large, the bottom of the pier may yield. Therefore, the ECC material is proposed in this study as an alternative to the conventional concrete material for the pier. The ultimate drift ratio α_u can be calculated by Δ_u/H, as shown in Figure 2d. When the designated ultimate drift ratio (i.e., 2.0%) reaches, the SMA washer sets at one side of the pier is fully flattened. If the drift ratio of the bridge exceeds 2.0% during earthquakes, the pier will yield. The proposed resilient rocking bridge system is capable of self-locking, which is a unique property over the conventional one.

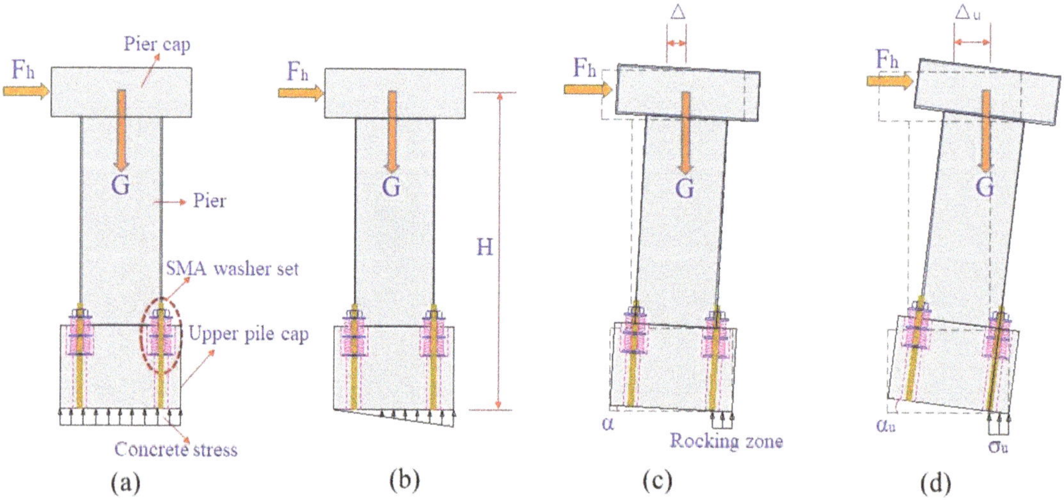

Figure 2. Rocking mechanism of the SMA washer-based pier. (**a**) Original state; (**b**) Onset of uplift; (**c**) Rocking state; (**d**) Self-locking state.

3. Verification of ECC Material Constitutive Models

3.1. Constitutive Models of ECC Material

In order to study the seismic response of the ECC-reinforced structures, a constitutive model [45,46], which is capable of considering the computational efficiency and accuracy, is essential for establishing finite element (FE) model, e.g., OpenSees model. The envelop curves of the constitutive model is composed of two parts, of which one is related to the tension state (i.e., *O-A-F-I*) and the other part is associated with the compression state (i.e., *O-J-P*), as shown in Figure 3. The envelop curve in tension is multilinear curves which are expressed as Equation (1), and the envelop curve in compression is also the multilinear curves as written in Equation (2).

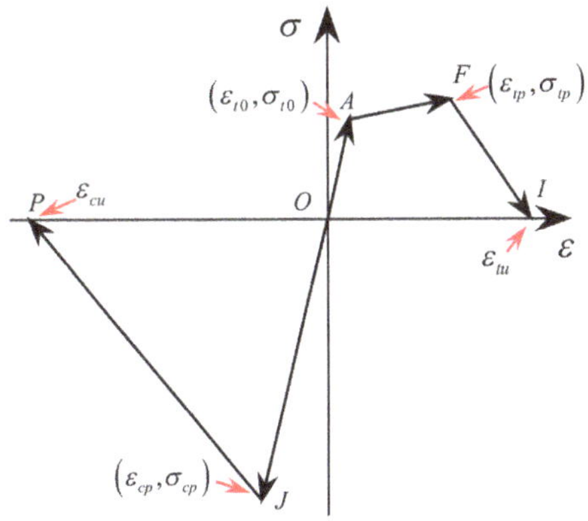

Figure 3. Envelop curves of the ECC constitutive model [46].

Segment *A-F* in the tension region is a strain hardening stage, which indicates that the ECC material is more resilient than the normal concrete. The three segments on the envelope curve within the tensile region (i.e., segments *O-A-F-I*) are given in

$$F_t = \begin{cases} E\varepsilon & 0 \le \varepsilon \le \varepsilon_{t0} \\ \sigma_{t0} + (\sigma_{tp} - \sigma_{t0})\frac{(\varepsilon-\varepsilon_{t0})}{(\varepsilon_{tp}-\varepsilon_{t0})} & \varepsilon_{t0} \le \varepsilon < \varepsilon_{tp} \\ \sigma_{tp}\frac{(\varepsilon_{tu}-\varepsilon)}{(\varepsilon_{tu}-\varepsilon_{tp})} & \varepsilon_{tp} \le \varepsilon < \varepsilon_{tu} \\ 0 & \varepsilon_{tu} \le \varepsilon \end{cases} \quad (1)$$

where E is Young's modulus. The first microcrack will occur if the strain ε is larger than the cracking strain ε_{t0} at point *A*. The loading path further follows the segment *A-F* until the strain reaches the peak tensile strain ε_{tp} at point *F*, where the stress simultaneously reaches the peak tensile stress σ_{tp}. However, a soft stiffness (i.e., segment *F-I*) will occur if the tensile strain exceeds ε_{tp} but is smaller than the ultimate tensile strain ε_{tu} at point *I*, where the corresponding stress becomes zero. It is worth noting that the loading path will continuously move forward along the positive abscissa axis if the strain is larger than ε_{tu}.

The envelope curve consists of multilinear curves in the compressive region (i.e., segments *O-J-P*) which can be expressed as

$$F_c = \begin{cases} E\varepsilon & \varepsilon_{tp} \le \varepsilon < 0 \\ \sigma_{cp}\frac{(\varepsilon_{cu}-\varepsilon)}{(\varepsilon_{cu}-\varepsilon_{cp})} & \varepsilon_{cp} \le \varepsilon < \varepsilon_{cu} \\ 0 & \varepsilon_{cu} \le \varepsilon \end{cases} \quad (2)$$

where σ_{cp} and ε_{cp} are the peak compressive stress and corresponding strain, respectively. ε_{cu} is the ultimate compressive strain. It is noted that loading path will further move forward along the negative abscissa axis if the strain is smaller than ε_{cu}.

The loading, the unloading and the reloading rules of the ECC model in the tension region (see Figure 4a) are given by

$$F_t = \begin{cases} E\varepsilon & 0 \le \varepsilon_{tm} < \varepsilon_{t0} & (3) \\ \sigma'_{tl}\left(\frac{\varepsilon-\varepsilon_{tl}}{\varepsilon'_{tm}-\varepsilon_{tl}}\right)^{\alpha_t} & \varepsilon_{t0} \le \varepsilon_{tm} < \varepsilon_{tp}, \dot{\varepsilon} < 0 & (4) \\ \sigma'_{tl} + (\sigma_{tm} - \sigma'_{tl})\left(\frac{\varepsilon-\varepsilon'_{tl}}{\varepsilon'_{tm}-\varepsilon'_{tl}}\right) & \varepsilon_{t0} \le \varepsilon_{tm} < \varepsilon_{tp}, \dot{\varepsilon} \ge 0 & (5) \\ \sigma_{tm}\left(\frac{\varepsilon-\varepsilon_{tl}}{\varepsilon_{tm}-\varepsilon_{tl}}\right) & \varepsilon_{tp} \le \varepsilon_{tm} < \varepsilon_{tu} & (6) \end{cases}$$

where α_t is a constant that is larger than or equal to 1. It can be calibrated using the experimental data. The segment *B-C-E* is the initial unloading path expressed by the Equation (4). ε_{tm} is the maximum strain reached in history on the envelop curve where the unloading is triggered. ε_{tl} is the strain corresponding to the stress vanishing on the initial unloading path (i.e., the strain associated with the point *E* or *H*). The value $\varepsilon_{tl} = \beta_t \cdot \varepsilon_{tm}$, where β_t is a constant. Segment *C-D* is a typically partial reloading path, where the stress at point C is not zero, expressed by Equation (5), which ensures the extension of the segment *C-D* passing through the historically reached maximum strain point on the envelop curve (i.e., point *B*). The ε_{tr} and ε_{tu} are strains at points C and *D*, respectively. The unloading path starting from *D* is controlled by Equation (4). The subscript on the parameter ε'_{tm} denotes that ε'_{tm} should be set as ε_{tm} or ε_{tr} when they are used to define the initial unloading path or partially reloading path, respectively. The same specification applies to the parameters ε'_{tl} and ε'_{tu}.

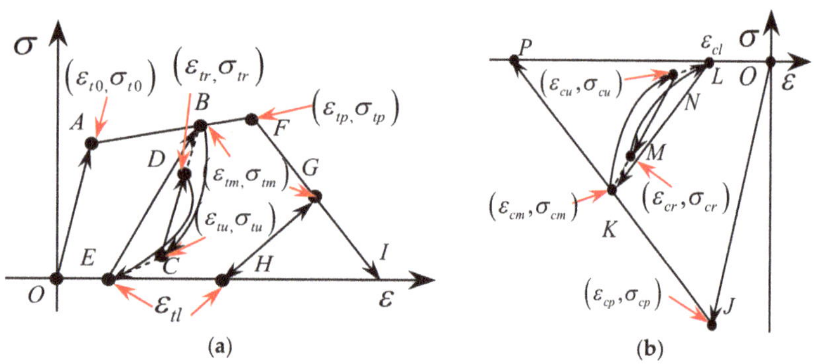

Figure 4. Unloading and reloading rules of the ECC model (**a**) Tension region; (**b**) Compression region [46].

The unloading and reloading paths after point *F* (i.e., *F-G*) are a linear curve given by Equation (6). The loading, the unloading and the reloading paths of the ECC model in the compression region (see Figure 4b) are expressed as:

$$F_c = \begin{cases} E\varepsilon & \varepsilon_{cp} \le \varepsilon_{cm} < 0 & (7) \\ \sigma'_{cm}\left(\frac{\varepsilon-\varepsilon_{cl}}{\varepsilon'_{cm}-\varepsilon_{cl}}\right)^{\alpha_c} & \varepsilon_{cu} \le \varepsilon_{cm} < \varepsilon_{cp}, \dot{\varepsilon} > 0 & (8) \\ \sigma'_{cu} + (\sigma_{cm} - \sigma'_{cm})\left(\frac{\varepsilon-\varepsilon'_{cl}}{\varepsilon'_{cm}-\varepsilon'_{cl}}\right) & \varepsilon_{cu} \le \varepsilon_{cm} < \varepsilon_{cp}, \dot{\varepsilon} \le 0 & (9) \end{cases}$$

The segment *K-N*, as presented in Figure 4b, is formulated by Equation (8). ε_{cm} is the strain on the envelope curve where unloading is triggered (i.e., the strain at point *K*) and the ε_{cl} is the strain on the initial unloading path corresponding to zero stress (i.e., the strain at point *L*). The value $\varepsilon_{cl} = \beta_c \cdot \varepsilon_{cm}$, where β_c is a constant. Segment *N-M* is formulated by Equation (9). ε_{cr} and ε_{cu} are strains of the points *M* and *N*, respectively. The unloading path from *N* is given by Equation (8). The parameter ε'_{cm} should be set to ε_{cm} or ε_{cr} when they are employed to define the initial unloading path or partially reloading path.

3.2. ECC-Reinforced Column and Numerical Verification

A 1/5 scaled ECC-reinforced column [47,48], as shown in Figure 5a, is used to verify the effectiveness of the ECC constitutive model proposed in Section 3.1. To provide lateral load on the top of the cantilever column, a rigid ECC transverse beam is monolithically casted with the cantilever base. This loading configuration is chosen to promote a flexural deformation mode in the specimen and to investigate the effect of ECC material properties on the expected plastic hinge region in particular. Longitudinal steel reinforcement was bent at a 90° angle at the bottom of the transverse beam and further extended to provide sufficient anchorage. A total of four reinforcing steels with a diameter of 10 mm are arranged at the four corners of the cross section of the pier, resulting in a reinforcement ratio of 3.14%. The compressive strength of the ECC material is 80.0 MPa at a strain of 1.2%. The tensile strength is 6.0 MPa at a strain of 6.0%. The Young's modulus of the ECC is 16,000.0 MPa. The Passion's ratio of the ECC material is 0.15. The yielding strength of the reinforcing steel is 410.0 MPa at a strain of 0.02% and the ultimate uniaxial strength is 640.0 MPa at a strain of 14.0%. The lateral load protocol for the quasi-static test is shown in Figure 5b.

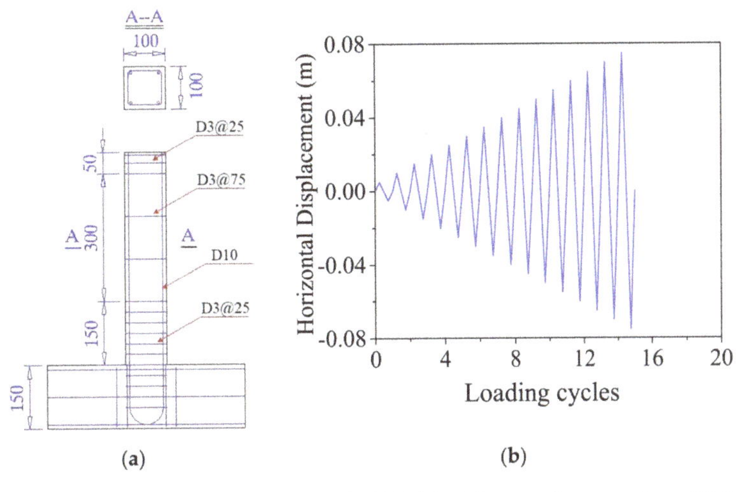

Figure 5. Specimen and loading protocol. (**a**) ECC-reinforced column specimen (Unit: mm); (**b**) Loading protocol for quasi-static test.

The FE model of the ECC-reinforced column is established in OpenSees (Version 2.4.1). A total of 5 displacement-based beam-column elements are used to model the ECC column. The *ECC02 material model* that has been developed and implemented into OpenSees is employed to capture the response of the column. The analysis results regarding the horizontal force against drift ratio together with the test results are displayed in Figure 6. It indicates that the numerical simulation results agree well with the test results. It confirms that the ECC constitutive model proposed in this study is sufficiently accurate for further dynamic analysis.

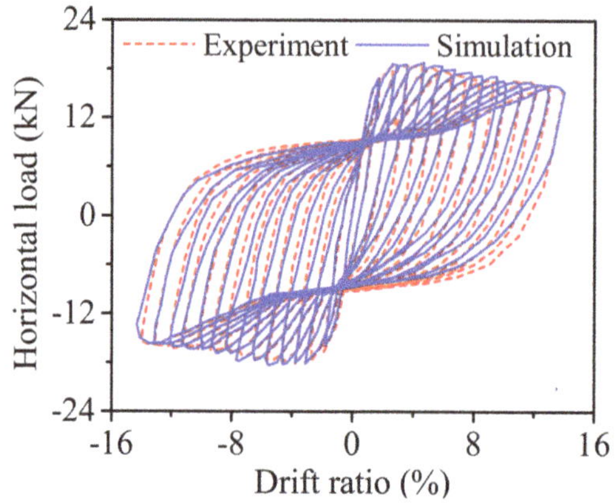

Figure 6. Numerical simulation results against test results.

3.3. Constitutive Models of SMA Washer

The chemical composition of the SMA washer is 55.87% nickel and 44.13% titanium alloys by weight and it is supposed to exhibit super elasticity at room temperature (austenite finish temperature is 4.5 °C). The constitutive model of the SMA washer has been developed in [40], which has a flag-shaped envelop curves, including loading stage (i.e., O-A-B-E) and unloading stage (i.e., E-B-C-E-O), as shown in Figure 7, where F_2 and δ_2 represent the "yield" force and the corresponding deformation, respectively; F_3 and δ_3 are the force and deformation when the SMA washer is fully flattened. Similarly, F_4, δ_4, F_1, and δ_1 represent the characteristic forces and deformations during the reverse plateau. Once the SMA washer reaches its maximum compressive deformation, the axial stiffness increases abruptly (i.e., BE).

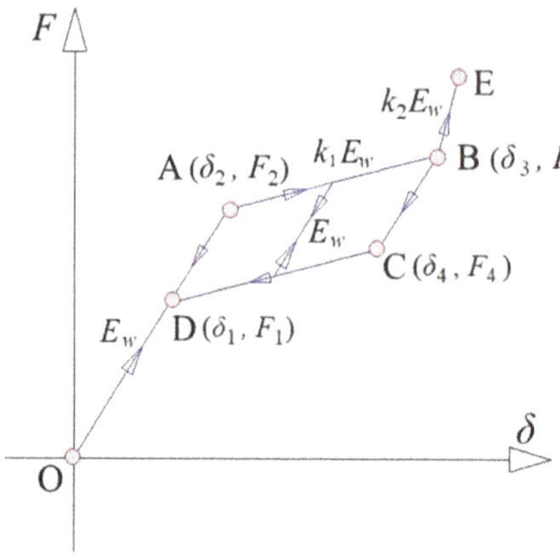

Figure 7. Idealized constitutive model for SMA washer.

The loading stag on the envelop curve is composed of three segments, which are expressed as:

$$F = \begin{cases} E_w\delta & 0 \leq \delta < \delta_2 \\ F_2 + k_1E_w(\delta - \delta_2) & \delta_2 \leq \delta < \varepsilon_3 \\ F_3 + k_2E_w(\delta - \delta_3) & \varepsilon_3 \leq \delta \end{cases} \quad (10)$$

where E_w is the elastic stiffness of the single SMA washer; k_1 and k_2 are constants which can be determined from test.

The unloading stage on the envelop curve consists of four segments, which are given as follows:

$$F = \begin{cases} E_w\delta & 0 \le \delta < \delta_1 \\ F_4 - k_1 E_w(\delta_4 - \delta) & \delta_1 \le \delta < \delta_4 \\ F_3 - E_w(\delta_3 - \delta) & \delta_4 \le \delta < \delta_3 \\ F_3 + k_2 E_w(\delta - \delta_3) & \delta \ge \delta_3 \end{cases} \tag{11}$$

The uploading and reloading stiffness away from the envelop curve is E_w.

The key parameters for a single SMA washer are shown in Figure 7.

The diagram of a typical SMA washer is presented in Figure 8, where the outer and inner diameters of the SMA washer are D and d, respectively. The thickness and the height of the SMA washer are h and t, respectively. These parameters of a SMA washer can be appropriately adjusted according to the seismic design objectives of rocking bridges. Once all the parameters are designated, the maximum compressive deformation and the restoring force provided by a SMA washer can be calculated.

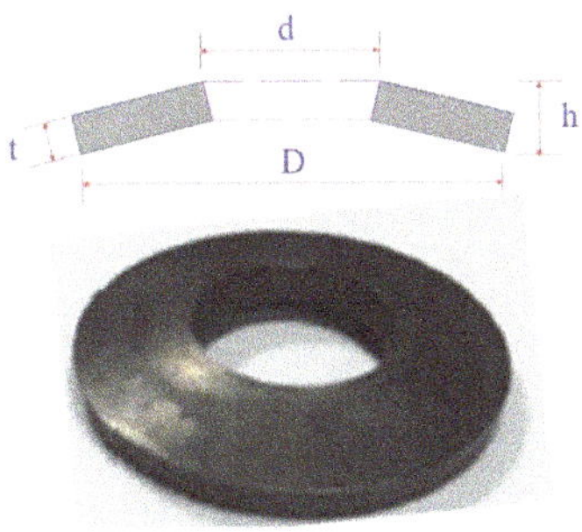

Figure 8. Configuration of a typical SMA washer.

4. Validation of the Numerical Simulation Method for Capturing Rocking Behavior

A 1/4 scale RC rocking pier specimen equipped with SMA washer sets [29] is employed to validate the numerical simulation method used in this study for capturing the overall rocking behavior. The configuration of the rocking pier is shown in Figure 9. The test specimen includes one bent cap, the main body of a pier, two separated pile caps, four SMA washer sets, and a tendon. The diameter of the RC pier is 0.3 m. The strength of the normal concrete was 39.0 MPa at the test day. A total of 12 longitudinal HRB400 rebars with a diameter of 16 mm were uniformly arranged along the perimeter, resulting in a reinforcement ratio of 0.3%. The average yielding strength of the longitudinal rebar was 400.4 MPa. The diameter of the transverse reinforcement was 10 mm and the space between two adjacent stirrups was 75 mm. A total of four SMA washer sets were installed on the top of the upper pile cap and each SMA washer set composed of the loading protocol is shown in Figure 10. Other information regarding this rocking pier can be find elsewhere [29].

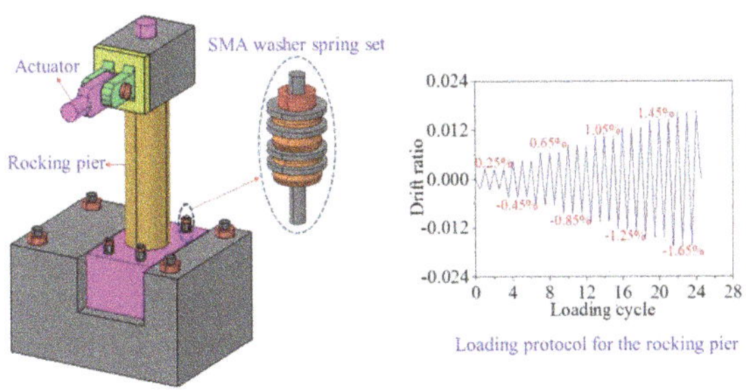

Figure 9. Configuration of a rocking pier specimen and loading protocol.

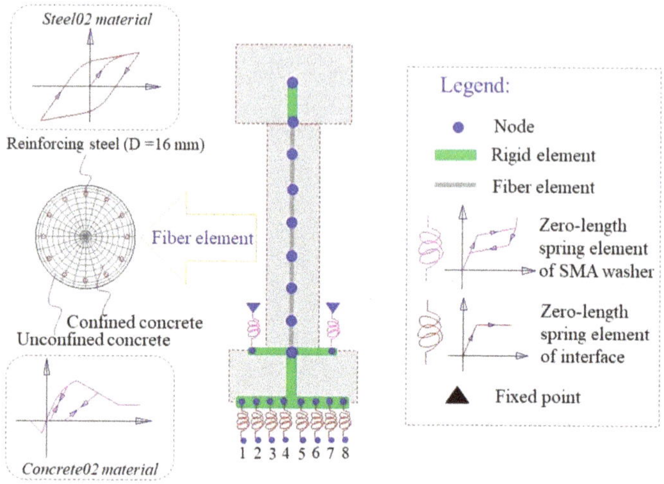

Figure 10. FE model of the rocking pier specimen.

To validate the numerical simulation method proposed in this study for capturing the overall rocking behavior, the experimental results of the rocking pier specimen are used to make comparison with the numerical analysis results. The RC pier was divided into seven displacement-based fiber beam-column elements, of which the normal concrete fiber was assigned with *Concrete02 material model* and the reinforcement fiber was assigned with *Steel02 material model*. The connection between the pier and the bent cap or the pile cap is modeled by rigid elements. The interface between upper and lower pile caps was simulated by eight zero-length spring elements, of which the tension strength is ignored. Each SMA washer set was modeled by a zero-length spring element, of which the assigned material was the *Self-centering material model*. The FE model is shown in Figure 10. It is worth noting that the nonlinear material properties of the normal concrete and the reinforcement can be considered by the *Concrete02 material model* and *Steel02 material model*, respectively. The geometric nonlinearity of the RC pier can also be accounted for by the co-rotational geometric transformations [39]. The numerical simulation results and the test ones are presented in Figure 11, which shows that the analysis results match well with the test regarding the amplitude of the lateral load and the trend of the hysteretic curve. It confirms that the numerical simulation method proposed in this study can be used to capture the overall rocking behavior of the rocking bridge system.

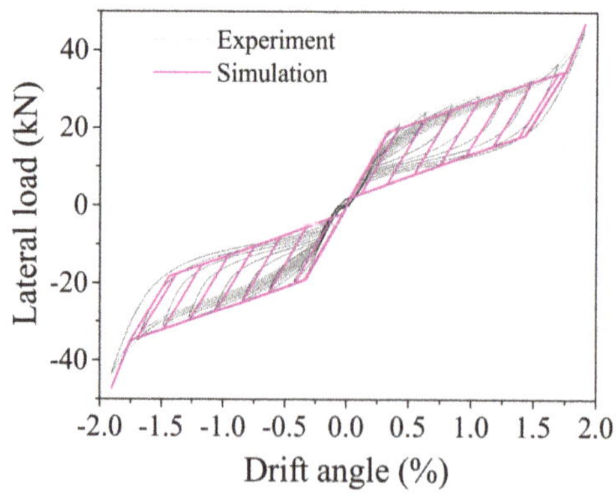

Figure 11. Comparison between numerical simulation and test results.

5. Seismic Responses of the Resilient Bridges with SMA-Based Rocking Piers

In order to demonstrate the superiorities of the proposed bridge system in resilience enhancement, four different bridge systems are studied. The first bridge system is a conventional bridge with a RC fixed pier. The second one is a conventional bridge with an ECC-reinforced fixed pier. The third one is a resilient bridge with a RC rocking pier with SMA washers. The last one is a resilient bridge with an ECC-reinforced rocking pier with SMA washers.

5.1. FE Model of Resilient Bridge Systems

A comprehensive FE model of the resilient continuous RC box girder bridge is established in OpenSees, as demonstrated in Figure 12. The RC girder (i.e., 20 + 20 m) of the bridge is modeled by 35 elastic beam-column elements. The rocking RC or ECC-reinforced pier is 8.0 m high which is modeled by six displacement-based beam-column fiber elements. The *Concrete02 material model* is used to model the properties of the normal concrete, of which the tension strength is considered. The *ECC02 material model* is employed to simulate the ECC material. The *Steel02 material model* is used for modeling the reinforcing steel. The heights of the two separated pile caps are both 2.0 m, which are modeled using rigid elements. The width of each expansion joint is 0.25 m, which is modeled by a zero-length element. Each bent cap and abutment are modeled by two rigid beam elements along the transverse direction of the bridge. Similarly, the transverse beams in the girder are simulated by two rigid beam elements. Two conventional sliding bearings are placed on each abutment and two fixed bearings are installed on the bent cap above the middle pier. The frictional coefficient of the sliding bearing is 0.02. Each bearing is simulated by a zero-length element connecting the girder and the bent caps or the abutments. The soil and structural interaction of each pile foundation has been considered by six zero-length elements [49,50]. The compressive strength of the conventional concrete for the box girder is 50.0 MPa. The maximum compressive strength and the crush strain of the ECC material in the pier and the rocking pile caps are 80.0 MPa and 1.2%, respectively. The maximum compressive strength and the crush strain of the conventional concrete material in the pier and the rocking pile caps are 40.0 MPa and 1.0%, respectively. The diameters of the longitudinal reinforcement and stirrup used in the pier are 32 mm and 16 mm, of which the corresponding yielding strengths are 440.0 MPa and 300.0 MPa, respectively. A total of 72 reinforcing steels are uniformly arranged around the perimeter of the ECC-reinforced pier, resulting in a reinforcement ratio of 2.28%. The net thickness of the cover concrete or ECC material is 0.05 m. The stirrup interval is 0.1 m at the plastic hinge region and 0.15 m at elsewhere. The SMA washer set is modeled by a compression-only element with a *self-centering material model* and *an Elastic-perfectly plastic material model* in parallel. The *self-centering material model* is used to model the superelasticity of the SMA washer set and the *elastic-perfectly plastic material model* is employed to simulate the precompression by designated an appropriate initial deformation. For instance, the maximum compressive deformation of each SMA washer set is 0.06 m, half of which is consumed to impose prestressed force on the rocking pier.

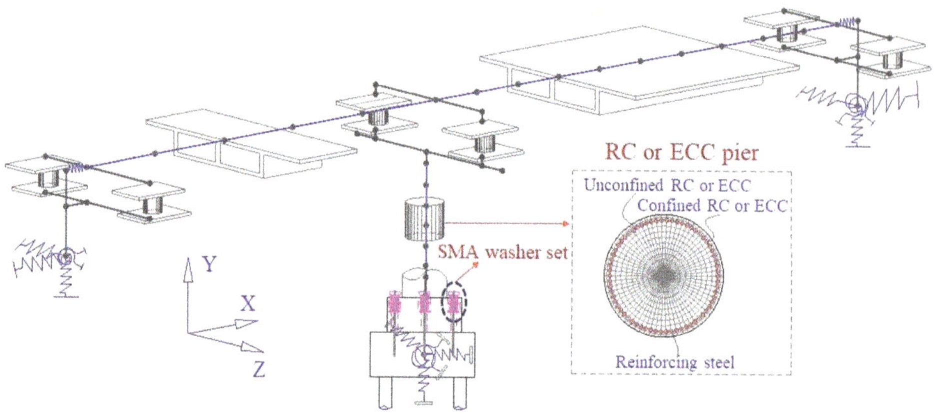

Figure 12. FE model of the resilient bridge with SMA-based rocking pier.

Two additional conventional bridge systems and a resilient bridge system with a RC rocking pier are also considered to demonstrate the super resilience of the innovative bridge with an ECC-reinforced rocking pier. The main difference between the conventional bridge and the resilient bridge is that the 10 m high fixed pier in the conventional bridge is replaced by an 8 m high pier and a 2 m rocking upper pile cap in height. The FE models of the other three bridge systems can be easily adjusted from that shown in Figure 12 and are therefore not elaborated.

5.2. Earthquake Ground Motions

A suite of ground motions being compatible to the acceleration spectra (i.e., E1 and E2 levels) [51,52] are generated and each suite includes seven earthquake ground motions, as shown in Figure 13a,b, respectively. The exceedance probabilities of E1 and E2 level earthquakes in a recurrence interval of 50 years are 10% and 2.5%, respectively. The damping ratio of the acceleration spectra is 5%.

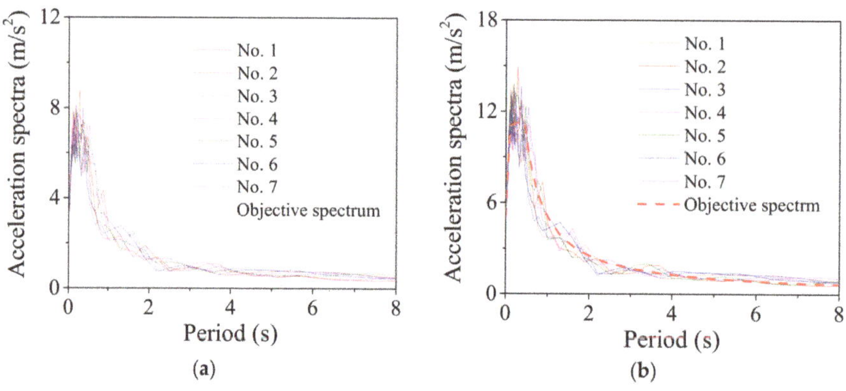

Figure 13. Earthquake spectra together with earthquake motions. (**a**) E1 level; (**b**) E2 level.

5.3. Comparison of Seismic Responses between the Conventional and Resilient Bridges

To evaluate the damage state of the RC and ECC-reinforced pier after earthquake excitation, sectional characteristic analyses are conducted prior to analysis of bridge system. The equivalent bending moments versus curvatures of the sections are shown in

Figure 14, where the equivalent bending moments of the RC and ECC-reinforced sections are 19,230.0 kNm and 28,023.0 kNm, respectively, and the associated curvatures are 0.003 1/m and 0.0044 1/m, respectively. The figure indicates that the elastic stiffness and the yield strength of the ECC-reinforced section are both much larger than those of the RC section.

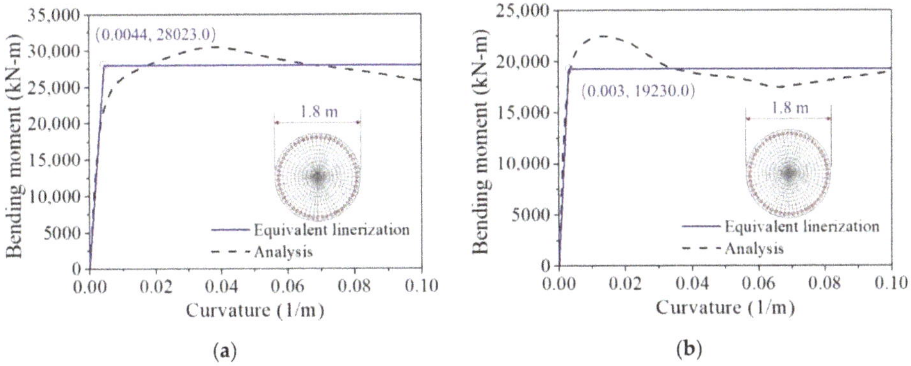

Figure 14. Sectional analyses of bending moment vs. curvature. (**a**) RC; (**b**) ECC.

The curvature ductility and the drift ratio of the pier as well as the shear deformation of the bearing are usually selected as the damage indicators for seismic performance assessment of bridges with rocking piers. Therefore, these indicators are used to make a comparison between the conventional and the resilient bridge systems under the considered earthquakes. The curvature ductility indicates the damage state of the plastic hinge region of the pier during earthquakes. Drift ratio is employed to assess the lateral deformation of the bridge system. The uplift ratio indicates the rocking amplitude of the rocking pier, which is expressed as $c/(a + b)$ shown in Figure 15, where c is the maximum uplift distance of the pier; a is the distance between the outmost edge of the upper pile cap and the left side of the stress block; b is the width of the compressive stress block (i.e., rocking zone). The maximum seismic responses regarding the aforementioned damage indicators of the conventional (fixed base) and the resilient (rocking), bridges with the normal RC piers under E1 and E2 earthquakes are summarized in Tables 1 and 2, respectively.

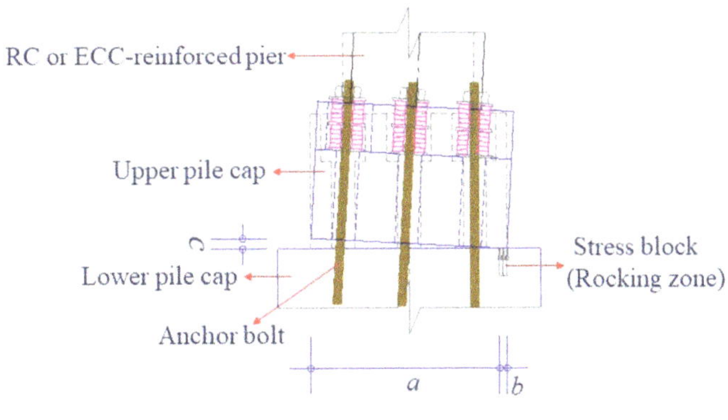

Figure 15. Schematic diagram of the uplift ratio.

Table 1. Maximum seismic responses of RC bridges at E1 level.

Earthquake	Curvature Ductility		Bearing Deformation (cm)		Drift Ratio (Residual Drift Ratio) (%)		Uplift Ratio (%)
No.	Conventional	Resilient	Conventional	Resilient	Conventional	Resilient	Resilient
1	1.15	0.76	11.99	7.34	1.01 (0.005)	0.62 (0.008)	0.07
2	1.15	0.92	11.91	8.70	1.00 (0.029)	0.72 (0.008)	0.08
3	0.80	0.82	9.30	7.68	0.79 (0.002)	0.65 (0.001)	0.07
4	0.89	1.17	10.20	11.75	0.86 (0.011)	0.99 (0.013)	0.30
5	0.74	0.70	8.77	6.95	0.74 (0.014)	0.57 (0.009)	0.06
6	0.83	1.09	9.86	9.80	0.81 (0.027)	0.83 (0.006)	0.17
7	0.99	0.68	10.81	6.54	0.91 (0.002)	0.56 (0.008)	0.05
Avg. value	0.94	0.88	10.41	8.40	0.88 (0.013)	0.70 (0.008)	0.11

Table 2. Maximum seismic responses of RC bridges at E2 level.

Earthquake	Curvature Ductility		Bearing Deformation (cm)		Drift Ratio (Residual Drift Ratio) (%)		Uplift Ratio (%)
No.	Conventional	Resilient	Conventional	Resilient	Conventional	Resilient	Resilient
1	3.24	1.47	20.33	19.43	1.71 (0.039)	1.63 (0.009)	1.01
2	2.70	1.73	18.38	20.77	1.54 (0.022)	1.74 (0.002)	1.12
3	3.00	1.47	18.96	16.42	1.59 (0.057)	1.38 (0.020)	0.70
4	3.23	1.56	19.02	17.58	1.58 (0.054)	1.47 (0.009)	0.86
5	2.95	1.31	18.37	16.17	1.54 (0.060)	1.36 (0.005)	0.73
6	2.68	1.47	18.97	16.68	1.57 (0.046)	1.39 (0.011)	0.72
7	3.15	1.49	19.52	17.89	1.64 (0.040)	1.50 (0.001)	0.86
Avg. value	2.99	1.50	19.08	17.85	1.60 (0.045)	1.49 (0.008)	0.86

The results summarized in Table 1 reveal that the average maximum values of the curvature ductility of the pier, bearing deformation, the drift ratio, and the residual drift ratio of the conventional RC bridge are all larger than those of the resilient RC bridge under E1 level earthquakes. The curvature ductility responses of the RC piers of the conventional bridge and the resilient bridge are both less than 1.0, which means that the RC piers in two bridge systems keep linear state under E1 level earthquakes. The drift ratio of the resilient bridge is 0.88%, which satisfied the principle of the seismic design objective. All the seismic responses confirm that the two bridge systems are both safe under E1 level earthquakes. The average maximum uplift ratio is 0.11%. Even though the earthquake intensity (i.e., E1 level) is not large, the unique property, such as rocking, of the resilient bridge is well exhibited.

When the earthquake intensity increases from E1 to E2 level, the average maximum curvature ductility of the conventional RC bridge sharply increases from 0.94 to 2.99. It implies that the RC pier experiences severe damage when the bridge is subjected to E2 level earthquakes although it performs linearly under E1 level earthquakes. A similar trend also can be found from the seismic performance of the resilient RC bridge system even though the increasing amplitude is not so large compared with the conventional bridge. To gain a thorough understanding about the rocking mechanism of the pier, the typical shear force against drift ratio of the bridge, and the bending moment versus curvature ductility of the section at the bottom of the RC pier under a typical ground motion (i.e., Earthquake No. 1 at E2 level) are shown in Figures 16 and 17, respectively. From these two figures, it can be concluded that the drift ratio of the conventional RC bridge (i.e., 1.71) is slightly larger than that of the resilient RC bridge (i.e., 1.63), but the maximum curvature ductility of the RC pier and the residual drift ratio of the conventional bridge (i.e., 3.24 and 0.039%) are much larger than those of the resilient bridge (i.e., 1.47 and 0.009%). It confirms that the resilient rocking bridge system can significantly alleviate the seismic damage to the RC pier because the dominate period of the rocking bridge is elongated. The seismic response of a SMA washer set is displayed in Figure 18, where the compressive deformation is 0.021 m that is less the uplift threshold of the rocking pier (i.e., 0.03 m). The hysteretic loop response as shown in Figure 18 demonstrates that the SMA washer sets dissipates a lot of energy during earthquakes.

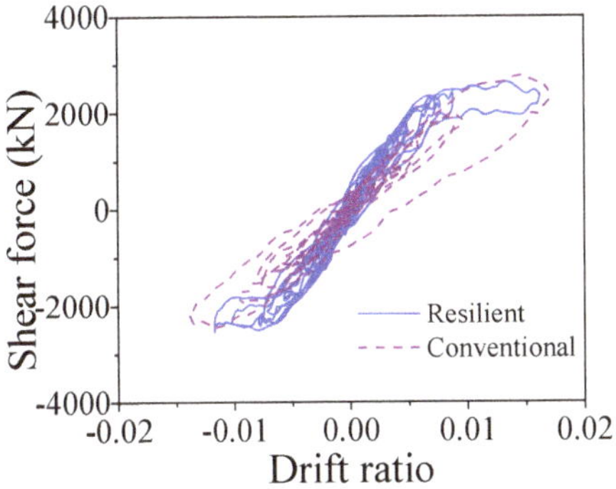

Figure 16. Shear force vs. drift ratio.

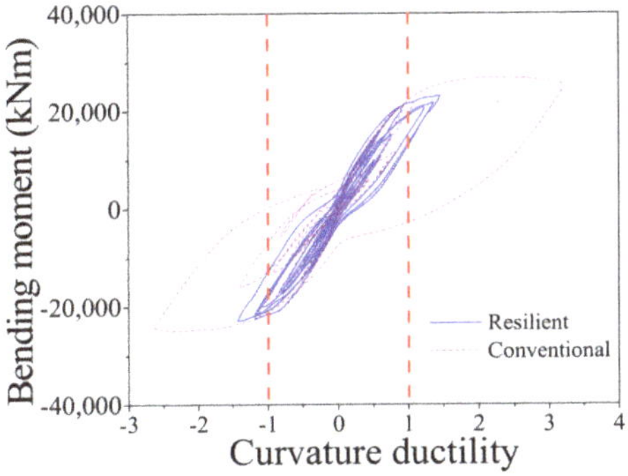

Figure 17. Bending moment vs. curvature ductility.

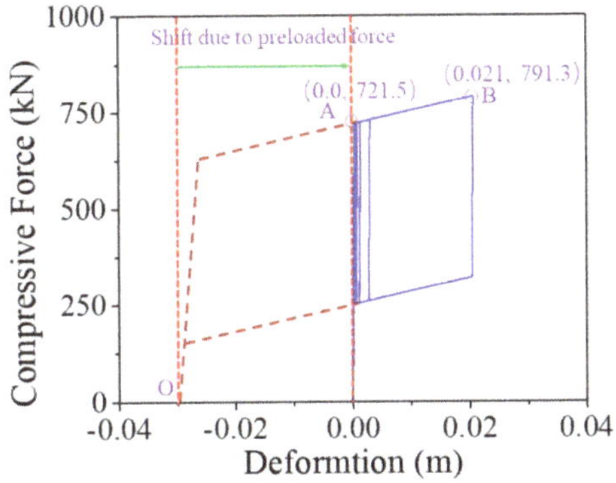

Figure 18. Compressive force vs. deformation of SMA washer set.

The maximum and average maximum seismic responses regarding the aforementioned damage indicators of the conventional and the resilient bridges with the ECC-reinforced piers under E1 and E2 level earthquakes are summarized in Tables 3 and 4, respectively.

Table 3. Maximum seismic responses of ECC-reinforced bridges at E1 level.

Earthquake	Curvature Ductility		Bearing Deformation (cm)		Drift Ratio (Residual Drift Ratio) (%)		Uplift Ratio (%)
No.	Conventional	Resilient	Conventional	Resilient	Conventional	Resilient	Resilient
1	0.67	0.76	11.93	10.23	0.98 (0.013)	0.87 (0.014)	0.08
2	0.62	0.93	11.17	15.03	0.94 (0.026)	1.26 (0.005)	0.49
3	0.68	0.57	11.30	8.35	1.00 (0.035)	0.69 (0.016)	0.06
4	0.65	0.87	11.74	13.63	0.97 (0.021)	1.10 (0.017)	0.32
5	0.63	0.62	11.13	8.66	0.94 (0.011)	0.73 (0.011)	0.06
6	0.71	0.72	12.60	10.06	1.04 (0.035)	0.83 (0.018)	0.08
7	0.61	0.74	10.91	10.18	0.92 (0.003)	0.86 (0.014)	0.08
Avg. value	0.65	0.74	11.54	10.88	0.97 (0.021)	0.91 (0.014)	0.17

Table 4. Maximum seismic responses of ECC-reinforced bridges at E2 level.

Earthquake	Curvature Ductility		Bearing Deformation (cm)		Drift ratio (Residual Drift Ratio) (%)		Uplift Ratio (%)
No.	Conventional	Resilient	Conventional	Resilient	Conventional	Resilient	Resilient
1	1.84	1.07	22.17	24.66	1.84 (0.137)	2.07 (0.007)	1.48
2	1.66	0.99	21.26	20.89	1.78 (0.040)	1.75 (0.006)	1.08
3	1.88	0.93	22.15	19.31	1.86 (0.065)	1.62 (0.006)	0.93
4	2.21	0.98	24.30	18.45	2.02 (0.080)	1.32 (0.029)	0.83
5	1.51	0.99	19.90	19.96	1.67 (0.131)	1.67 (0.011)	0.98
6	2.04	0.92	23.24	18.54	1.93 (0.022)	1.54 (0.011)	0.84
7	1.47	0.98	19.69	23.32	1.66 (0.037)	1.95 (0.009)	1.37
Avg. value	1.80	0.98	21.81	20.73	1.82 (0.073)	1.70 (0.011)	1.07

The results presented in Table 3 show that all the average maximum responses of the resilient bridge except for the curvature ductility of the ECC-reinforced pier are smaller than those of the conventional bridge under E1 earthquakes. Even though the maximum average curvature ductility of the resilient bridge is larger than that of the conventional one, it is still less than 1.0. In other words, the ECC-reinforced pier in the resilient bridge always experiences elastic state under E1 earthquakes.

When the earthquake intensity increased from E1 to E2 level, all the average maximum responses sharply enhanced except the curvature ductility of the ECC-reinforced pier of the resilient bridge increases slightly (i.e., from 0.74 to 0.98). The average maximum responses of the bearing deformation of the conventional and the resilient ECC-reinforced bridges are similar under E2 level earthquakes, but the average maximum curvature ductility of the ECC-reinforced pier and the residual drift ratio of the resilient bridge (i.e., 0.98 and 0.011%) are much smaller than those of the conventional bridge (i.e., 1.80 and 0.073%). The ECC-reinforced pier in the resilient bridge system can always keep in linear performance under E2 earthquakes but it may yield in the conventional bridge system under the same level earthquakes. The average maximum drift ratio of the conventional bridge is larger in comparison with the resilient bridge because the ECC-reinforced pier in the conventional bridge yields but the ECC-reinforced pier in resilient bridge keeps a linear behavior. To take a close look at the seismic response, the typical time history responses of the lateral seismic force versus drift ratio of the ECC-reinforced conventional and the ECC-reinforced resilient bridges under a typical earthquake (i.e., Earthquake No.1 at E2 level) are illustrated in Figure 19. It can be confirmed that although the two bridges experienced similar lateral displacement, the rocking effect contributes to the most part of the total lateral displacement but the fixed pier mainly depended on the flexural deformation of the pier itself. The seismic responses of the section at the bottom of the plastic hinge region in two bridge systems are presented in Figure 20, which implies that the fixed base ECC-reinforced pier suffered severe damage but the rocking pier stayed elastic. The seismic response of one SMA washer set is displayed in Figure 21, where the self-locking effect is triggered when the compressive deformation reached 0.03 m. Once the maximum drift ratio of the resilient bridge reaches 2.0%, the self-locking effect will act and the ECC-reinforced pier will yield simultaneously. In other words, if the drift ratio of the resilient bridge exceeds 2.0%, the incremental lateral displacement will completely depend on the yielding deformation of the ECC-reinforced pier.

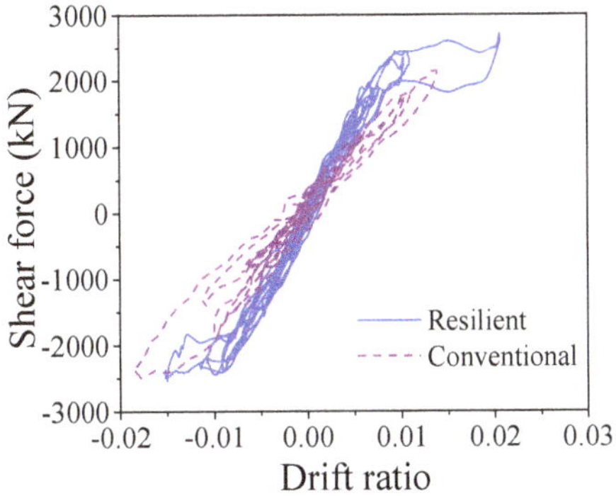

Figure 19. Shear force vs. drift ratio.

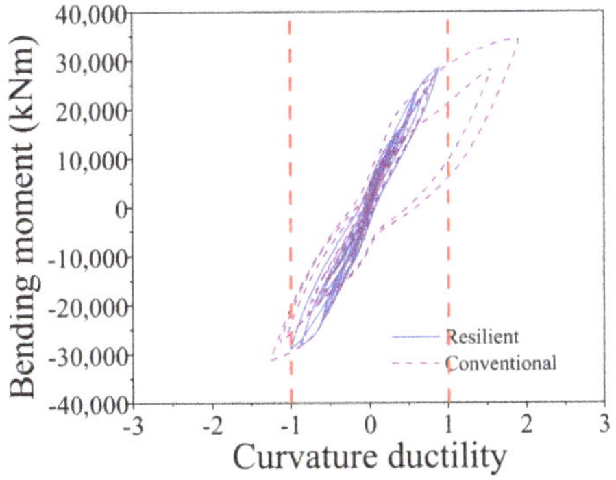

Figure 20. Bending moment vs. curvature ductility.

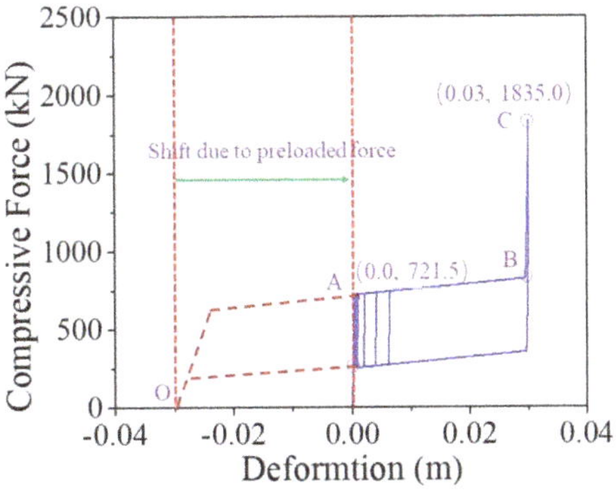

Figure 21. Compressive force vs. deformation of SMA washer set.

5.4. Comparison of Seismic Responses between the RC and ECC-Reinforced Resilient Bridges

The aforementioned comparisons demonstrates that the ECC-reinforced bridge systems are more resilient than the conventional bridge systems. The following part will make a comparison between the two resilient bridge systems with the RC rocking pier and the ECC-reinforced rocking pier. The average maximum drift ratio of the resilient bridge with RC rocking bridge under E2 level earthquakes is 1.49 but the value of the ECC-reinforced resilient rocking bridge is 1.70. The reason is that the yielding strength of the RC pier is smaller than the ECC-reinforced pier so that it cannot sustain large rocking amplitude. It can reconfirm by the response of the average maximum curvature ductility of two bridges. For instance, the average maximum curvature ductility of the resilient bridge with RC rocking bridge under E2 level earthquake is 1.50, whereas the value of the ECC-reinforced resilient rocking bridge is 0.98. A case is selected for further investigation: the drift ratio versus lateral seismic force responses of two resilient bridges subjected to a typical earthquake (i.e., Earthquake No. 1 at E2 level) is shown in Figure 22. From Figure 23, it can be recognized that the maximum drift ratio of the resilient bridge with the ECC-reinforced pier is 2.07, but the corresponding value is only 1.63. The curvature ductility versus bending moment responses of two resilient bridges under a typical earthquake (i.e., Earthquake No. 1 at E2 level) is shown in Figure 23. The maximum curvature ductility of the RC pier is 1.47 but the counterpart of the ECC-reinforced pier is 1.07. The earthquake-induced damage of the ECC-reinforced pier is so small (i.e., 1.07), which needs no repair but the RC pier should be retrofitted after earthquake.

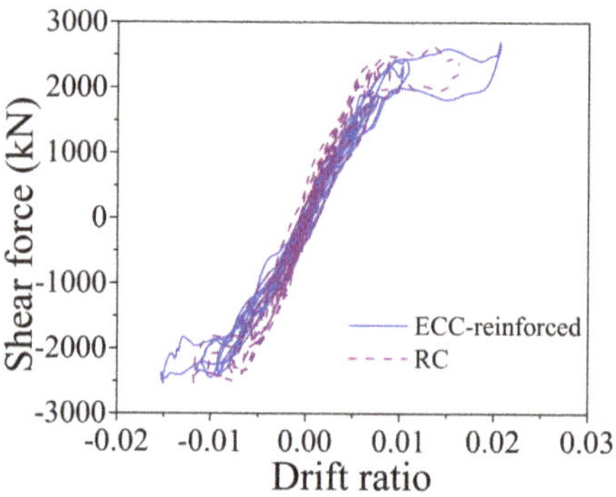

Figure 22. Shear force vs. drift ratio.

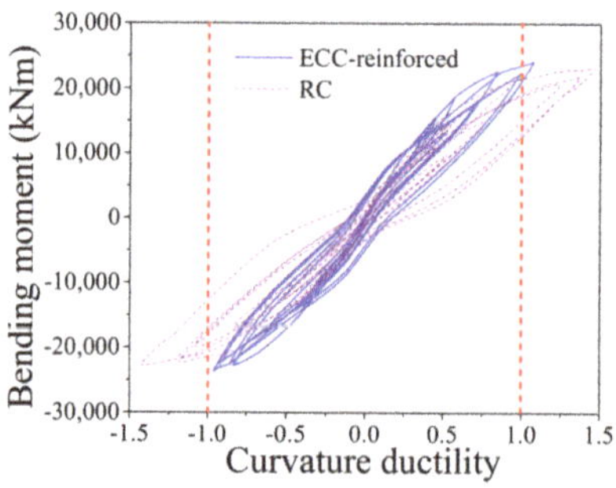

Figure 23. Bending moment vs. curvature ductility.

6. Conclusions

This study presented a resilient self-centering bridge system with the ECC-reinforced rocking pier. An experimental study on the ECC-reinforced column was employed to demonstrate the accuracy of the proposed constitutive model of the ECC material implemented in OpenSees. The advantages of the resilient ECC-reinforced rocking bridge system, such as low damage and negligible residual deformation, were verified by conducting a series of nonlinear seismic analyses on both resilient and conventional bridges with RC or ECC-reinforced piers. Several comments and conclusions are summarized as follows:

1. All the conventional and the resilient rocking bridge systems with the RC pier or the ECC-reinforced pier can satisfy the seismic design objectives under E1 level earthquakes;
2. The proposed resilient rocking bridge system with the RC pier or the ECC-reinforced pier shows superior seismic performance over the conventional bridge in terms of the response such as the curvature ductility of the pier, bearing deformation, drift ratio, and residual drift ratio of the bridge under E2 level earthquakes;
3. The resilient rocking bridge with the ECC-reinforced pier can achieve superior damage retrofitting capacity than the resilient rocking bridge with RC pier. The average maximum curvature ductility of the ECC-reinforced pier was only 0.98 but the counterpart of the RC pier was 1.5 which indicates the RC experienced yielding state during E2 level earthquakes;
4. Due to the protection mechanism (i.e., self-locking effect) of the SMA washer spring device against overload and the super resilient property of the ECC material, the resilient rocking bridge system shows the excellent damage control capacity;
5. The SMA washer spring device cannot only provide restoring force for the resilient bridge system but also can dissipate moderate earthquake energy input.

Author Contributions: Conceptualization, K.C.; Data curation, D.Y.; Formal analysis, X.L.; Investigation J.C., Y.L. and D.Y.; Validation, Y.L.; Writing—original draft, X.L.; Writing—review & editing, K.C. All authors have read and agreed to the published version of the manuscript.

Funding: This research was funded by National Key R&D Program of China, grant number 2018YFC0809606 and the Fundamental Research Funds for the Central Public Research Institute, grant number 2018-9025.

Institutional Review Board Statement: Not applicable.

Informed Consent Statement: Not applicable.

Data Availability Statement: The data presented in this study are available on request from the corresponding author.

Conflicts of Interest: The authors declare no conflict of interest.

References

1. Kawashima, K.; Macrae, G.A.; Hoshikuma, J.-I.; Nagaya, K. Residual Displacement Response Spectrum. *J. Struct. Eng.* **1998**, *124*, 523–530. [CrossRef]
2. Wang, W.; Fang, C.; Shen, D.; Zhang, R.; Ding, J.; Wu, H. Performance assessment of disc spring-based self-centering braces for seismic hazard mitigation. *Eng. Struct.* **2021**, *242*, 112527. [CrossRef]
3. Fang, C.; Wang, W.; Shen, D. Development and Experimental Study of Disc Spring–Based Self-Centering Devices for Seismic Resilience. *J. Struct. Eng.* **2021**, *147*, 04021094. [CrossRef]
4. Wang, W.; Fang, C.; Zhao, Y.; Sause, R.; Hu, S.; Ricles, J. Self-centering friction spring dampers for seismic resilience. *Earthq. Eng. Struct. Dyn.* **2019**, *48*, 1045–1065. [CrossRef]
5. Buckle, I. Overview of seismic design methods for bridges in different countries and future directions. In Proceedings of the Eleventh World Conference on Earthquake Engineering, Acapulco, Mexico, 23–28 June 1996.
6. Priestley, M.J.N.; Tao, J.R. Seismic Response of Precast Prestressed Concrete Frames With Partially Debonded Tendons. *PCI J.* **1993**, *38*, 58–69. [CrossRef]
7. Mander, J.B.; Cheng, C.-T. *Seismic Resistance of Bridge Piers Based on Damage Avoidance Design*; Report No. NCEER-97-0014; University at Buffalo: Buffalo, NY, USA, 1997.
8. Roha, H.; Reinhorn, A.M. Hysteretic behavior of precast segmental bridge piers with superelastic shape memory alloy bars. *Eng. Struct.* **2010**, *32*, 3394–3403. [CrossRef]
9. Roh, H.; Reinhorn, A.M. Analytical modeling of rocking elements. *Eng. Struct.* **2009**, *31*, 1179–1189. [CrossRef]

10. Roh, H.; Reinhorn, A.M. Nonlinear Static Analysis of Structures with Rocking Columns. *J. Struct. Eng.* **2010**, *136*, 532–542. [CrossRef]
11. Cheng, C.-T. Shaking table tests of a self-centering designed bridge substructure. *Eng. Struct.* **2008**, *30*, 3426–3433. [CrossRef]
12. Palermo, A.; Pampanin, S.; Calvi, G.M. Concept and development of hybrid solutions for seismic resistant bridge systems. *J. Earthq. Eng.* **2005**, *9*, 899–921. [CrossRef]
13. Palermo, A.; Pampanin, S.; Marriott, D. Design, Modeling, and Experimental Response of Seismic Resistant Bridge Piers with Posttensioned Dissipating Connections. *J. Struct. Eng.* **2007**, *133*, 1648–1661. [CrossRef]
14. Sakai, J.; Mahin, S.A. Mitigation of residual displacements of circular reinforced concrete bridge columns. In Proceedings of the 13th World Conference on Earthquake Engineering, Vancouver, BC, Canada, 1–6 August 2004.
15. Pollino, M.; Bruneau, M. Seismic Retrofit of Bridge Steel Truss Piers Using a Controlled Rocking Approach. *J. Bridg. Eng.* **2007**, *12*, 600–610. [CrossRef]
16. Marriott, D.; Pampanin, S.; Palermo, A. Quasi-static and pseudo-dynamic testing of unbonded post-tensioned rocking bridge piers with external replaceable dissipators. *Earthq. Eng. Struct. Dyn.* **2009**, *38*, 331–354. [CrossRef]
17. Kam, W.Y.; Pampanin, S.; Palermo, A.; Carr, A.J. Self-centering structural systems with combination of hysteretic and viscous energy dissipations. *Earthq. Eng. Struct. Dyn.* **2010**, *39*, 1083–1108. [CrossRef]
18. White, S.; Palermo, A. Quasi-Static Testing of Posttensioned Nonemulative Column-Footing Connections for Bridge Piers. *J. Bridg. Eng.* **2016**, *21*, 04016025. [CrossRef]
19. MacRae, G.A.; Priestley, M.J.N. *Precast Post-Tensioned Ungrouted Concrete Beam–Column Subassemblage Tests*; Report No. SSRP-94/10; University of California: San Diego, CA, USA, 1994.
20. Fang, C.; Zheng, Y.; Chen, J.; Yam, M.C.; Wang, W. Superelastic NiTi SMA cables: Thermal-mechanical behavior, hysteretic modelling and seismic application. *Eng. Struct.* **2019**, *183*, 533–549. [CrossRef]
21. Billington, S.L.; Yoon, J.K. Cyclic response of unbonded post-tensioned precast concrete columns with ductile fiber-reinforced concrete. *J. Bridge Eng.* **2004**, *9*, 353–363. [CrossRef]
22. Desroches, R.; McCormick, J.; Delemont, M. Cyclic Properties of Superelastic Shape Memory Alloy Wires and Bars. *J. Struct. Eng.* **2004**, *130*, 38–46. [CrossRef]
23. Wilson, J.C.; Wesolowsky, M.J. Shape Memory Alloys for Seismic Response Modification: A State-of-the-Art Review. *Earthq. Spectra* **2005**, *21*, 569–601. [CrossRef]
24. Saiidi, M.S.; O'Brien, M.; Sadrossadat-Zadeh, M. Cyclic Response of Concrete Bridge Columns Using Superelastic Nitinol and Bendable Concrete. *ACI Struct. J.* **2009**, *106*, 69–77. [CrossRef]
25. Youssef, M.A.; Alam, M.S.; Nehdi, M. Experimental Investigation on the Seismic Behavior of Beam-Column Joints Reinforced with Superelastic Shape Memory Alloys. *J. Earthq. Eng.* **2008**, *12*, 1205–1222. [CrossRef]
26. Alam, M.; Youssef, M.; Nehdi, M. Analytical prediction of the seismic behaviour of superelastic shape memory alloy reinforced concrete elements. *Eng. Struct.* **2008**, *30*, 3399–3411. [CrossRef]
27. Zheng, Y.; Dong, Y.; Chen, B.; Ghazanfar, A.A. Seismic damage mitigation of bridges with self-adaptive SMA-cable-based bearings. *Smart Struct. Syst.* **2019**, *24*, 127–139.
28. Zheng, Y.; Dong, Y.; Li, Y. Resilience and life-cycle performance of smart bridges with shape memory alloy (SMA)-cable-based bearings. *Constr. Build. Mater.* **2018**, *158*, 389–400. [CrossRef]
29. Zheng, Y.; Dong, Y. Performance-based assessment of bridges with steel-SMA reinforced piers in a life-cycle context by numerical approach. *Bull. Earthq. Eng.* **2018**, *17*, 1667–1688. [CrossRef]
30. Fang, C.; Liang, D.; Zheng, Y.; Yam, M.C.; Sun, R. Rocking bridge piers equipped with shape memory alloy (SMA) washer springs. *Eng. Struct.* **2020**, *214*, 110651. [CrossRef]
31. Dong, Y.; Frangopol, D.M.; Saydam, D. Time-variant sustainability assessment of seismically vulnerable bridges subjected to multiple hazards. *Earthq. Eng. Struct. Dyn.* **2013**, *42*, 1451–1467. [CrossRef]
32. Fang, C.; Yam, M.C.H.; Lam, C.C.; Xie, L. Cyclic performance of extended end-plate connections equipped with shape memory alloy bolts. *J. Constr. Steel Res.* **2014**, *94*, 122–136. [CrossRef]
33. Roh, H.; Reinhorn, A.M. Modeling and seismic response of structures with concrete rocking columns and viscous dampers. *Eng. Struct.* **2010**, *32*, 2096–2107. [CrossRef]
34. Sharabash, A.M.; Andrawes, B.O. Application of shape memory alloy dampers in the seismic control of cable-stayed bridges. *Eng. Struct.* **2009**, *31*, 607–616. [CrossRef]
35. Yam, M.C.H.; Fang, C.; Lam, C.C.; Zhang, Y. Numerical study and practical design of beam-to-column connections with shape memory alloys. *J. Constr. Steel Res.* **2015**, *104*, 177–192. [CrossRef]
36. Fang, C.; Wang, W.; He, C.; Chen, Y. Self-centring behaviour of steel and steel-concrete composite connections equipped with NiTi SMA bolts. *Eng. Struct.* **2017**, *150*, 390–408. [CrossRef]
37. Wang, W.; Fang, C.; Yang, X.; Chen, Y.; Ricles, J.; Sause, R. Innovative use of a shape memory alloy ring spring system for self-centering connections. *Eng. Struct.* **2017**, *153*, 503–515. [CrossRef]
38. Fang, C.; Yam, M.C.H.; Chan, T.M.; Wang, W.; Yang, X.; Lin, X. A study of hybrid self-centering connections equipped with shape memory alloy washers and bolts. *Eng. Struct.* **2018**, *164*, 155–168. [CrossRef]
39. Wang, W.; Fang, C.; Zhang, A.; Liu, X. Manufacturing and performance of a novel self-centring damper with SMA ring springs for seismic resilience. *Struct. Control. Health Monit.* **2019**, *26*, e2337. [CrossRef]

40. Zheng, Y.; Fang, C.; Liang, D.; Sun, R. An innovative seismic-resilient bridge with shape memory alloy (SMA)-washer-based footing rocking RC piers. *J. Intell. Mater. Syst. Struct.* **2021**, *32*, 549–567. [CrossRef]
41. Varela, S.; Saiidi, M. A bridge column with superelastic NiTi SMA and replaceable rubber hinge for earthquake damage mitigation. *Smart Mater. Struct.* **2016**, *25*, 075012. [CrossRef]
42. Wang, W.; Fang, C.; Liu, J. Self-Centering Beam-to-Column Connections with Combined Superelastic SMA Bolts and Steel Angles. *J. Struct. Eng.* **2017**, *143*, 04016175. [CrossRef]
43. Crisfield, M.A. A consistent co-rotational formulation for non-linear, three-dimensional, beam-elements. *Comput. Methods Appl. Mech. Eng.* **1990**, *1*, 131–150. [CrossRef]
44. Wang, W.; Fang, C.; Liu, J. Large size superelastic SMA bars: Heat treatment strategy, mechanical property and seismic application. *Smart Mater. Struct.* **2016**, *25*, 075001. [CrossRef]
45. Mazzoni, S.; McKenna, F.; Scott, M.H.; Fenves, G.L.; Jeremic, B. *Open System for Earthquake Engineering Simulation (Opensees) Command Language Manual*; Pacific Earthquake Engineering Research Center, University of California at Berkeley: Berkeley, CA, USA, 2006.
46. Zhang, N.; Gu, Q.; Dong, Y.; Qian, J.; Zheng, Y. Seismic performance of bridges with ECC-reinforced piers. *Soil Dyn. Earthq. Eng.* **2021**, *146*, 106753. [CrossRef]
47. Han, T.S.; Feenstra, P.H.; Billington, S.L. Simulation of highly ductile fiber-reinforced cement-based composite components under cyclic loading. *ACI Struct. J.* **2003**, *100*, 749–757.
48. Fischer, G.; Li, V.C. Effect of matrix ductility on deformation behavior of steel reinforced ECC flexural members under reversed cyclic loading conditions. *ACI Struct. J.* **2002**, *99*, 781–790.
49. Zheng, Y.; Chen, B.; Chen, W. Elasto-Plastic Seismic Response of RC Continuous Bridge with Foundation-Pier Dynamic Interaction. *Adv. Struct. Eng.* **2015**, *18*, 817–836. [CrossRef]
50. Zheng, Y.; Xiao, X.; Zhi, L.; Wang, G. Evaluation on impact interaction between abutment and steel girder subjected to non-uniform seismic excitation. *Shock. Vib.* **2015**, *2015*, 981804. [CrossRef]
51. *Specifications for Seismic Design of Highway Bridges (JTG/T 2231-01—2020)*; Ministry of Transportation of China: Beijing, China, 2020.
52. Fang, C.; Liang, D.; Zheng, Y.; Lu, S. Seismic performance of bridges with novel SMA cable-restrained high damping rubber bearings against near-fault ground motions. *Earthq. Eng. Struct. Dyn.* **2021**, 1–22. [CrossRef]

materials

Article

Thermomechanical Fatigue Testing on Fe-Mn-Si Shape Memory Alloys in Prestress Conditions

Eva Marinopoulou and **Konstantinos Katakalos** *

Laboratory of Strength of Materials and Structures, Department of Civil Engineering, Faculty of Engineering, Aristotle University of Thessaloniki, 541 24 Thessaloniki, Greece
* Correspondence: kkatakal@civil.auth.gr

Abstract: Active materials have gained increasing momentum during the last decades due to their ability to act as sensors and actuators without the need for an external controlling system or an electronic signal. Shape memory alloys (SMAs), which are a subcategory of active materials, are slowly being introduced in the civil engineering sector in applications that refer to prestressing and strengthening of various structural elements. Low-cost iron-based SMAs are a good alternative to the Ni-Ti SMAs for such uses since the cost of large-scale civil engineering applications would otherwise be prohibitive. The scope of this study is the investigation of the thermomechanical response of the Fe-17Mn-5Si-10Cr-4Ni-1(V,C) ferrous SMA. In particular, this study focuses on the application of prestress, and on the alloy's behavior under fatigue loadings. In addition, the effect of loading frequency on the recovery stress of the material is thoroughly investigated. Four dog-bone specimens were prepared and tested in low-cycle fatigue. All the experiments aimed at the simulation of prestress. The recovery stress was monitored after pre-straining and heating applied under strain–control conditions. The experimental results are promising in terms of the is situ prestress feasibility since the measured recovery stress values are satisfactory high.

Keywords: ferrous shape memory alloys; prestress; recovery stress; relaxation; thermomechanical behavior; fatigue; active materials; low-cost SMAs; civil engineering applications

Citation: Marinopoulou, E.; Katakalos, K. Thermomechanical Fatigue Testing on Fe-Mn-Si Shape Memory Alloys in Prestress Conditions. *Materials* **2023**, *16*, 237. https://doi.org/10.3390/ma16010237

Academic Editors: Andrey Belyakov, Cheng Fang, Canxing Qiu and Yue Zheng

Received: 26 September 2022
Revised: 24 November 2022
Accepted: 16 December 2022
Published: 27 December 2022

1. Introduction

According to Lagoudas [1], shape memory alloys (SMAs) are a unique class of shape memory materials with the ability to recover their shape when the temperature is increased or, under specific conditions, upon cyclic loading. SMAs are categorized as active materials due to their ability to recover strain under stress. Other active materials include piezoelectric and magnetostrictive materials [2]. Due to these attributes, SMAs have been employed as actuators in multiple fields, such as the aerospace and automotive industry; however, they have also been used for biomedical applications, for instance, in orthodontic wires. Ni-Ti alloys are the most popular in these fields. However, their cost is restrictive for civil engineering applications. Iron-based SMAs are an alternative to this issue. Some civil engineering applications include prestressing, vibration dampening, strengthening and actuation.

1.1. Shape Memory Effect

Shape recovery takes place upon transformation between two distinct crystal phases that exist on the microstructure of SMAs: austenite and martensite. These two phases differ in both properties and microstructure.

Austenite has typically a cubic structure, while martensite has a structure of lower symmetry. Martensitic transformation is a diffusionless and displacive first-order phase transition driven by stress and/or temperature. The transformation from austenite to martensite is called forward, while the opposite from martensite to austenite is known as reverse transformation.

Martensite is present between Mf and Ms, which are typically low, while austenite can be found between As and Af, which are relatively higher temperatures. These four temperature thresholds create the thermal boundaries of the two phases. Martensite formation can be caused by either thermal or mechanical loading. Thus, it can be categorized into two groups, thermal martensite and stress-induced martensite (SIM), respectively [3]. The phase transformation process during monotonic loading is presented in Figure 1, as obtained from Zhang et al. [4].

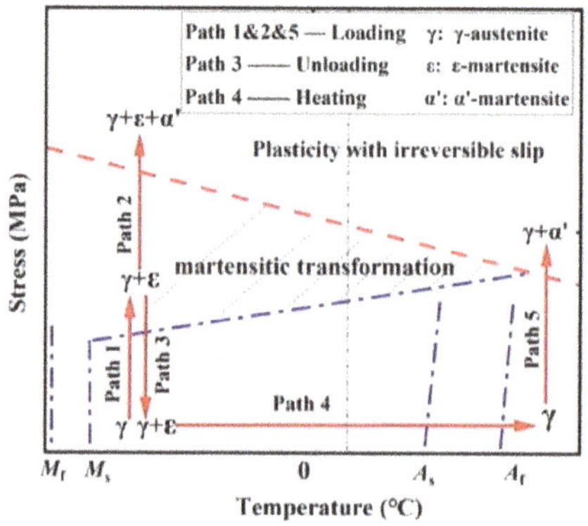

Figure 1. Stress–temperature diagram for an iron-based SMA, Zhang et al. [4].

1.2. *Brief Introduction to the Iron-Based SMAs (FeSMAs)*

Since Sato et al. discovered that a small Si addition to the chemical composition of an Fe-Mn alloy enhances the shape memory effect [5,6], many iron-based alloys have been developed. Various chemical compositions have been formed, such as Fe-32Mn-6Si by Murakami et al. [7], Fe-28-Mn-6Si-5Cr, Fe-20-Mn-5Si-8Cr-5Ni and Fe-16-Mn-5Si-12Cr-5Ni by Otsuka et al. [8,9].

Tsuzaki et al. [10] proved that adding carbon on an Fe-Mn-Si system strongly enhances the shape memory effect. Many researchers have also demonstrated that austenite can be strengthened through the addition of microparticles, such as niobium and nitrogen [11–13]. The shape recovery percentage of iron-based SMAs is significantly affected by both the addition of carbon and the subsequent aging process, as Wen et al. [14] and Farjami et al. [15] indicate.

Aside from these additions in the chemical composition of the alloys, thermomechanical training is also employed to achieve an optimal shape memory effect. The main essence of training an alloy, through various thermomechanical processes, is the enhancement of its mechanical properties. Otsuka et al. [16,17] introduced the term "training" and many researchers continued to examine the topic, proving that these thermomechanical sequences vastly expand the alloys' shape memory potential [18–21].

These are all very significant findings that have been thoroughly investigated throughout the years and have made the industrial production of iron-based SMAs possible.

1.3. *Fatigue Behavior of Iron-Based SMAs under Mechanical Loading*

According to Zhang et al. [4], Sawaguchi et al. [22,23] were the first to discover the material's energy dissipation prospects in 2006. Following these findings, Nikulin et al. [24] discovered that chemical composition plays a major role in fatigue life. It was then demonstrated that silicon presence has a significant impact on fatigue life [25]. In 2021, Fang et al. [26] investigated iron-based SMAs in low-cycle fatigue for strain rates that ranged from ±1% to ±9% and proved that the alloys' behavior under fatigue could profoundly outperform the steel alloys' response [26].

A significant amount of research has been conducted since 2006 referring to Fe-Mn-Si systems' fatigue response in mechanical loading and energy dissipation potential. Among many researchers, particularly Nikulin, Sawaguchi et al. [27–30], have covered a wide variety of topics, which include microstructural investigations, strain amplitude effect, deformation temperature effect, composition percentages impacts and many more. A comprehensive literature review that covers many studies on fatigue is provided by Zhang et al. [4]. Nonetheless, thermomechanical tests and phase transformation sequences have not been extensively investigated on iron-based SMAs. The present study caters to this exact need, attempting to add more material to the existing literature referring to the implications of low-cycle fatigue on the recovery stress that is required during prestress applications.

1.4. *Civil Engineering Applications*

Iron-based SMAs are used in civil engineering applications during the past decade, in uses relevant to prestressing and strengthening; however, they are also used as energy-dissipating components in seismic damping [26]. Recent advances in additive manufacturing, generally regarding SMAs [31] and particularly focusing on iron-based SMAs [32], could enhance the material's response and facilitate its incorporation on the field.

In 2016, Shahverdi et al. [33,34] investigated the behavior of concrete beams reinforced with prestressed iron-based SMAs. Following this investigation, further research has confirmed the alloys' successful use as a prestressing or strengthening component [35–37]. On the topic of finite element modeling, few researchers have dealt with the structural simulation of iron-based SMAs. Although there has been extensive investigation on modeling NiTi SMAs, these constitutive models are not quite applicable on Fe-SMAs, since their thermomechanical behavior differs substantially. Khalil et al. [38] have developed a constitutive model, based on previous works of the relevant literature, that described Fe-SMAs' nonlinearities and phase transformation process, which was also implemented into Abaqus software [38]. Following this attempt, some researchers further examined the material's modeling towards finite element structural analysis, among which are Abouali et al. [39], who modeled the behavior of prestressed Fe-SMAs used as a reinforcement in a concrete structural element.

2. Materials and Methods

The iron-based shape memory alloy that was used for this study is the Fe-17Mn-5Si-10Cr-4Ni-1(V,C) (mass%), as-received, without any additional treatments or thermomechanical training. The specimen geometry was chosen to be of dog-bone shape, and the dimensions are displayed in Figures 2 and 3. All the specimens were cut from an SMA strip.

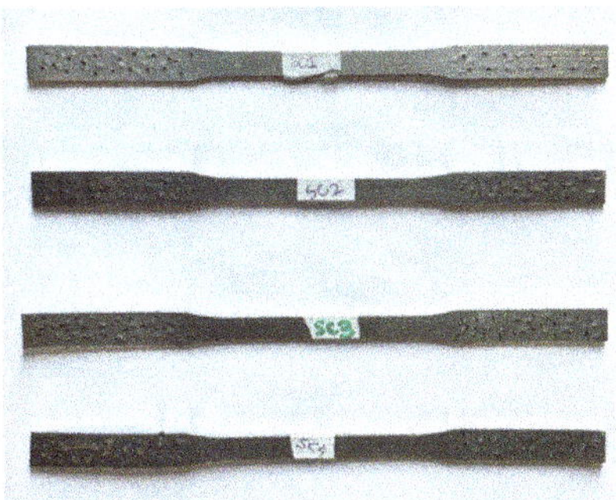

Figure 2. Dog-bone specimens used for the fatigue tests.

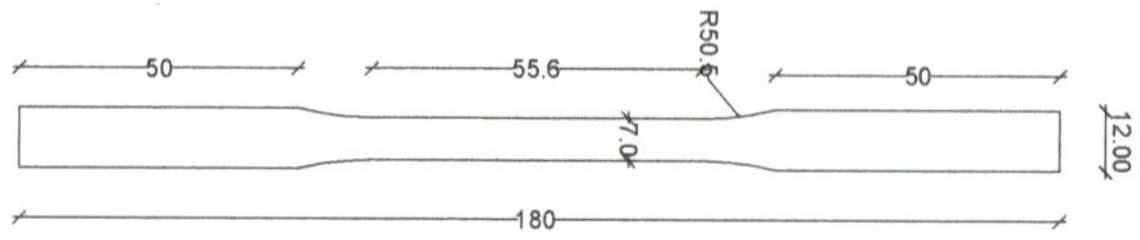

Figure 3. Specimen dimensions (mm).

The austenite start (As) and finish (Af) threshold values for each phase of the alloy were obtained from experiments conducted by the research group of the Laboratory of Experimental Strength of Materials and Structures, as depicted in Figure 4.

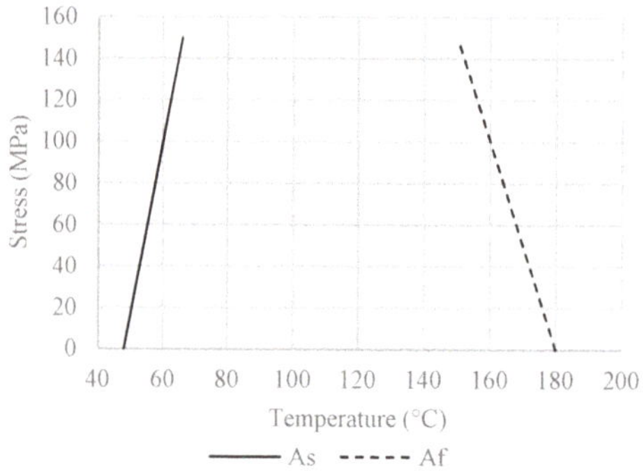

Figure 4. Austenite start and finish range.

The Instron 596 tensile testing machine was employed for the thermomechanical tests, as shown in Figure 5. The Instron machine contains an integrated dynamic load cell with a capacity of 50 kN, while its operating system is the Bluehill software data acquisition system. During the mechanical loading tests, the strain evolution was monitored using a clip-on extensometer with a gauge length of 50 mm. The thermomechanical tests, which employed displacement control, were monitored through the Bluehill system. The thermal cycles were conducted using a custom-made induction heating and air-cooling machine, which is also shown in Figure 5. The machine was programmed and manufactured internally in the laboratory for the scope of the thermomechanical experiments performed by the research team.

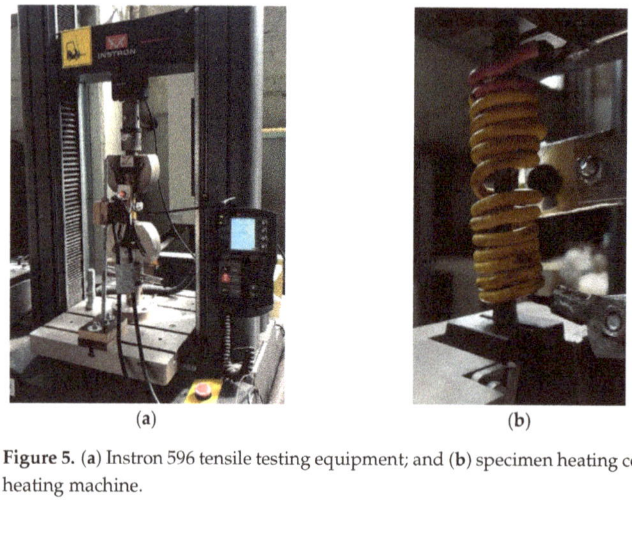

(**a**) (**b**)

Figure 5. (**a**) Instron 596 tensile testing equipment; and (**b**) specimen heating coil, part of the induction heating machine.

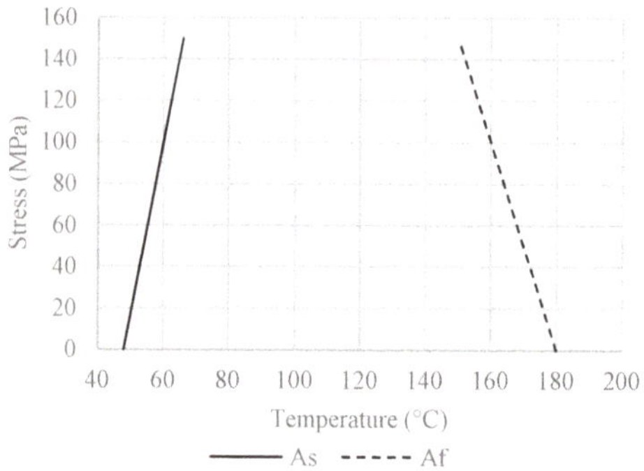

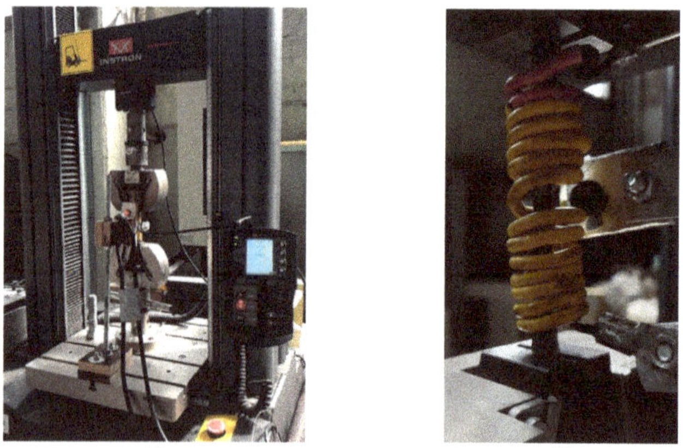

2.1. Tensile Test Procedure

During prestressing, SMA tendons are expected to generate compressive stresses, adequate for the purpose of balancing out the impact tensile stresses have on concrete members. SMA tendons can vastly facilitate on-field pre-stressing: being already in a prestressed state, the sole remaining process to be conducted in situ would be heating, which would enable the activation of the shape memory effect; thus, producing the required recovery stresses. SMAs would therefore eliminate many of the common disadvantages that emerge from the mechanical pre-tensioning process.

The test profiles were based on the tensile and fatigue tests conducted by Ghafoori et al. [40]. The test profile is presented in Figure 6. Three initial experiments were performed in order to characterize the mechanical properties of the alloy. The yield and ultimate strength and strain were acquired through this process. Strain was monitored using an extensometer and the displacement values were recorded directly through Instron's software. Simultaneously measuring strain and displacement was crucial in order to record the correlation between the two values and subsequently use it for calibration (this is explained more thoroughly in the following sections).

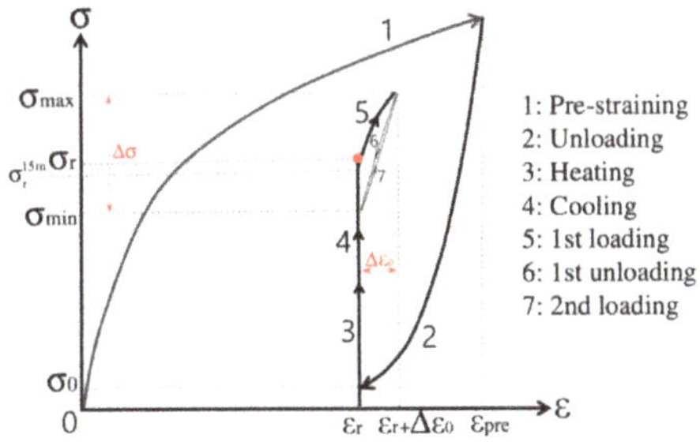

Figure 6. Cyclic loading path [40].

One standard tensile test (S1), along with two loading–unloading sequences with target strains of 2% and 1.6% (S2 and S3), were conducted. Figure 7 presents the tensile tests results. The experiments were performed using strain–control under a strain rate of 0.15%/s, as obtained from the literature [40].

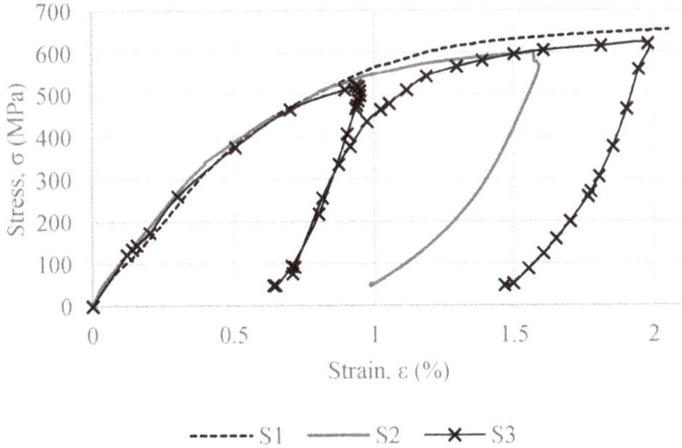

Figure 7. Stress vs. strain curve, specimens S1–S3.

Different target strains were employed throughout the tests, aiming to examine how the thermomechanical response of the alloys varies in relation to this change. Altering the pre-strain values has a practical purpose in civil engineering applications. During on-site prestressing, it is challenging to acquire a high level of precision. Thus, a target pre-strain of 2% might eventually turn out to be 1.6%, 1.8% or 2.2%. Given this fact, the research team decided to employ varying pre-strain values. The concept of the unstable test parameters is further illustrated in the following sections.

S1 was the first specimen to be tested. Figure 8 presents the stress vs. strain curve derived from the uniaxial tensile test of S1, up to 9.5% strain. Following this experiment, S2 was tested up to 1.6% strain; and S3 was tested in two increments, firstly up to 1% and subsequently up to 2% strain (see Figure 7).

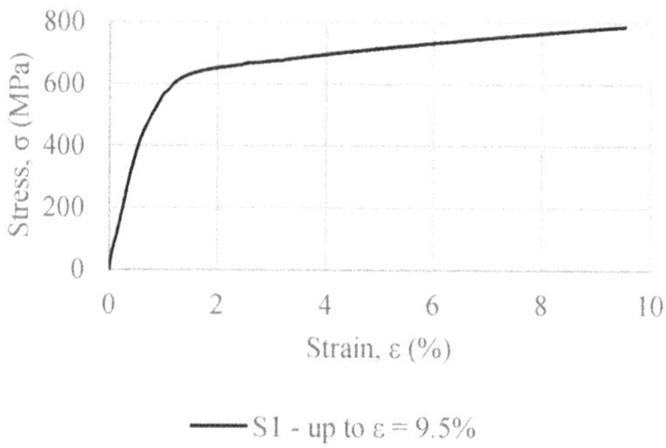

Figure 8. Stress vs. strain curve up to 9.5%, specimen S1.

2.2. *Low-Cycle Fatigue Test Procedure*

The fatigue test profiles and loading protocols were also based on the literature [40]. The aim of this test setup is to investigate the recovery stress generation, while strain is monitored. The applied loading path is presented in Figure 6.

Various researchers have already dealt with strain recovery formation; however, there are few studies referring to the recovery stress. In civil engineering applications, the recovery stress is highly significant since when prestressing a member, the emphasis is given to the maximum stress that a tendon can produce. SMAs offer the possibility of a new prestressing technique that is simpler and easier to conduct on field.

In addition to calculating the amount of recovery stress that can be generated, estimating how well SMAs tendon can perform over time and under cyclic loadings is equally substantial for prestressing applications. The tendon will most likely experience fatigue since it is expected to receive various additional live loads and potentially even seismic loads. Measuring the relaxation that occurs during low-cycle fatigue loadings is, therefore, essential to determine the applicability of SMAs in this configuration.

Three variables that directly influence material behavior on fatigue are considered in the tests: loading frequency, level of pre-strain and limit stress target value. These parameters are variable in each test. The deviations aim to simulate the conditions of on-field prestressing, where the level of precision is most likely lower than in a laboratory setup. Each parameter is briefly highlighted below:

Loading frequency: Four different frequencies are examined: 0.5, 1, 2 and 4 Hz, which are imposed on the samples C1, C2, C3 and C4, respectively.

Level of pre-strain: Pre-strain values are denoted as ε_r in Figure 8. For the present study, pre-strain values of 1.85%, 1.80%, 1.90% and 2% were imposed on the specimens C1, C2, C3 and C4, respectively.

Limit stress target value: Thermal expansion, which occurs at the beginning of the heating process, induces stress drop [40]. In order to avoid the formation of compressive stresses during that phase, setting a lower stress limit is practical. While Ghafoori et al. [40] employ a stable lower limit of 50 MPa, the present study uses four different ones: 125, 70, 120 and 150 MPa on samples C1, C2, C3 and C4, respectively.

Due to testing configuration limitations, strain control was not feasible. As can be seen in Figure 5b, the induction heating coil that is used for the thermal cycles poses a practical restriction for the positioning of the extensometer. Similarly, a strain gauge would be useless in such an application since the high temperatures would affect the measurements. Thus, after careful strain-to-displacement calibration was conducted through the tensile tests, displacement control was performed. However, all the results were carefully converted and presented in strain values for the purpose of comparability.

The detailed cyclic test profile loading pattern for all four of the specimens is presented in Figure 8, as obtained from Ghafoori et al. [40]. Both tensile loading and unloading were performed at a strain rate of ε = 0.075%/s. After reaching the target pre-strain, the specimen is unloaded. Both loading and unloading, paths 1 and 2 of Figure 8, respectively, take place at room temperature. The subsequent step (path 3 of Figure 6) is heating. The specimen is heated at 160 °C and then, cooled back to room temperature (path 4). After reaching room temperature, the strain is kept constant for 15 min in order to monitor any potential stress relaxation. According to findings in the literature, [40], this hold time is sufficient for the most part of the relaxation to occur. Stress relaxation is attributed to creep phenomena [40,41].

Following the end of the hold time, 10 low-frequency cycles were performed at the same strain range, 0.04%; however, they were performed at different frequencies. After the cyclic loading, all the specimens were subjected to an increasing tensile load (again, through displacement control), with the ultimate target strains being the initial pre-strain values of each specimen.

3. Results

3.1. Tensile Tests

The results of the three initial experiments are presented in Figure 7. All three of these tests were performed at room temperature. According to the recorded data, the standard 0.01% and 0.2% strength of the material are 175 and 510 MPa, respectively. For the purposes of this study and in accordance with the literature [40], two limits are investigated: the 0.01% stress, as a limit of proportionality; and the limit stress that corresponds to 0.2% strain. This consideration is further explained in the following section.

S1 was tested first and produced an ultimate tensile strength of 1066 MPa. Specimen S2 was loaded until a pre-strain of 1.6%, which will be a target strain for the cyclic tests. It was observed that the stress recovers from 1.6% to almost 1%. Specimen S3 was loaded in two increments in order to determine if the loading–unloading and reloading sequence influences the stress that corresponds to 2% strain. According to the experiments, there is only a slight deviance of approximately 5%, which is considered negligible. From this experiment, it was also concluded that for a target strain of ε = 1% and ε = 2%, the recovered strain is equal to 0.65% and 1.5%, respectively. It can also be observed that the slope of the curve changes at approximately 300 MPa due to phase transformation.

3.2. Low-Cycle Fatigue Tests

The results of the fatigue tests are displayed in Figure 9a–c. Figure 9a is the stress–strain graph for each specimen, plotted with the stress–strain curve of S1, which serves as a reference. After the completion of the loading–unloading path, significant recovery stresses are generated, ranging from 265 to 275 MPa. However, this is attributed to both SME and thermal contraction. Upon heating and until 100 °C, thermal expansion is prevalent; therefore, the monitored stress drops at the beginning of the thermal cycle. Once this temperature is exceeded, SME becomes the governing factor; thus, stress starts to increase. As soon as the cooling phase is initiated, thermal contraction further increases the stress.

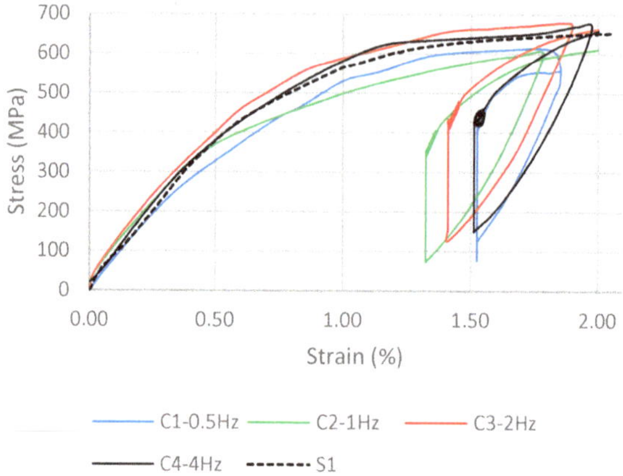

(**a**)

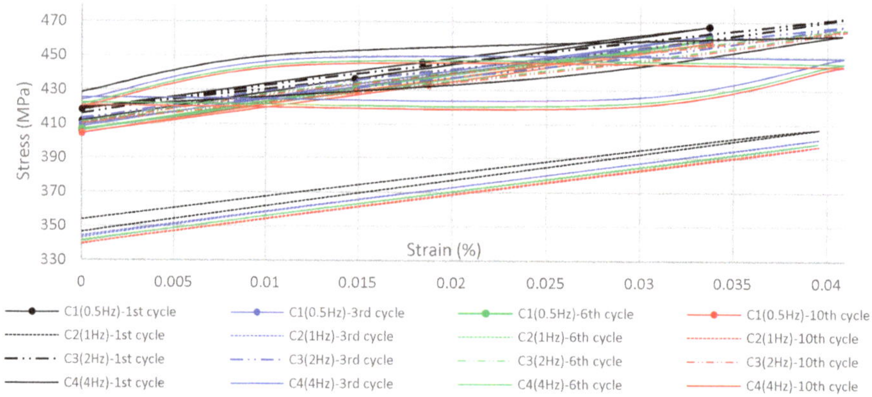

(**b**)

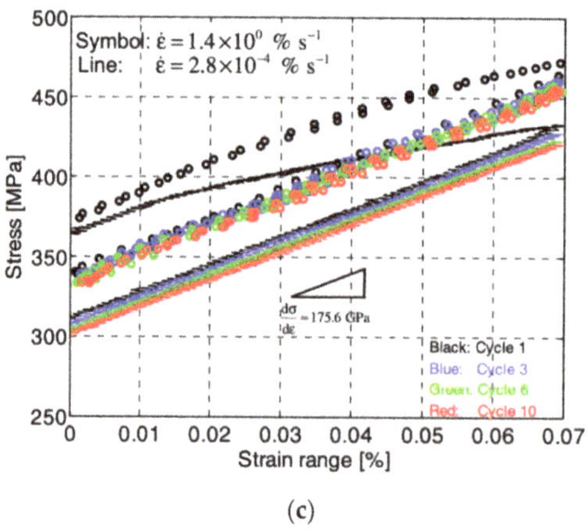

(**c**)

Figure 9. (**a**) Fatigue tests results. (**b**) Detailed graph of the alloy's cyclic behavior at the programmed strain range. (**c**) Ghafoori et al. [40]: detailed graph of the alloy's cyclic behavior at the programmed strain range.

After the thermal cycle is finished, an initial recovery stress is recorded, which is denoted as σ_r in Figure 5. The hold time between the thermal and the mechanical loading cycles ensures that stress relaxation is allowed to take place. The subsequent mechanical loading begins from the reduced recovery stress, σ_{min} (see Figure 5). In Figure 9b, it is evident that each frequency has a different effect on the SMA's behavior.

4. Discussion

4.1. Tensile Tests

Detecting an exact yield point is a challenging and uncertain procedure that has already been a matter of discussion in the relevant literature. According to Lee et al. [42] and Khalil et al. [43], there are two mechanisms taking place upon loading that complicate the detection of a clear yield stress. Khalil et al. [43] denote two different yield limits, one that occurs due to the activation of stress-induced martensite and another that relates to plasticity. The interaction of these two mechanisms complicates the identification of the precise point where plasticity begins to take effect.

Temperature should also be considered since these two limits are not stable to temperature variations. Additionally, the stress–strain curve's slope varies at different temperatures. Khalil et al. [43] have produced a series of explanatory graphs and tables that clarify the interaction between the two mechanisms at various temperatures. They report which mechanism is prevalent for each temperature.

According to the literature, the material's unloading behavior is not linear, as would be expected according to Hooke's law. The nonlinearity of the unloading curve demonstrates an amount of pseudo-elasticity. The present study's outcomes are in good agreement with the literature. This behavior is demonstrated by the change of slope in the unloading curves of specimens S2 and S3, presented in Figure 6.

The following paragraphs discuss the major findings of the present research. The adopted loading protocol is chosen from the literature for comparison purposes, taking under consideration the fact that there is a lack of an official testing standard (ASTM or EN) for this type of test (thermomechanical fatigue tests for Fe-SMAs). The investigated parameters are different from the ones reported in the past; thus, they increase the density of the experimental results concerning Fe-SMAs. The different heating rate, loading rate and pre-stress range that are applied are the basic parameters that are discussed hereafter and compared with the literature.

4.2. Low-Cycle Fatigue Tests

Table 1 includes all the parameters investigated in each test. It contains details on the pre-strain level, unloading target value, cyclic loading frequency and recovery stress for each specimen. It is expected that different pre-strain and unloading target stresses result in different residual strains and consequently, different recovery stresses. However, the stress drops due to relaxation do not significantly deviate from each other, all ranging between 7 and 9 MPa. The corresponding stress drop percentages are all below 2.10%, which means that the SMA tendons can withstand the anticipated cyclic loads that will be applied on the structure without experiencing major relaxation. After the cyclic loading–unloading sequences, the final recovery stresses are quite high, with increases from the corresponding unloading stresses ranging from 270 to 290 MPa.

Figure 9b displays each specimen's detailed response on the mechanical cyclic loading–unloading sequence. Figure 9b is presented similarly to the literature [40]. It is evident that between the first and tenth cycle, some amount of the recovery stress was lost. This stress loss is attributed to plasticity. The maximum stress drop is found to occur between the first and the second cycle. This is apparently the stage where plasticity begins to have an effect on the material's mechanical behavior. However, the stress drop due to plasticity that takes place throughout this process is considered a small fraction of the total stress-drop.

Table 1. Low-cycle fatigue test results.

Specimen	Pre-Strain (%)	Unloading Target Value (MPa)	Residual Strain ε (%)	Recovery Stress σ_r (MPa)	Recovery Stress after 15′ σ_r^{15min} (MPa)
C1	1.85	125	1.50	427	418
C2	1.80	70	1.30	354	347
C3	1.90	120	1.40	424	416
C4	2.00	150	1.50	437	428
Specimen	**Stress Drop Due to Relaxation (MPa)**		**Stress Drop Percentage (%)**	**Cyclic Loading Frequency (Hz)**	**Recovery Stress after Cyclic Loading σ_r (MPa)**
C1	9		2.10	0.5	404
C2	7		2.00	1	340
C3	8		1.90	2	409
C4	9		2.10	4	420

Specimen C4 demonstrates a partly different response to the fatigue loadings compared to C1–C3: It exhibits a clear hysteresis loop for each cycle, as can be seen in both Figure 9b,c. The recorded hysteretic behavior can not entirely be attributed to the increase of frequency since it only occurs for C4 and not for C2 or C3. It can be assumed that the microstructural response of C4 was different from the rest; however, in order to reach a conclusion on the factors that drive this change in behavior, microstructural investigation needs to be conducted. However, the hysteretic behavior of C4 is in good agreement with the literature covering fatigue [22–30] since it has been reported that iron-based SMAs have energy-dissipating potential. The hysteresis loop on 4 Hz, thus, verifies the material's damping capacity. Setting that aside, similarly to the rest of the specimens, C4 demonstrates plasticity mostly during the first cycle; After that, its response is stabilized with each cycle.

Figure 10a presents the stress–temperature curves for the specimens used in the present study. Figure 10b is a comparison between the findings of this study and the literature. The shape of the curves and, therefore, the thermomechanical behavior of the alloy, depends on the heating and cooling rates. It is evident from Figure 10b that the more rapid heating rate induces thermal expansion more intensely than the slow rate does. In this study, the heating and cooling rate is equal to 1.2 °C/s; a very quick heating rate, trying to simulate realistic in situ civil engineering applications (e.g., the increase of temperature that occurs with the use of an inductive current or with a simple flamethrower). On the other hand, the literature adopts a slower heating rate that reaches 2 °C/min [40]. In Figures 10 and 11, the rapid heating rate results in a small initial increase of the recorded stress at the range of 50–80 MPa, followed by an equivalent decrease in stress. During cooling, the shape memory effect (SME) is initialized and coupled with the thermal contraction, it results in the increase of the recorded stress at the range of 450 MPa. Similar findings of the SME initiation and the thermal contraction effect are also mentioned in the literature (see Figures 10b and 11b).

Figure 11a shows the stress–time and temperature–time curves that correspond to the thermal protocol, without the hold time for temperature equalization. The left vertical axis depicts the recorded stress, whereas the right vertical axis the applied temperature. Two curves are presented: the stress curve (continuous line) and the temperature curve (dashed line). The time record upon heating and cooling is different since heating was performed using a standard rate, whereas cooling was completed naturally. In combination with Figure 4, the interaction between the SME and the thermal expansion effect can be better understood: firstly, Figure 4 shows that the austenite start ranges between 50–80 °C. From Figure 10a, it can be seen that stress increase takes place around 40 °C and is continued until 70 °C. Therefore, during the austenite start, stress increase occurs at an almost constant rate and it can be considered that at this point, the SME is dominant. After 70 °C, a rapid decrease in stress is observed; thus, for the temperature range of 70–110 °C, thermal expansion appears to be prevalent. From 110 °C to about 160 °C, an increase is observed at a significantly lower rate. Therefore, coupling of the two effects takes place at this point, with the SME prevailing. During cooling, SME is still present since the quick heating rate prevents the phase transformation from martensite to austenite, to be entirely completed before cooling begins. Namely, it takes some time for the specimen to distribute the temperature changes across its whole volume, which means that the initiation of cooling does not signify the end of the SME. A slower heating rate would allow the temperature to be more uniformly distributed and such peaks would be avoided; an overall smoother curve (similar to the literature's) would most likely be formed. After 110 °C, a rapid increase in stress is observed due to the thermal contraction effect.

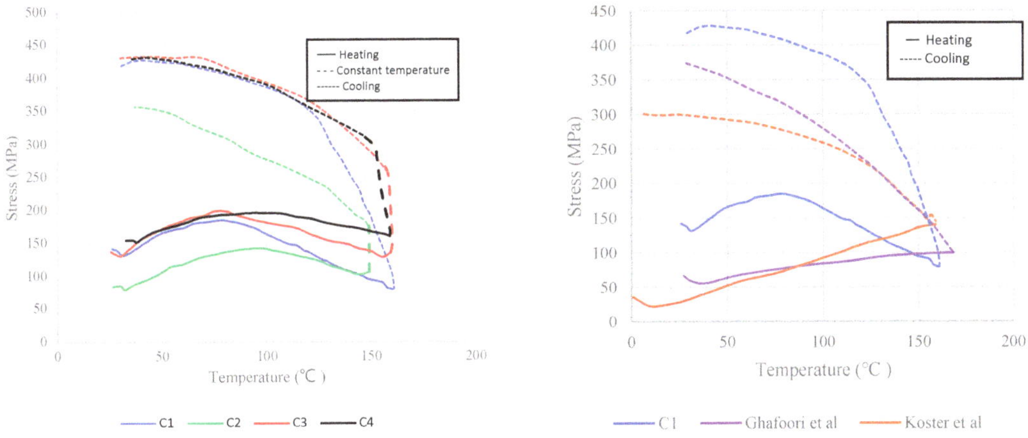

(a) (b)

Figure 10. (**a**) Stress vs. temperature graph for all specimens, C1–C4; and (**b**) comparison with the literature [40,44].

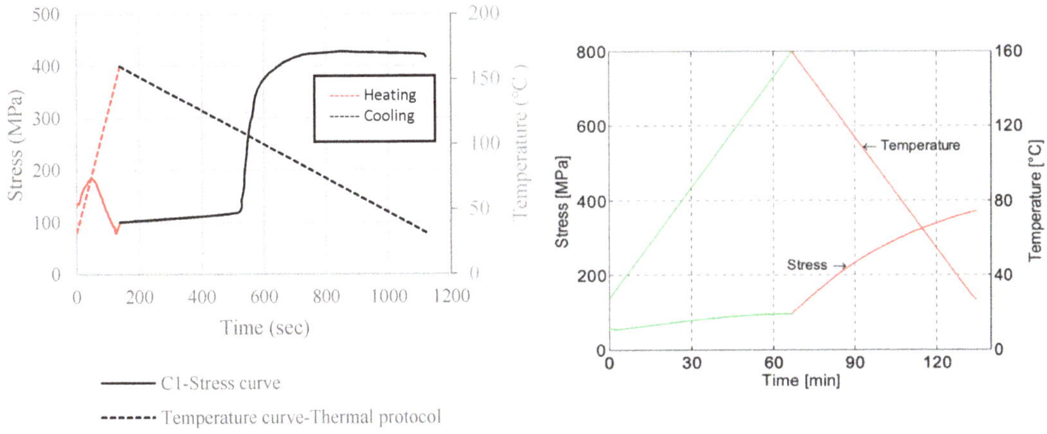

(a) (b)

Figure 11. (**a**) Stress–time–temperature graph, specimen C1; and (**b**) stress–time–temperature graph, Ghafoori et al. [40].

Figure 11a,b demonstrate that the heating–cooling rate plays an important role in the thermomechanical behavior of the alloy, especially in terms of the thermal expansion effect and the alloys' temperature distribution across its whole volume. The specimens used by Ghafoori et al. [40] were heated at a rate of 2 °C/min, while the samples of the present study were heated at an average rate of 1.2 °C/s. The specimens are left to cool naturally. According to the monitored data, cooling occurs at an average rate of 8 °C/min, (~0.13 °C/s), which is evidently slower than heating.

5. Conclusions

The present study focused on the thermomechanical behavior of an iron-based shape memory alloy subjected to tensile and low-cycle fatigue tests. A set of specimens was subjected to a certain target of pre-strain values, thermal and then, mechanical cyclic loading for different frequencies. Under constant strain conditions, the recovered stress was monitored. The following conclusions were drawn from the experimental investigation:

1. Cyclic loading was applied with four different frequencies (i.e., different strain rates) that were missing from the literature. It was shown that while at 0.5, 1 and 2 Hz, the behavior of the alloy was similar, at 4 Hz, a hysteresis loop was observed. This finding confirms the alloy's energy-dissipating capacity.
2. The studied alloy demonstrated a significant recovery stress, which was largely maintained after the fatigue tests. After the loading cycles, a small drop in the recovery stress, around 2–3% was observed. Moreover, the measured recovery stress decrease, during the 15 min hold time, was relatively small.
3. The experimental results are promising in terms of the on-field prestress feasibility. Recovery stress values are high enough, although accuracy can be an issue that needs careful handling.
4. The heating and cooling rate has a profound effect, significantly changing the thermomechanical behavior of the material.

Author Contributions: Conceptualization, E.M. and K.K.; Methodology, E.M. and K.K.; Software, K.K.; Formal analysis, E.M.; Investigation, E.M. and K.K.; Resources, K.K.; Data curation, E.M.; Writing—original draft, E.M.; Writing—review & editing, K.K.; Supervision, K.K.; Project administration, K.K.; Funding acquisition, K.K. All authors have read and agreed to the published version of the manuscript.

Funding: This research received no external funding.

Data Availability Statement: The data can be requested by the authors.

Acknowledgments: This study would not have been possible without the contribution of the entire research group of the Laboratory of Strength of Materials and Structures. Special thanks shall be given to Konstantinos Chrysopoulos, who produced the austenite range diagram used in this study under the guidance of Konstantinos Katakalos; Tasos Vlachos, who developed the induction heating and air-cooling machine; Lazaros Melidis and Thomas Koukouftopoulos.

Conflicts of Interest: The authors declare no conflict of interest.

References

1. Lagoudas, D.C. *Shape Memory Alloys, Modeling and Engineering Applications*; Springer: New York, NY, USA, 2008. [CrossRef]
2. Hartl, D.J.; Lagoudas, D. Aerospace applications of shape memory alloys. *Proc. Inst. Mech. Eng. Part G J. Aerosp. Eng.* **2007**, *4*, 535–552. [CrossRef]
3. Cladera, A.; Weber, B.; Leinenbach, C.; Czaderski, C.; Shahverdi, M.; Motavalli, M. Iron-based shape memory alloys for civil engineering structures: An overview. *Constr. Build. Mater.* **2014**, *63*, 281–293. [CrossRef]
4. Zhang, Z.-X.; Zhang, J.; Wu, H.; Ji, Y.; Kumar, D.D. Iron-Based Shape Memory Alloys in Construction: Research, Applications and Opportunities. *Materials* **2022**, *15*, 1723. [CrossRef] [PubMed]
5. Sato, A.; Chishima, E.; Soma, K.; Mori, T. Shape memory effect in $\gamma \rightleftarrows \epsilon$ transformation in Fe-30Mn-1Si alloy single crystals. *Acta Metall.* **1984**, *30*, 1177–1183. [CrossRef]
6. Sato, A.; Chishima, E.; Yamaji, Y.; Mori, T. Orientation and composition dependencies of shape memory effect IN Fe-Mn-Si alloys. *Acta Met.* **1984**, *32*, 539–547. [CrossRef]

7. Murakami, M.; Otsuka, H.; Suzuki, H.; Matsuda, S. Complete shape memory effect in polycrystalline Fe-Mn-Si alloys. *Trans-Actions Iron Steel Inst. Jpn.* **1986**, *27*, 985–990.
8. Otsuka, H.; Yamada, H.; Maruyama, T.; Tanahashi, H.; Matsuda, S.; Murakami, M. Effects of alloying additions on Fe-Mn-Si shape memory alloys. *ISIJ Int.* **1990**, *30*, 674–679. [CrossRef]
9. Otsuka, H. Fe-Mn-Si Based Shape Memory Alloys. *MRS Proc.* **1991**, *246*, 309–320. [CrossRef]
10. Tsuzaki, K.; Natsume, Y.; Kurokawa, Y.; Maki, T. Improvement of the shape memory effect in Fe-Mn-Si alloys by the addition of carbon. *Scr. Met. Mater.* **1992**, *27*, 471–473. [CrossRef]
11. Jian, L.; Wayman, C. Shape memory effect and related phenomena in a microalloyed Fe-Mn-Si alloy. *Mater. Charact.* **1994**, *32*, 215–227. [CrossRef]
12. Kajiwara, S.; Liu, D.; Kikuchi, T.; Shinya, N. Remarkable improvement of shape memory effect in Fe-Mn-Si based shape memory alloys by producing NbC precipitates. *Scr. Mater.* **2001**, *44*, 2809–2814. [CrossRef]
13. Ariapour, A.; Perovic, D.D.; Yakubtsov, I. Shape-memory effect and strengthening mechanism in a Nb and N-doped Fe-Mn-Si-based alloy. *Met. Mater. Trans. A* **2001**, *32*, 1621–1628. [CrossRef]
14. Wen, Y.; Yan, M.; Li, N. Effects of carbon addition and aging on the shape memory effect of Fe–Mn–Si–Cr–Ni alloys. *Scr. Mater.* **2004**, *50*, 441–444. [CrossRef]
15. Farjami, S.; Hiraga, K.; Kubo, H. Shape Memory Effect and Crystallographic Investigation in VN Containing Fe-Mn-Si-Cr Alloys. *Mater. Trans.* **2004**, *45*, 930–935. [CrossRef]
16. Otsuka, K.; Shimizu, K. MRS International Meeting on Advanced Materials. In Proceeding of the MRS International Meeting on Advanced Materials, Tokyo, Japan, 31 May–3 June 1988; Volume 9.
17. Otsuka, K.; Wayman, C.M. *Shape Memory Materials*; Cambridge University Press: Cambridge, MA, USA, 1998.
18. Watanabe, Y.; Mori, Y.; Sato, A. Training effect in Fe-Mn-Si shape-memory alloys. *J. Mater. Sci.* **1993**, *28*, 1509–1514. [CrossRef]
19. Wang, D.F.; Chen, Y.R.; Gong, F.Y.; Liu, D.Z.; Liu, W.X. The Effect of Thermomechanical Training on the Microstructures of Fe-Mn-Si Shape Memory Alloy. *J. Phys. Colloq.* **1995**, *05*, 527–530. [CrossRef]
20. Chung, C.; Shuchuan, C.; Zuyao), T.H. Thermomechanical training behavior and its dynamic mechanical analysis in an Fe-Mn-Si shape memory alloy. *Mater. Charact.* **1996**, *37*, 227–236. [CrossRef]
21. Liu, D.Z.; Kajiwara, S.; Kikuchi, T.; Shinya, N. Atomic force microscopy study on microstructural changes by 'training' in Fe-Mn-Si-based shape memory alloys. *Philos. Mag.* **2003**, *83*, 2875–2897. [CrossRef]
22. Sawaguchi, T.; Kikuchi, T.; Ogawa, K.; Yin, F.X.; Kajiwara, S.; Kushibe, A.; Ogawa, T. Internal Friction of Fe-Mn-Si-Based Shape Memory Alloys Containing Nb and C and Their Application as a Seismic Damping Material. *Key Eng. Mater.* **2006**, *319*, 53–58. [CrossRef]
23. Sawaguchi, T.; Sahu, P.; Kikuchi, T.; Ogawa, K.; Kajiwara, S.; Kushibe, A.; Higashino, M.; Ogawa, T. Vibration mitigation by the reversible fcc/hcp martensitic transformation during cyclic tension–compression loading of an Fe–Mn–Si-based shape memory alloy. *Scr. Mater.* **2006**, *54*, 1885–1890. [CrossRef]
24. Nikulin, I.; Sawaguchi, T.; Tsuzaki, K. Effect of alloying composition on low-cycle fatigue properties and microstructure of Fe–30Mn–(6−x)Si–xAl TRIP/TWIP alloys. *Mater. Sci. Eng. A* **2013**, *587*, 192–200. [CrossRef]
25. Nikulin, I.; Sawaguchi, T.; Ogawa, K.; Tsuzaki, K. Effect of γ to ε martensitic transformation on low-cycle fatigue behaviour and fatigue microstructure of Fe–15Mn–10Cr–8Ni–xSi austenitic alloys. *Acta Mater.* **2016**, *105*, 207–218. [CrossRef]
26. Fang, C.; Wang, W.; Ji, Y.; Yam, M. Superior low-cycle fatigue performance of iron-based SMA for seismic damping application. *J. Constr. Steel Res.* **2021**, *184*, 106817. [CrossRef]
27. Sawaguchi, T.; Nikulin, I.; Ogawa, K.; Sekido, K.; Takamori, S.; Maruyama, T.; Chiba, Y.; Kushibe, A.; Inoue, Y.; Tsuzaki, K. Designing Fe–Mn–Si alloys with improved low-cycle fatigue lives. *Scr. Mater.* **2015**, *99*, 49–52. [CrossRef]
28. Nikulin, I.; Sawaguchi, T.; Ogawa, K.; Tsuzaki, K. Microstructure Evolution Associated with a Superior Low-Cycle Fatigue Resistance of the Fe-30Mn-4Si-2Al Alloy. *Met. Mater. Trans. A* **2015**, *46*, 5103–5113. [CrossRef]
29. Nikulin, I.; Sawaguchi, T.; Kushibe, A.; Inoue, Y.; Otsuka, H.; Tsuzaki, K. Effect of strain amplitude on the low-cycle fatigue behavior of a new Fe–15Mn–10Cr–8Ni–4Si seismic damping alloy. *Int. J. Fatigue* **2016**, *88*, 132–141. [CrossRef]
30. Tasaki, W.; Sawaguchi, T.; Nikulin, I.; Sekido, K.; Tsuchiya, K. Effect of Deformation Temperature on Low-Cycle Fatigue Properties of Fe-28Mn-6Si-5Cr Shape Memory Alloy. *Mater. Trans.* **2016**, *57*, 639–646. [CrossRef]
31. Alagha, A.N.; Hussain, S.; Zaki, W. Additive manufacturing of shape memory alloys: A review with emphasis on powder bed systems. *Mater. Des.* **2021**, *204*, 109654. [CrossRef]
32. Ferretto, I.; Kim, D.; Della Ventura, N.; Shahverdi, M.; Lee, W.; Leinenbach, C. Laser powder bed fusion of a Fe–Mn–Si shape memory alloy. *Addit. Manuf.* **2021**, *46*, 102071. [CrossRef]
33. Shahverdi, M.; Czaderski, C.; Motavalli, M. Iron-based shape memory alloys for prestressed near-surface mounted strengthening of reinforced concrete beams. *Constr. Build. Mater.* **2016**, *112*, 28–38. [CrossRef]
34. Shahverdi, M.; Czaderski, C.; Annen, P.; Motavalli, M. Strengthening of RC beams by iron-based shape memory alloy bars embedded in a shotcrete layer. *Eng. Struct.* **2016**, *117*, 263–273. [CrossRef]
35. Izadi, M.; Ghafoori, E.; Shahverdi, M.; Motavalli, M.; Maalek, S. Development of an iron-based shape memory alloy (Fe-SMA) strengthening system for steel plates. *Eng. Struct.* **2018**, *174*, 433–446. [CrossRef]

36. Strieder, E.; Aigner, C.; Petautschnig, G.; Horn, S.; Marcon, M.; Schwenn, M.; Zeman, O.; Castillo, P.; Wan-Wendner, R.; Bergmeister, K. Strengthening of Reinforced Concrete Beams with Externally Mounted Sequentially Activated Iron-Based Shape Memory Alloys. *Materials* **2019**, *12*, 345. [CrossRef] [PubMed]
37. Rezapour, M.; Ghassemieh, M.; Motavalli, M.; Shahverdi, M. Numerical Modeling of Unreinforced Masonry Walls Strengthened with Fe-Based Shape Memory Alloy Strips. *Materials* **2021**, *14*, 2961. [CrossRef]
38. Khalil, W.; Mikolajczak, A.; Bouby, C.; Ben Zineb, T. A constitutive model for Fe-based shape memory alloy considering martensitic transformation and plastic sliding coupling: Application to a finite element structural analysis. *J. Intell. Mater. Syst. Struct.* **2012**, *23*, 1143–1160. [CrossRef]
39. Abouali, S.; Shahverdi, M.; Ghassemieh, M.; Motavalli, M. Nonlinear simulation of reinforced concrete beams retrofitted by near-surface mounted iron-based shape memory alloys. *Eng. Struct.* **2019**, *187*, 133–148. [CrossRef]
40. Ghafoori, E.; Hosseini, E.; Leinenbach, C.; Michels, J.; Motavalli, M. Fatigue behavior of a Fe-Mn-Si shape memory alloy used for prestressed strengthening. *Mater. Des.* **2017**, *133*, 349–362. [CrossRef]
41. Leinenbach, C.; Lee, W.; Lis, A.; Arabi-Hashemi, A.; Cayron, C.; Weber, B. Creep and stress relaxation of a FeMnSi-based shape memory alloy at low temperatures. *Mater. Sci. Eng. A* **2016**, *677*, 106–115. [CrossRef]
42. Lee, W.; Weber, B.; Feltrin, G.; Czaderski, C.; Motavalli, M.; Leinenbach, C. Phase transformation behavior under uniaxial deformation of an Fe–Mn–Si–Cr–Ni–VC shape memory alloy. *Mater. Sci. Eng. A* **2013**, *581*, 1–7. [CrossRef]
43. Khalil, W.; Saint-Sulpice, L.; Chirani, S.A.; Bouby, C.; Mikolajczak, A.; Ben Zineb, T. Experimental analysis of Fe-based shape memory alloy behavior under thermomechanical cyclic loading. *Mech. Mater.* **2013**, *63*, 1–11. [CrossRef]
44. Koster, M.; Lee, W.; Schwarzenberger, M.; Leinenbach, C. Cyclic deformation and structural fatigue behavior of an FE–Mn–Si shape memory alloy. *Mater. Sci. Eng. A* **2015**, *637*, 29–39. [CrossRef]

materials

MDPI

Review

Iron-Based Shape Memory Alloys in Construction: Research, Applications and Opportunities

Zhe-Xi Zhang [1], Jie Zhang [1], Honglei Wu [1,2,*], Yuezhen Ji [1,3] and Dheeraj D. Kumar [1]

1 Department of Structural Engineering, College of Civil Engineering, Tongji University, Shanghai 200092, China; zzx888@tongji.edu.cn (Z.-X.Z.); 2032316@tongji.edu.cn (J.Z.); 1832499@tongji.edu.cn (Y.J.); 2093369@tongji.edu.cn (D.D.K.)
2 Tongji Architectural Design (Group) Co., Ltd., Shanghai 200092, China
3 China Railway ERYUAN Engineering (Group) Co., Ltd., Chengdu 610031, China
* Correspondence: 8whl@tjad.cn; Tel.: +86-021-3537-7165

Abstract: As a promising candidate in the construction industry, iron-based shape memory alloy (Fe-SMA) has attracted lots of attention in the engineering and metallography communities because of its foreseeable benefits including corrosion resistance, shape recovery capability, excellent plastic deformability, and outstanding fatigue resistance. Pilot applications have proved the feasibility of Fe-SMA as a highly efficient functional material in the construction sector. This paper provides a review of recent developments in research and design practice related to Fe-SMA. The basic mechanical properties are presented and compared with conventional structural steel, and some necessary explanations are given on the metallographic transformation mechanism. Newly emerged applications, such as Fe-SMA-based prestressing/strengthening techniques and seismic-resistant components/devices, are discussed. It is believed that Fe-SMA offers a wide range of applications in the construction industry but there still remains problems to be addressed and areas to be further explored. Some research needs at material-level, component-level, and system-level are highlighted in this paper. With the systematic information provided, this paper not only benefits professionals and researchers who have been working in this area for a long time and wanting to gain an in-depth understanding of the state-of-the-art, but also helps enlighten a wider audience intending to get acquainted with this exciting topic.

Keywords: iron-based shape memory alloy (Fe-SMA); shape memory effect; martensitic transformation; prestressing; low cycle fatigue; seismic; damping

Citation: Zhang, Z.-X.; Zhang, J.; Wu, H.; Ji, Y.; Kumar, D.D. Iron-Based Shape Memory Alloys in Construction: Research, Applications and Opportunities. *Materials* **2022**, *15*, 1723. https://doi.org/10.3390/ma15051723

Academic Editors: Francesco Iacoviello and Eunsoo Choi

Received: 20 December 2021
Accepted: 21 January 2022
Published: 25 February 2022

1. Introduction

Iron-based shape memory alloy (Fe-SMA, especially referring to Fe-Mn-Si class shape memory alloy) possesses shape memory effect (SME) [1,2], outstanding low-cycle fatigue (LCF) resistance [3], and some other desirable characteristics by which the material has proven its potential in the field of the construction industry. Fe-SMA was traditionally regarded as an ideal material used in fasteners and tie systems, e.g., pipe joints, rail couplings, and crane rail joint plates, where constrained stress is required [4]. The main purpose of using such materials is to simplify the construction process by its SME property which induces prestress conveniently. The so-called SME-induced prestress, also known as 'recovery stress', is associated with its unique deformation-induced martensitic transformation and subsequent heating–cooling process, where an approximate stress of 200–400 MPa can be generated.

Apart from its desirable prestressing capability, Fe-SMA also has excellent low-cycle fatigue resistance, which was first recognized by Sawaguchi in 2006 [3]. A series of studies have been carried out on this front and considerable achievements have been made. The engineering community has gained particular confidence with the completion of the 196-m skyscraper 'JP Tower Nagoya', where Fe-SMA seismic dampers were first employed in an actual project in seismic-prone cities. These works further expand the application boundary of Fe-SMA and inspire the interests of seismic engineers.

Fe-SMA has some extra benefits. It is reported that the corrosion-resistance of Fe-SMA is close to that of stainless steel due to the addition of Nickel and Chromium elements [5]. This makes Fe-SMA well suited to chloride environments, e.g., coastal/offshore engineering construction. In addition, in contrast to Nitinol (another popular class of SMA) which is less easily produced in large scale because of the demanding metallurgical process [6–8], Fe-SMA can be mass produced with conventional metallurgical equipment [9], and, more encouragingly, the cost of the raw materials is inherently low [10]. This facilitates practical use of Fe-SMA in the civil engineering sector, where the necessary size of elements/members is often large and the budget is often controlled.

This paper seeks to summarize the recent technological advances in the research and application of Fe-SMA in construction, covering diverse aspects including structural retrofitting and seismic damping. The paper is also keen to highlight the authors' unique reflections on the present issue and future needs in this field, in addition to the latest solutions. An overview of the basic mechanical properties of Fe-SMA is first presented, followed by the possible application scenarios emerging over the past 10 years. Challenges arising from the application of the new material are described, and where further studies are emphasized that are required to respond to the identified issues. Research needs and new application opportunities of Fe-SMA are also presented. While this paper is of scientific interest to the mechanical and material science community, much emphasis is placed on making the contents easier to learn by civil engineers. Therefore, in most cases, the presentation is structured following the civil engineering custom and terminology.

2. Basic Properties

Prior to the introduction of the application of Fe-SMA, it is necessary to have a comprehensive understanding of its basic characteristics. Typical physical properties of Fe-SMA together with other important steel types are listed in Table 1. The Fe-SMA presented in the table is Fe-17Mn-5Si-10Cr-4Ni-1(V,C) (mass%), which is one of the most typical classes. Other classes will be expressed in the form of their compositions when they are discussed.

Table 1. Physical properties for SMAs and conventional structural steels [5,11–13].

Property	Units	Value				
		Fe-SMA	NiTinol		Q235	SUS304
			(Martensite)	(Austenite)		
Density	g/cm^3	7.2–7.5	6.45–6.5		7.86	7.93
Young's modulus	GPa	170	28–41	83	201	193
Electrical resistivity	$\mu\Omega\cdot cm$	100–130	76–80	82–100	29.3	73
Specific Heat Capacity	$J/kg\cdot °C$	540	836.8	836.8	745	500
Thermal conductivity	$W/(m\cdot °C)$	8.4	8.6–10	18	61.1	16.3
Thermal expansion coefficient	$(\times 10^{-6})\ °C^{-1}$	16.5	6.6	11	12.6	17.2
Melting point	°C	1320–1350	1240–1310		1468	1398–1454
Strain recovery limit	%	2	10		-	-
Poisson's ratio	-	0.359	0.33		0.294	0.25

2.1. Metallographic Transformation in Fe-SMA

When a metallic substance is subjected to an external force, metallographic transformation process is triggered. At the micro-level, the crystal lattice matrix is rearranged and the atoms move in a particular way. The accumulated movements of the atoms on the microscopic scale directly leads to deformation of the metal on a macroscopic scale [5]. Fe-SMA has three types of metallographic phases, i.e., γ-austenite (face-centered cubic structure, fcc), ε-martensite (hexagonal close-packed structure, hcp), and α'-martensite (body-centered tetragonal structure, bct). Schematic illustrations of these crystal lattices are shown in Figure 1. The metallographic transformation between γ-austenite and ε-martensite (so-called 'martensitic transformation') occurs when the material is under an applied force and/or a temperature change, and thus the martensitic transformation of Fe-SMA can be categorized into stress-induced martensitic transformation and thermal-induced martensitic transformation.

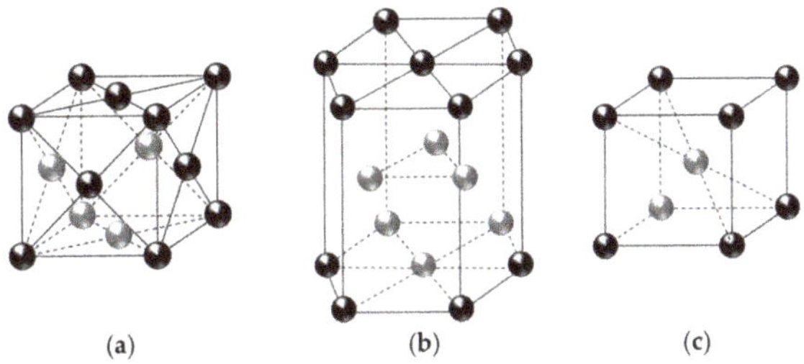

Figure 1. Micro-structure of crystal lattices: (**a**) γ-austenite; (**b**) ε-martensite; (**c**) α′-martensite.

The principle of thermal-induced martensitic transformation and its reverse process can be readily understood by the following description. As seen in Figure 2, four typical phase-transformation temperatures, i.e., martensite start temperature (M_s), martensite finish temperature (M_f), austenite start temperature (A_s), and austenite finish temperature (A_f), are representative ones related to the start and finish of martensitic transformation or its reverse process. When the initial temperature is high and it decreases to M_s, the transformation from γ-austenite to ε-martensite starts and then the ε-martensite fraction in Fe-SMA increases gradually. When the temperature drops below M_f, the transformation process is completed and the ε-martensite fraction in Fe-SMA reaches its maximum value (i.e., ζ in Figure 2). For the reverse transformation, when the rising temperature reaches A_s, the ε-martensite starts to transform into the γ-austenite phase, and this process continues until the temperature increases beyond A_f, where the ε-martensite is transformed into γ-austenite completely. Table 2 gives the measured results of the critical phase-transformation temperatures of typical SMAs.

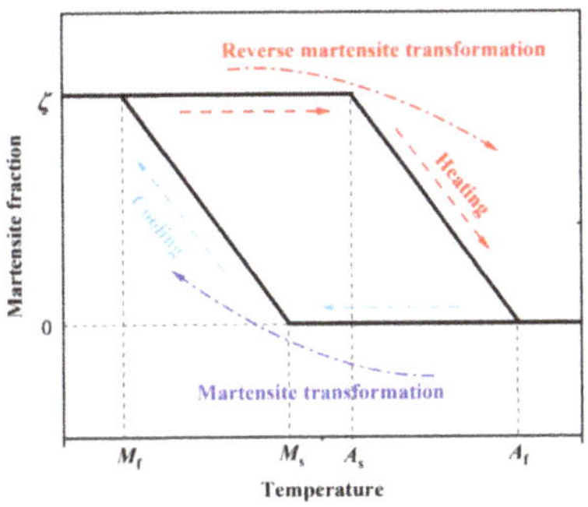

Figure 2. Schematic definition of thermal-induced martensitic transformation of Fe-SMA. ζ denotes the maximum martensite fraction, and $0 < ζ < 1$.

Table 2. Phase-transformation temperatures of SMAs.

Material	M_f (°C)	M_s (°C)	A_s (°C)	A_f (°C)	Ref.
Fe-SMA	−90	−75	85	110	[14]
Fe-SMA	−64	−60	103	162	[15]
NiTinol	−70–55	15–21	−22–2	17–30	[16]

The martensitic transformation of Fe-SMA has two characteristics which directly promote its unique mechanical behavior: (1) The stacking fault energy required for martensitic transformation is low, which makes the martensitic transformation easy to occur [5], (2) The martensitic transformation is a diffusionless phase transformation process (i.e., it creates a new crystal structure without introducing any compositional change) [5], making its reversible martensitic transformation possible. The former gives the basis on why the transformation between γ-austenite and ε-martensite would take place first when Fe-SMA is exposed to external force and the latter explains the basic mechanism of pseudo-elasticity & shape memory effect (this part will be further described in Section 2.3). More detailed descriptions of the martensitic transformation from a perspective of crystallography can be found elsewhere [5,17].

2.2. Monotonic Loading Property

Figure 3 compares the typical monotonic stress-strain curves of Fe-SMA with those of other structural steel, i.e., austenitic stainless steel S30408, low point steel LYP100, mild steel Q355, high-strength steel Q690 and aluminum alloy 7A04-T6. The stress-strain responses of Fe-SMAs seem to be similar to that of stainless steel, although the strength of the former is higher. Since there is no yield plateau in Fe-SMA, 0.2% proof stress is considered as an equivalent yield strength. The comparison of the basic mechanical properties between Fe-SMA and other metals is summarized in Table 3. The Young's moduli of the Fe-SMAs are generally comparable to, and may be slightly lower than, those of steel (except for aluminum alloy). High ultimate strengths (f_u > 700 MPa) and remarkable ductility are observed in the Fe-SMAs, characteristics which are encouraging for seismic application. The yield to ultimate strength ratio of Fe-SMA is around 0.4, indicating a pronounced strain hardening behavior. Fracture of Fe-SMA occurs soon after reaching the ultimate strength, indicating an inadequate necking process. This feature can be more clearly reflected when compared with conventional structural steel (such as mild steel Q355) where a significant localized shrinkage appears before fracture, as shown in Figure 4. This phenomenon also indicates that the ductility of Fe-SMA is mainly derived from its evenly distributed plasticity rather than localized necking. Apart from these characteristics, Fe-SMA is also reported to possess a higher yield strength at higher strain rates [18], which may make Fe-SMA an ideal material for blast-resistant structures, but no such study is currently available.

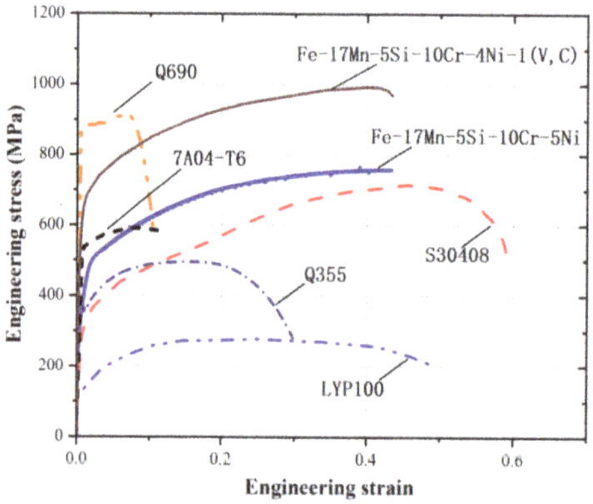

Figure 3. Typical full stress-strain curves of Fe-SMA and other typical structural steels [19–24].

Table 3. Mechanical properties of different metallic materials under quasi-static monotonic loading tests.

Material	E_0/GPa	f_y/MPa	f_u/MPa	f_y//f_u	ε_u	f_{u2}/MPa	ε_{u2}	*EL*(%)	Ref.
S30408	249	273	710	0.38	0.48	520	0.59	-	[22]
7A04-T6	72	537	594	0.91	0.08	590	0.10	-	[20]
LYP100	200	100	279	0.36	0.25	205	0.50	-	[19]
LY100	199	128	252	0.51	0.27	-	-	47.3	[25]
LY160	194	186	294	0.63	0.24	-	-	44.5	[25]
LY225	202	191	295	0.65	0.23	-	-	44.0	[25]
Q235	208	282	467	0.60	0.25	354	0.38	-	[24]
Q355	206	385	533	0.72	0.25	390	0.36	-	[24]
Q690	218	876	909	0.96	0.07	809	0.20	-	[24]
Fe-SMA	184	450	950	0.47	-	-	0.54	-	[26]
Fe-SMA	200	310	993	0.31	-	-	-	-	[27]
Fe-SMA	160	534	992	0.54	0.40	969	0.44	-	[21]
Fe-17Mn-5Si-10Cr-5Ni	172	297	774	0.38	0.45	698	0.48	-	[23]
Fe-28Mn-6Si-5Cr	170	250	800	0.31	0.40	-	-	-	[17]
Fe-15Mn-4Si-10Cr-8Ni	184	260	676	0.38	-	-	-	74.0	[28]

Notes: E_0 refers to Young's modulus; f_y refers to yield strength; f_u refers to ultimate tensile strength (UTS); f_{u2} refers to fracture stress; ε_u refers to the strain corresponds to UTS; ε_{u2} refers to the strain till fracture; *EL* refers to elongation after fracture.

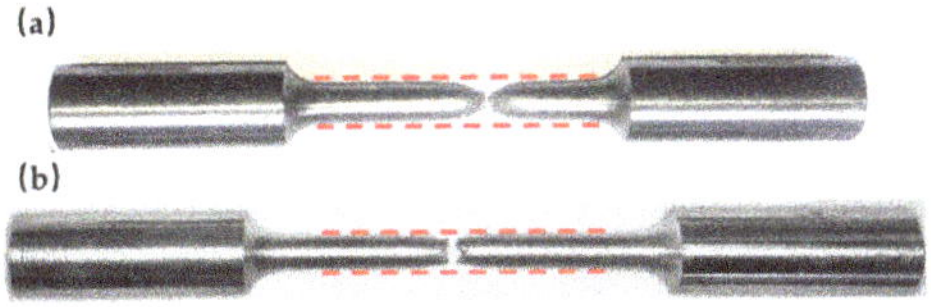

Figure 4. Macroscopic fracture behavior of (**a**) Q355 and (**b**) Fe-SMA under monotonic loading.

The evolution law of metallographic transformation during monotonic loading process [29–34] can be described as follows: the initial phase of Fe-SMA specimen can be considered pure γ-austenite since it is annealed during manufacturing, a process which is equivalent to the reverse transformation mentioned above. As the stress (or strain) increases, part of the parent γ-austenite gradually transforms into ε-martensite (see Path 1 in Figure 5), and this leads to a deviation of the monotonic stress-strain curve from the linear relationship [5,35]. As the stress further increases (see Path 2 in Figure 5), α′-martensite, whose structure is formed through the lattice expansion of fcc and hcp structures, is discovered in the field of γ-austenite and ε-martensite [5,9]. This transformation process continues until fracture. It is highlighted that when the material is loaded under a high service temperature, the formation of stress-induced ε-martensite may be strongly hindered, and this process is replaced by a direct transformation from parent γ-austenite to α′-martensite (see Path 5 in Figure 5).

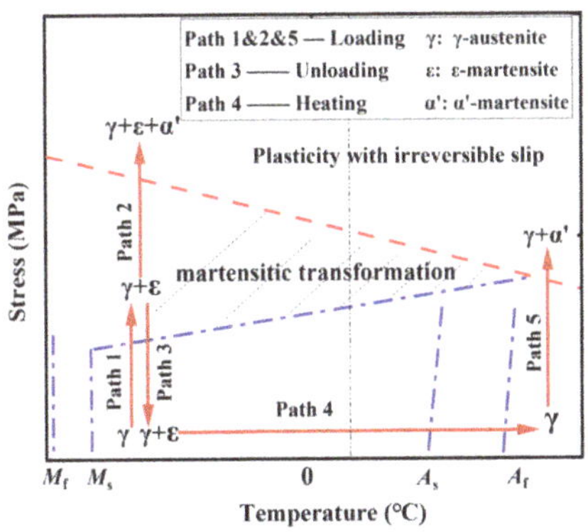

Figure 5. Evolution law of metallographic transformation for Fe-SMA under different thermal-mechanical states.

2.3. *Pseudo-Elasticity and Shape Recovery Property*

Hooke's law holds true for normal steel during both the loading and unloading stages. However, a nonlinear spring-back curve deviating from the linear path exists in Fe-SMA upon unloading (see Figures 6 and 7a). This unique phenomenon is called pseudo-elasticity [36–42], which is associated with the partial reverse transformation of the previously formed stress-induced ε-martensite. However, due to the limited fraction of stress-induced ε-martensite, the phenomenon of pseudo-elasticity is limited, i.e., much less significant than Nitinol [21,23,43,44]. In any case, the moderate pseudo-elasticity could still benefit residual deformation control during dynamic shakedown, as discussed later in Section 2.4.

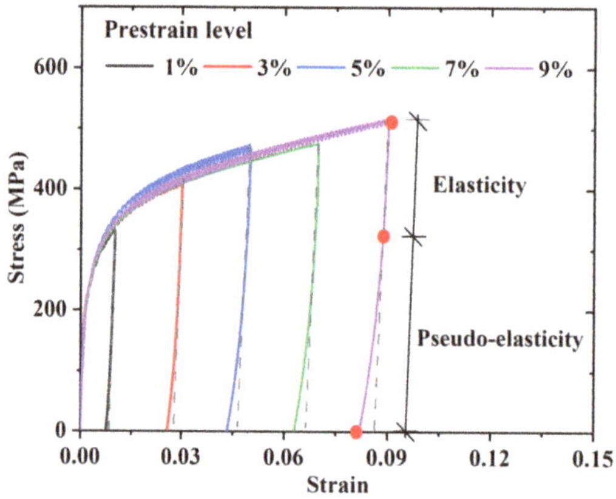

Figure 6. Pseudo-elasticity phenomenon of Fe-SMA.

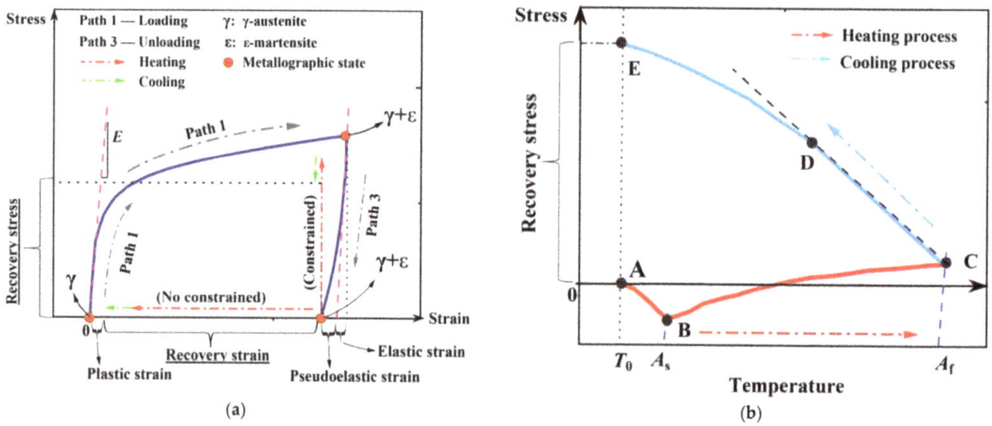

Figure 7. Schematic diagram of (**a**) activation process of shape recovery property, and (**b**) generation of recovery stress.

Shape recovery property, also known as shape memory effect (SME), is activated by heating the deformed material. The aforementioned thermal-induced ε→γ transformation contributes to the SME of Fe-SMA elements. Considering a Fe-SMA bar which is subjected to axial elongation and subsequently unloaded, residual deformation occurs, like normal steel. If the residual deformation of this bar is constrained, recovery stress (pre-stress) is generated after heating and cooling. Figure 7b illustrates the development of recovery stress (corresponding to path 4 in Figure 5). In the initial heating stage, stress relaxation is observed due to thermal expansion (path A→B). When the temperature reaches A_s, thermal-induced ε→γ transformation is triggered and the recovery stress starts to counteract thermal expansion. This process continues until the temperature increases to the predefined maximum temperature (path B→C). During the cooling stage (e.g., air cooling), a tensile stress is generated due to contraction and thus further recovery stress is induced, which first increases linearly with decreasing temperature (path C→D). As the tensile stress increases to a threshold of the minimum value for triggering the stress-induced martensite transformation, Fe-SMA again enters into the plastic stage. Consequently, a 'yield' phenomenon is observed (point D) and the stress-temperature curve becomes nonlinear (path D→E). Point E is the final recovery stress produced in this process. It is noteworthy that the preload level (i.e., path 1 in Figure 5) should be controlled within a moderate range to avoid the formation of α′-martensite, since the reverse transformation process cannot be realized when the material is in the α′-martensite state [9]. In other words, α′-martensite is responsible for the irreversible plasticity, i.e., unrecoverable plastic strain or permanent deformation.

When Fe-SMA is employed as a measure of prestressing, the amount of prestress is the property of most concern to engineers. With the aim of increasing the shape recovery properties, many research groups across the world have been devoted to developing improved production and manufacturing processes, including but not limited to thermal mechanical training, adjusting the chemical compositions and heat treatment [45–71]. These efforts greatly enhance the reliability of Fe-SMA as an emerging prestressing strategy. For engineering practice, a satisfactory prestress level can be achieved by pre-loading and heating the Fe-SMA in an appropriate way. Table 4 summarizes the reported recovery stress (f_R) of Fe-SMA generated from different activation conditions. A pre-strain level of 2–4% seems to be an optimum range to achieve a satisfactory recovery stress (300–450 MPa) when the activation temperature is below 200 °C [14,21,41,72]. It is also interesting to find that concurrently applying a higher activation temperature (350 °C) and a larger pre-strain level (6–8%) is beneficial for producing a larger recovery stress (which is almost 30% higher than that activated at 200 °C) [73]. An increase in the activation temperature may cause problems to concrete but is acceptable for steel structures [74]. Instead of applying monotonic pre-strain, researchers have also examined the recovery stress of Fe-SMA upon heating after experiencing tension-compression strain cycles, where the shape recovery capability is decreased [75]. One possible reason is that the compression history worsens the micro-structural environment for the reverse martensitic transformation process.

2.4. Cyclic Behavior, Low Cycle Fatigue and Energy Dissipation Capacity

The potential of Fe-SMA as energy-dissipating material was not recognized until 2006, when its stable hysteretic behavior and excellent low cycle fatigue resistance were first identified by Sawaguchi et al. [77]. For seismic application, it is essential to clarify the mechanical behavior of Fe-SMA under cyclic loading [78–94]. In this section, some basic properties of Fe-SMA under cyclic loading are discussed.

2.4.1. Hysteretic Behavior

Obtaining the hysteretic behavior under symmetrical cyclic loading is often a first and standard procedure to understand the basic performance of a member or material during earthquakes [95–100]. Figure 8a shows typical stabilized hysteretic loops (taken from half-life cycle response) of Fe-SMA and steel, where the curves are divided into the elastic (linear) part, transition part and hardening part. It is found that the loop shape of the Fe-SMA is slightly narrower than that of the mild steel. An early spring-back phenomenon, which results from the aforementioned pseudo-elasticity, is displayed during the unloading process, leading to a smaller elastic region. In addition, Fe-SMA shows a more obvious strain hardening response, whereas the mild steel shows little hardening with an almost flat stress-strain curve in the hardening part. It is reasonable to deduce that the residual deformation of structures with Fe-SMA-based components would be reduced during earthquake conditions, since the spring-back self-centering trend, together with a relatively large post-yield hardening, could effectively promote a self-centering capability of the system under dynamic shakedown [101,102]. Figure 8b further presents the full incremental stress-strain responses as well as the skeleton curves (obtained from connecting the peak stresses of the hysteretic loops) of Fe-SMA and mild steel.

Table 4. Recovery stresses for Fe-SMA under different activation conditions.

Ref.	Manufacturing Details	Specimen [1]	Pre-Strain [2]	Pre-Load	Activation Temperature	f_R/MPa [3]
[41]	Hot rolled at 1040 °C	SP-15 × 150	4%	50 MPa	160 °C	444
	Hot rolled at 1000 °C	SP-15 × 150	4%	50 MPa	160 °C	448
[14]	Hot rolled, solution treated at 1100 °C for 5 h, aging at 850 °C for 2 h	SP-0.7 × 1	4%	50 MPa	160 °C	330
		SP-0.7 × 1	4%	50 MPa	225 °C	380
[76]	Hot press at 1100 °C and cold rolled, solution treated at 1100 °C for 5 h and water quenched, aging at 850 °C for 2 h followed by air cooling	SP-0.7 × 1	4%	50 MPa	160 °C	565
[72]	Hot press at 1100 °C and cold rolled, solution treated at 1100 °C for 5 h and water quenched, aging at 850 °C for 2 h followed by air cooling	SP-0.8 × 2	2%	-	100 °C	290
		SP-0.8 × 2	4%	-	100 °C	303
		SP-0.8 × 2	2%	-	140 °C	317
		SP-0.8 × 2	4%	-	140 °C	355
[21]	Hot pressed and cold rolled	SP-1.5 × 15	0.5%	50 MPa	160 °C	293
		SP-1.5 × 15	2%	50 MPa	160 °C	346
		SP-1.5 × 15	2%	50 MPa	195 °C	388
		SP-1.5 × 15	4%	50 MPa	160 °C	334
		SP-0.5 × 15	4%	50 MPa	160 °C	331
		SP-1.5 × 15	6%	70 MPa	160 °C	298
		SP-1.5 × 15	8%	83 MPa	160 °C	334
[73]	Hot rolled at 1150 °C and cold rolled, solution treated at 1100 °C in H_2N_2-atmosphere for 5 h, aging at 850 °C for 2 h	SP-1.6 × 6	1%	50 MPa	200 °C	330
		SP-1.6 × 6	2%	50 MPa	200 °C	344
		SP-1.6 × 6	4%	50 MPa	200 °C	342
		SP-1.6 × 6	6%	50 MPa	200 °C	332
		SP-1.6 × 6	8%	50 MPa	200 °C	337
		SP-1.6 × 6	2%	50 MPa	349 °C	364
		SP-1.6 × 6	4%	50 MPa	350 °C	421
		SP-1.6 × 6	6%	50 MPa	350 °C	428
		SP-1.6 × 6	8%	50 MPa	350 °C	445

[1] 'SP' refers to the sheet plate specimen, and is expressed in the form of 'thickness × width'; [2] The pre-strain process is performed under room temperature environment; [3] 'f_R' refers to the recovery stress of Fe-SMA.

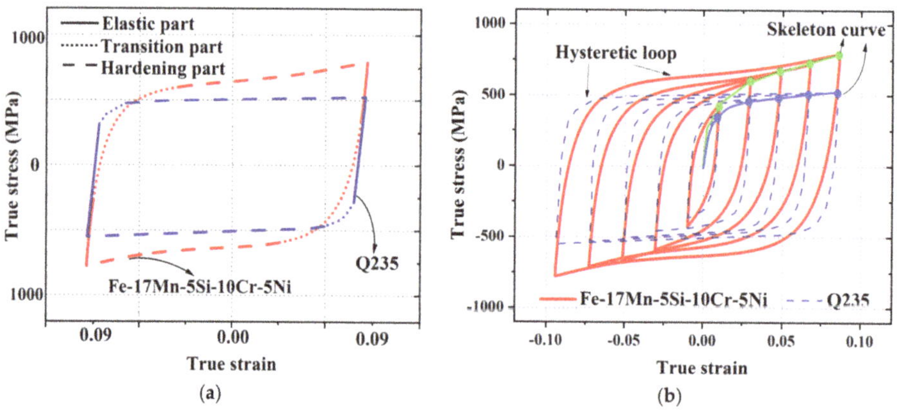

Figure 8. Comparison of (**a**) Decomposed hysteretic loops; and (**b**) Skeleton curves obtained from stabilized hysteretic loops of Fe-17Mn-5Si-10Cr-5Ni and mild steel (Q235).

Researchers have reported that the hysteretic loops of Fe-SMA can quickly saturate regardless of the strain range so that the peak stress is stable until failure [23]. However, according to the findings reported by Rosa et al. [26], a slight cyclic softening behavior (i.e., decrease in peak stress) appears when the strain rate reaches 0.08/s, which is an expected strain rate considering a real earthquake excitation [103]. Further research opportunity exists in comprehensive understanding of the rate effect of the material, especially in the context of seismic application.

Figure 9 presents the stabilized hysteretic loops at different strain amplitudes by moving their compressive corners to the same coordinate origin. If the ascending curves of these hysteresis loops coincide, i.e., a master curve can be drawn in accordance with the ascending curves, the material is deemed to possess Masing behavior. Previous research [84] reported that Masing behavior can be observed in Fe-SMA when the strain amplitudes are less than ±2%, as shown in Figure 9a. However, when the strain amplitude increases, the ascending branches of the saturated hysteresis loops deviate from the master curve and represent a non-Masing behavior (Figure 9b). Microstructural observations reveal that the adaptability of Fe-SMA to Masing behavior is related to its micro-deformation patterns. When the strain amplitude is moderate (e.g., not exceeding ±2%), the strain-induced martensitic transformation accompanied by a planar slip of Shockley partial dislocations in austenite is the main deformation mode [84]. However, when the strain amplitude advances to a larger range, the micro-deformation pattern is dominated by the formation of mechanical-twinning [90].

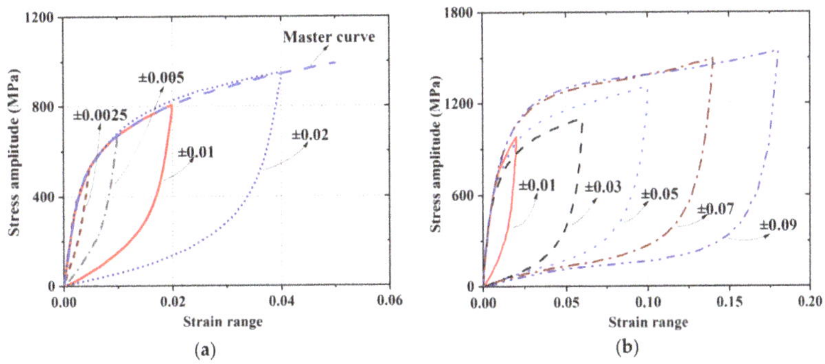

Figure 9. Adaptability of Fe-SMA to Masing behavior under (**a**) small strain ranges [84]; and (**b**) large strain ranges [23].

2.4.2. Low Cycle Fatigue (LCF) Behavior

Large-size Nitinol elements are often criticized for their brittle fracture behavior and poor LCF resistance [104–107]. Figure 10 and Table 5 summarize the LCF lives of some typical steels. It can be seen that Fe-SMA possesses significantly longer LCF life than conventional steels. This remarkable property results from the evolution of cyclically deformation-induced ε-martensitic transformation, in contrast to dislocation-based plasticity with irreversible slip in normal steels. As demonstrated in Figure 11, the parent γ-austenite phase partially transforms into a tension-induced ε-martensite during the loading process. During the unloading stage, the tension-induced martensite gradually diminishes and returns back to γ-austenite. When subjected to compression towards negative strain, compression-induced ε-martensite is generated in the field of γ-austenite [77,78]. The phase transformation of the re-loading stage is similar to that of the aforementioned unloading stage, and the repeated tension-compression cyclic loading process induces alternate formation and disappearance of stress-induced ε-martensite. The repeated phase transformation is beneficial in reducing the internal stress concentration caused by cyclic loading and inhibiting local accumulation of dislocations and hence the initiation and propagation of fatigue cracks [88].

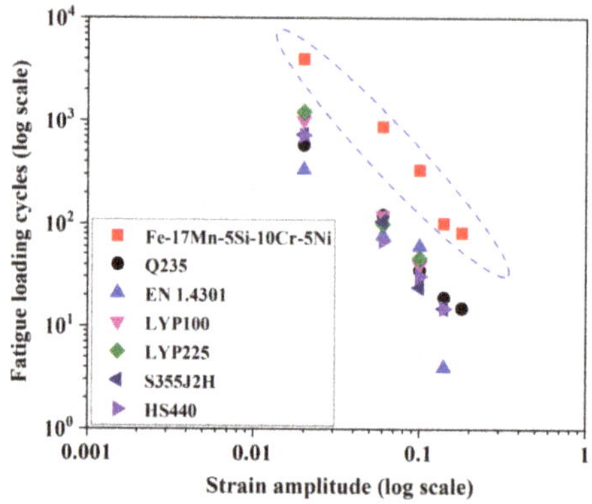

Figure 10. Comparison of fatigue lives between Fe-SMA and other typical metals [23,108,109].

Table 5. Numbers of cycles to failure of different metals.

Research Group	Material	Strain Amplitude					Ref.
		±1%	±3%	±5%	±7%	±9%	
Tongji University	Fe-17Mn-5Si-10Cr-5Ni	4007	880	210	102	83	[23]
Portland State University	GR345	536	69	27	16	-	[108]
	HPS485	400	51	21	13	-	[108]
	HS440	720	69	31	15	-	[108]
	LYP225	-	38	-	9	-	[108]
	LYP100	720	50	32	11	-	[108]
Imperial College	S355	495–732	53–107	22–24	9–15	-	[109]
	S235	439–521	16–21	8–20	3	-	[109]
	EN1.4301	266–335	27–78	7–61	2–4	-	[109]
Tsinghua University	LY100	512–694	82–103	-	-	-	[25]
	LY160	1008	121	40	-	-	[25]
	LY225	1220	101	46	-	-	[25]
Japanese National Institute for Materials Science	Fe-15Mn-(0-6)Si-10Cr-8Ni	2860–8500	-	-	-	-	[85]
	Fe-30Mn-(6-x)Si-xAl	2024–8070	-	-	-	-	[80]

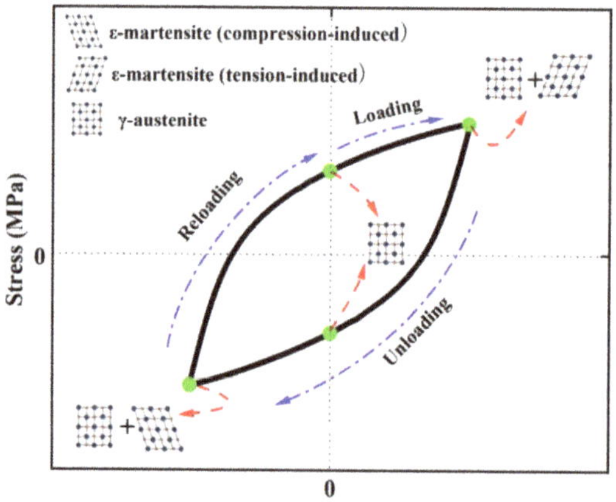

Figure 11. Phase transformation during cyclic tension-compression process for Fe-SMA.

2.4.3. Energy Dissipation

The equivalent viscous damping ratio (EVD), as defined in Equation (1), is considered to evaluate the energy dissipation capacity of Fe-SMA specimens under fully reversed cyclic loading.

$$\mathrm{EVD} = \frac{1}{2\pi}\frac{E_D}{E_S} \quad (1)$$

In the equation, E_D is the area within the inelastic force-displacement response curve, and E_S is the recoverable elastic strain energy stored in an equivalent linear elastic system [110]. The EVDs of different metals under half-life cycle are plotted in Figure 12a. It can be seen that the EVDs of Fe-SMA are lower (decreased by about 20%) than that of conventional steel. This is mainly attributed to its strain hardening behavior, resulting in a narrower hysteresis loop shape. The early spring-back behavior also decreases the E_S of Fe-SMA to some degree. On the other hand, the absolute energy dissipation W_D, which is calculated by the area of the stabilized stress-strain curves, shows that Fe-SMA could be have more energy dissipation, especially at larger strain amplitudes (see Figure 12b). More importantly, due to the excellent LCF resistance, the total energy dissipation (accumulation of E_D) of Fe-SMA is much larger than that of normal steel with the same geometric size.

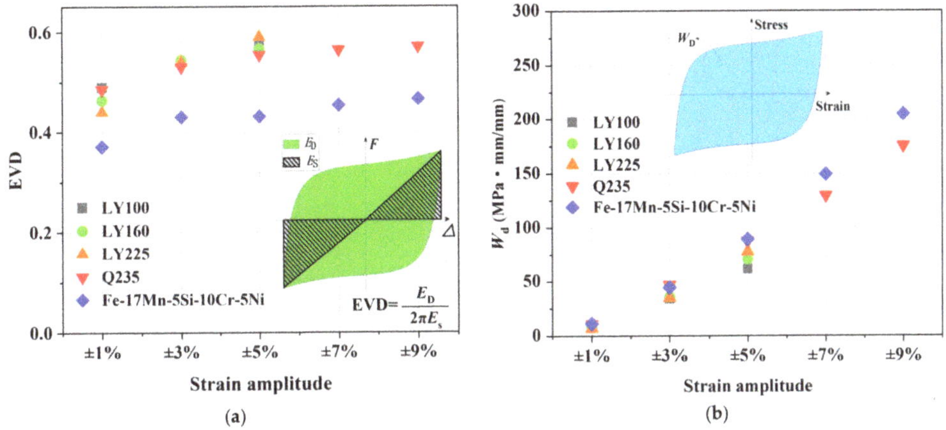

Figure 12. Parameters for evaluating energy dissipation capacity under different strain amplitudes: (**a**) EVDs; and (**b**) Absolute energy dissipation at half-life cycle. Results are re-produced from [23,25].

3. Research and Potential Engineering Applications

Early application of Fe-SMA in civil engineering began in the 1990s, where focus was mainly on special connections, such as railway fishplates, crane rail joint plates, and pipe connection devices for tunnel construction [5,9,17,111]. These connections are tightened via the SME of Fe-SMA. The SME-triggered tightening method greatly simplifies the construction/prestressing process.

Recent research and development activities enable a wider use of Fe-SMA in the construction industry. In particular, the research and application of Fe-SMA over the past 10 years can be mainly divided into two fields: (1) SME-induced prestressing technique for repairing and strengthening structures [112–139], and (2) seismic damping [23,28,75].

3.1. Novel Strengthening Solution Based on SME of Fe-SMA

High-cycle fatigue causes crack propagation in structural components, a case which in turn leads to the deterioration of stiffness/strength and shortens the service life of the structure. Practice has proven that through strengthening the cracked or damaged components, the service life can be effectively prolonged. Existing strengthening strategies include using external bonding reinforcing materials or applying prestressing. Overcoming some possible shortcomings such as the difficulty in construction for the traditional reinforcing strategies, a new method utilizing Fe-SMA has received great attention. The main procedure of the Fe-SMA-based strengthening solution is similar to that of the aforementioned SME-triggered tightening method and can be summarized as follows (see Figure 13):

(1) Pre-deform (for most cases, pre-tension) the Fe-SMA elements to a preset strain value or the expected shape;
(2) Connect the pre-deformed Fe-SMA elements to the base components (i.e., those ready for strengthening or connecting);
(3) Apply electric heating (or infrared heating) to the pre-deformed Fe-SMA elements to a preset temperature and keep it for a short period of time to guarantee fully activated SME;
(4) Wait until the Fe-SMA elements cool down to room temperature, and recovery stress is generated;
(5) The structures then return to normal service state.

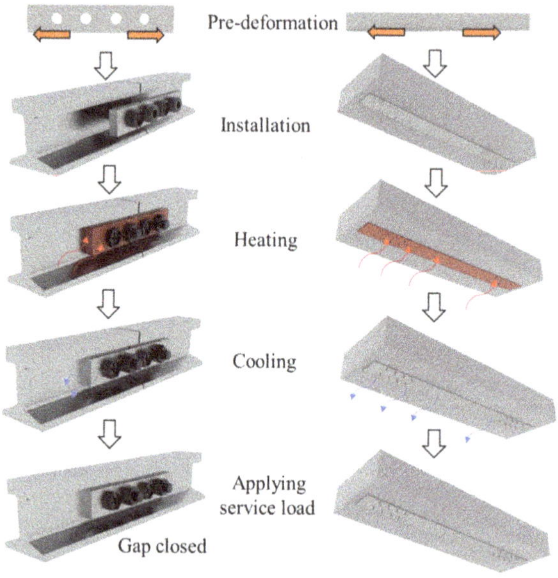

Figure 13. Schematic illustration of procedure for Fe-SMA-based strengthening/connecting solution.

3.1.1. Strengthening for Reinforced Concrete (RC) Structures

The earliest practical application of Fe-SMA in the field of prestressing can be traced back to 2001 where a bridge in Michigan, United States, experienced fatigue-induced cracking [120]. In this case, Fe-SMA tendons were installed perpendicularly to the shear cracks. After electric heating, a recovery stress of approximately 225 MPa was induced in the Fe-SMA tendons. Field measurements indicated that the generated recovery stress closed the width of the shear cracks to a large extent and the load-carrying capacity of this bridge was effectively recovered.

Since the initial success, laboratory research works have been conducted on concrete structures strengthened by Fe-SMA reinforcement, and suitable anchorage systems have been developed. For example, EMPA (Swiss Federal Laboratories for Materials Science and Technology) proposed an anchorage system for Fe-SMA tendons employed as near-surface mounted reinforcement (NSMR) in RC structures [121–123]. As shown in Figure 14a, Fe-SMA prestressing elements are embedded in pre-made grooves and covered with adhesive material such as cement-based mortar herein. Lap-shear experiments have been carried out to clarify the bonding behavior between the Fe-SMA strips and cement-based mortar [122].

Deeper embedment depth and ribbed surface for Fe-SMA strips are recommended for practical application [124]. It is also found that the current design guidelines would underestimate the necessary anchorage length for Fe-SMA bars [125]. The corresponding calculation methods are yet to be available.

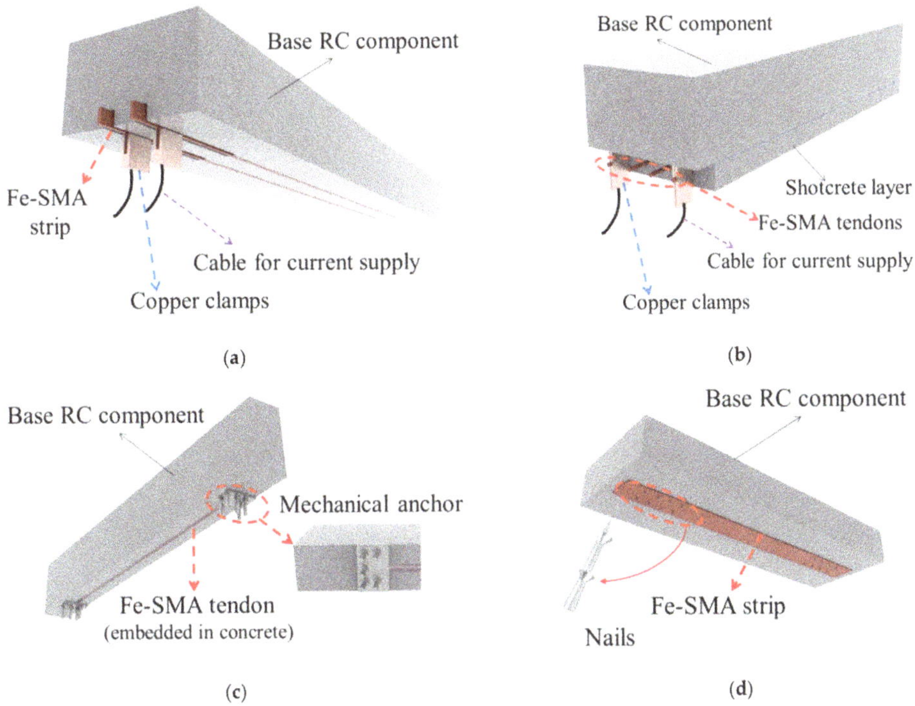

Figure 14. Schematic illustration of different anchoring methods for Fe-SMA tendon/strip strengthening RC components: (**a**) NSMR method; (**b**) shotcrete method; (**c**) NSMR-expansion anchor method; and (**d**) nail-based method.

With the aim of further simplifying the anchoring process and satisfying the objective of rapid recovery on site, a new anchoring method employing shotcrete is proposed (see Figure 14b) [126]. The previously tensioned Fe-SMA tendons are installed beneath the beam with an additional cementitious layer (shotcrete) sprayed on, covering the Fe-SMA tendons. After sufficient curing, current resistance heating is applied for activation and prestress is induced. Feasibility studies on flexural strengthening [126] and shear strengthening [127] of RC beams with this anchoring method have been conducted. Both the test results revealed that this strengthening system can efficiently increase the flexural/shear-resistant performance of RC beams. At the same time, beam deflections, number of cracks, and the widths of cracks were all reduced.

The flexural behavior of RC beams strengthened by the Fe-SMA NSMR system was investigated in [128], where ribbed Fe-SMA strips were longitudinally embedded at the bottom of the beam. Copper clamps were used to transmit the electric current. After current resistance heating to a target temperature of 160 °C, a permanent prestress of about 200 MPa was created in the Fe-SMA strips. Rojob et al. [129] conducted a comparative experimental study on the effect of strengthening through CFRP strips and Fe-SMA strips, and confirmed that the Fe-SMA strips lead to better ductility of the beam. Rojob et al. [130] further added expansion anchor to this system (see Figure 14c) and found that the ductility of the RC beam was further improved. This is because the additional expansion anchor provides an extra force transmission path, which maintains beam function after the Fe-SMA tendons are stripped from the adhesive material.

Nail-based mechanical anchorage system is an alternative method for the Fe-SMA NSMR system. As shown in Figure 14d, Fe-SMA strips can be easily fixed to the surface of the base concrete layer by the aid of direct-fasteners (e.g., nails) and nail-setting devices [131]. The total duration for installing and activating a 5-m Fe-SMA strip is within 20 min [132]. This method has been applied to some retrofitting cases in Switzerland [132]. However, nail-based anchorage systems may not be the best solution for bridges since the nails tend to loosen under HCF-loading conditions.

It should be noted that for RC structures, the target temperature during electric heating should be carefully controlled, since high temperatures may cause concrete cracking/damage and could be detrimental to the bond strength between Fe-SMA and concrete [73]. As reported in [125], a longitudinal splitting crack with a width of about 0.05 mm appeared in the mortar surface when the Fe-SMA was heated up to 190 °C. Most existing studies adopt a maximum activation temperature of around 160 °C, which can be regarded as a feasible target temperature.

3.1.2. Strengthening for Steel Structures

The friction-based mechanical anchorage system, which is feasible for strengthening steel structures with Fe-SMA prestressing strips, was first developed by EMPA, as shown in Figure 15a [133]. Glass-fiber-reinforced plastic (GFRP) laminates and friction foils are also involved in this anchorage system, along with clamping plates and bolts which are necessary for anchoring. The GFRP laminates electrically insulate the Fe-SMA strips from the steel plate during the activation procedure, thus avoiding energy waste and reduction of heating efficiency. Extra friction foils are often used to increase the static friction coefficient for this joint. The experiment was first conducted on simple steel plates, and it was shown that a 2% pre-strain of the Fe-SMA strips can produce a recovery stress of about 330–410 MPa after heating to 260 °C, resulting in a compressive stress of about 35–74 MPa in the base steel plates. A fatigue test was further conducted and the results proved that the fatigue life of these strengthened steel plates was evidently increased and the propagation of initial cracks was postponed and even arrested in some cases [134]. Appropriate modifications were subsequently made and the applications were extended to fatigue strengthening of metallic girders (see Figure 15b) [135,136] and connections (see Figure 15c) [137]. Similar conclusions were drawn from these works.

Recently, a novel fatigue strengthening solution for steel structures using adhesively bonded Fe-SMA strips was investigated by EMPA. The adhesive Sika1277 was used to bond the Fe-SMA strips to the steel plates [138]. It is reported that the bonding force is approximately twice the prestress achieved in the Fe-SMA strips. No softening behavior was observed during the activation process, which means that the adhesive can securely anchor the Fe-SMA strip throughout the whole strengthening process [139]. Due to the bridging mechanism of the adhesive anchorage, crack opening in the base structure was suppressed and stress singularity at the crack tip was also significantly reduced [139]. However, future studies are still needed to investigate the time-varying behavior of this bonding-based anchorage system during the entire service life.

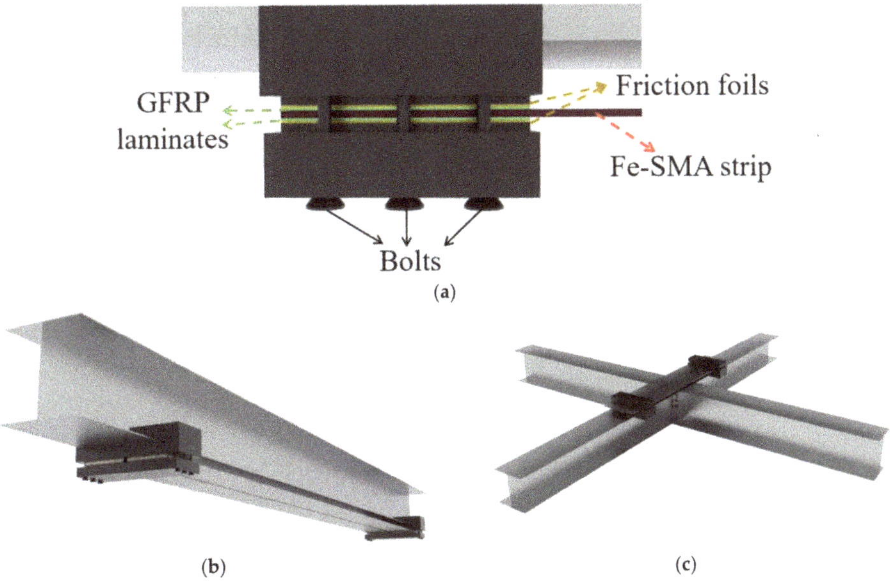

Figure 15. Schematic illustration of friction-based mechanical anchorage system for Fe-SMA strips in steel structures: (**a**) Configuration details; (**b**) Strengthening of metallic girders; and (**c**) strengthening of metallic connections.

3.2. Seismic Dampers

Conventional metal dampers are usually made of steel with a reasonably low yield strength, which encourages early participation in energy dissipation. Ductility and durability are also important characteristics, since many strong ground motions followed by a series of aftershocks have been recorded in the past decades [140]. A Japanese industry–academic–government joint research group had developed Fe-SMA-based (Fe-15Mn-4Si-10Cr-8Ni) buckling restrained shear dampers (BRS, see Figure 16a) and buckling-restrained braces (BRB, see Figure 16b), and used them in the JP Tower Nagoya and The Aichi International Convention & Exhibition Center, Tokoname, respectively [28]. Related experiments have been conducted [23,28,75] and the results confirmed that the Fe-SMA seismic dampers exhibit considerably longer fatigue life (around ten times) than conventional steel dampers. Loading tests with random seismic wave inputs were also performed and the results showed that the seismic dampers exhibit stable energy absorption behavior under a wide range of deformation angles, reflecting a reliable performance during earthquake sequences [28].

A more comprehensive experimental study on Fe-SMA (Fe-17Mn-5Si-10Cr-5Ni) BRS was conducted by the authors and co-workers recently [23] (Figure 17a). Loading protocols with constant and incremental symmetrical shear displacement amplitudes, marked as 'protocol I' and 'protocol II' in Figure 17b, respectively, were employed to investigate the hysteresis response of the Fe-SMA-based BRSs. Such loading protocols were also conducted on steel (Q235) BRSs with the same geometry, and the test results are compared in Figure 18. The hysteretic loops of Fe-SMA-based BRSs are slightly narrower than those of steel BRSs (half-life cycle EVD = 0.42 vs. 0.52 under constant displacement amplitude), which is consistent with the material-level observation described previously. Importantly, significantly enhanced fatigue resistance was achieved in the Fe-SMA-based BRSs, leading to a considerable increase in the total accumulated energy dissipation (E_T) (Figure 19). Figure 20 shows the final crack patterns of Fe-SMA- and Q235-based BRSs in this experiment. The cracks of the Fe-SMA-based BRSs tended to be initiated in the center region of the core plates, with a subsequent crack propagation to the arc-shaped edge, whereas the cracks of the Q235-based ones were initiated from the arc-shaped edge region. Research opportunities exist in further investigating the reasons behind the difference in the fracture mechanism between Fe-SMA and steel shear dampers.

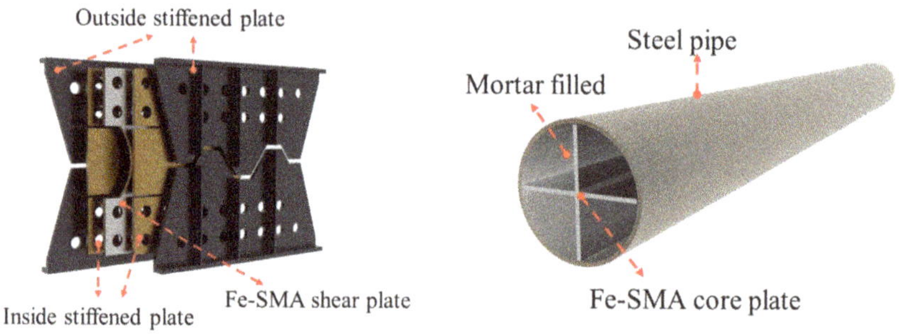

(a) (b)

Figure 16. Configuration of Fe-SMA-based (**a**) BRS; and (**b**) BRB developed by a Japanese industry–academic–government joint research group.

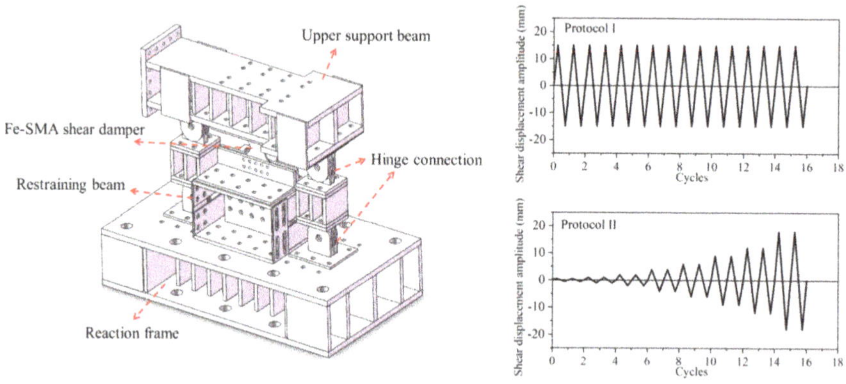

(a) (b)

Figure 17. Laboratory experiment of Fe-SMA-based BRS: (**a**) illustration of test setup; and (**b**) loading protocols.

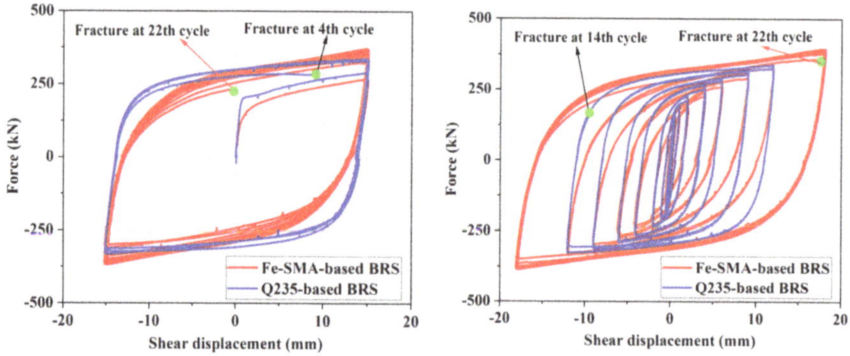

(a) (b)

Figure 18. Hysteretic behavior for Fe-SMA- and Q235-based BRS under: (**a**) Protocol 'I'; and (**b**) Protocol 'II'.

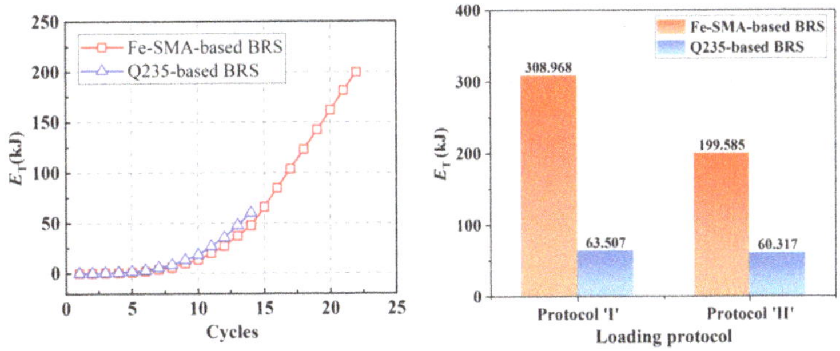

(**a**) (**b**)

Figure 19. The total amount of accumulated energy dissipation (E_T) for Fe-SMA- and Q235-based BRS: (**a**) E_T vs. Cycles under Protocol 'I'; and (**b**) a comparison of E_T between these two metal BRS under different loading protocols.

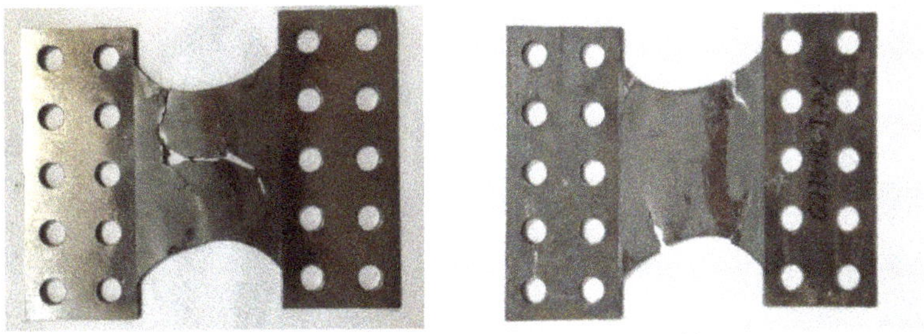

(**a**) (**b**)

Figure 20. Macroscopic fracture behavior of (**a**) Fe-SMA-based BRS; and (**b**) Q235-based BRS under loading protocol 'I'.

More recently, the authors and co-workers have completed a series of tests on Fe-SMA U-shaped dampers, as shown in Figure 21a. It was found that the LCF life of the Fe-SMA dampers is 5–7 times that of their steel counterparts. A representative Fe-SMA damper hysteretic curve is shown in Figure 21b, where full and stable hysteresis curves are observed. The details of this experimental program will be published in a separate paper.

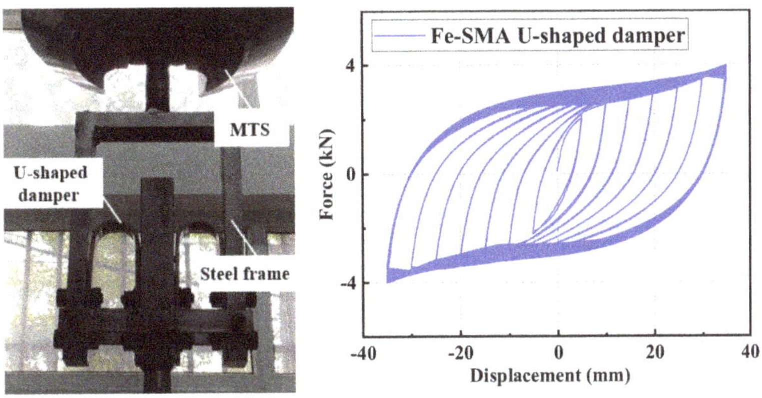

(**a**) (**b**)

Figure 21. Fe-SMA U-shaped damper tests: (**a**) test setup; and (**b**) hysteretic response under incremental loading protocol.

3.3. *Advantages Compared with Alternative Solutions*

Summarizing the above studies and applications, the main advantages of the Fe-SMA solutions are further elaborated here. In the field of retrofitting, the prestressing process of Fe-SMA strips is easier than that of the CFRP tendon-based reinforcement which currently prevails [133,136,139,141]. This is mainly because the former can be activated through electrical heating without any heavy hydraulic jacks or dedicated mechanical clamps. Moreover, the required fire protection for Fe-SMA strips can be less demanding than that required for CFRP strips. Studies have found that Fe-SMA strips have a positive effect in countering the relaxation behavior of structures at elevated temperatures [142]. Furthermore, corrosion, creep, and relaxation behavior under extreme environments (e.g., high-temperature and chlorine environments) have been investigated which confirmed the suitability and reliability of Fe-SMA [120,143–150]. Another attractive advantage of Fe-SMA-based strengthening strategy is its re-prestressing property, i.e., HCF-induced relaxation in recovery stress can be restored through repeated rounds of thermal activation [18,134,151]. This process is simple and can be implemented without the necessity for the time and labor-intensive disassembling procedure.

As for the Fe-SMA-based seismic dampers which have attracted great attention in the community of seismic designers, the fatigue-free and maintenance-free seismic design ambition may become possible. These significantly benefit society since the maintenance costs including those related to hazards have become an important part of governments' expenditure over the years (for example, about 400 billion Euros are paid for maintenance of buildings in Europe [152]).

Despite the fact that the price of Fe-SMA is currently higher than conventional structural steel, the whole-life costing of Fe-SMA-based applications may be comparable with the conventional technologies. It is noted that Fe-SMA is much cheaper than NiTi SMA which is often criticized for its high cost [153–157]. Existing studies also found that the total costs of Fe-SMA- and CFRP-based strengthening solutions are comparable when the cost of dedicated mechanical clamps for prestressing CFRP strips as well as the cost of labor were considered [141]. Moreover, long term costs can be saved to some degree due to the low maintenance requirements and the corrosion-resistance property of Fe-SMA. Importantly, since Fe-SMA shares similar production process to stainless steel [9], it is optimistic to predict that the price of Fe-SMA has the potential to approach that of stainless steel as long as the demand matches with production quantities.

4. Further Research Needs

There is still a gap to be filled for a comprehensive understanding of Fe-SMA's mechanical properties towards more confident application in practice. Metal plastic forming and heat treatment have a strong influence on the mechanical properties of Fe-SMA. Cold-worked Fe-SMA displays higher yield strength and higher shape recovery stress than that of hot-rolled ones [76], and the stress-strain curves may also show differences. The rationale behind the influence of these factors has not been comprehensively studied, and there are still limited constitutive models developed for Fe-SMA.

Welding is another important aspect for engineering application of Fe-SMA. Although the applicability of different welding technologies such as tungsten-inert gas, laser beam, and electron beam welding have been studied [158–162], some crucial welding properties and technologies are still under investigation. For example, in most of the existing experiments, Fe-SMA was welded to Fe-SMA or austenitic stainless steel in which the base metal shares the same metallography (γ-austenite). However, there has been little experimental research focusing on welding methods for connections between Fe-SMA and conventional structural steel. In addition, most of the existing studies focused on the effect of welding on the SME of Fe-SMA, whereas the mechanical properties of the weld itself and the heat-affected zone (HAZ) have been inadequately studied. Special attention should be paid to the fracture initiation mechanism of the weld, because the fatigue resistance as well as the plastic deformation capability of the weld and HAZ are usually inferior to the parent material of Fe-SMA, a case which leads to early fracture of the weld zone before the LCF failure of Fe-SMA. Further research opportunities exist in seeking for reliable weld techniques for Fe-SMA-based components, especially in the field of seismic engineering.

From a fracture mechanics' point of view, previous research works have found that the crack distribution pattern of Fe-SMA-based dampers is very different from steel dampers [23], which may be related to the unique reversible martensitic transformation of Fe-SMA. However, an accurate explanation for the difference in the crack initiation and propagation behavior is still unavailable. Research opportunities exist in revealing how the microscopic feature of Fe-SMA would affect the macroscopic fatigue failure mode.

In future, Fe-SMAs may be used as next-generation structural steel due to its high strength, high ductility, excellent energy dissipation capability and LCF resistance during earthquakes. To this end, a more in-depth understanding of the mechanical behavior of Fe-SMA elements (such as springs, tendons, cables, etc.) and structural components (such as beams, columns, shear plate walls, etc.) is required. Since differences in mechanical properties exist between Fe-SMA and conventional structural steel, the applicability of the existing construction and design methods for normal steel to Fe-SMA-based structural components is questionable. In particular, a more unique system-level behavior is anticipated when the Fe-SMA elements are employed because of the special cyclic hardening and unloading spring-back behavior. Fe-SMA may also be used together with super-elastic Nitinol to reach a hybrid design which enables a good balance among self-centering capability, energy dissipation, and cost, compared with a pure Nitinol solution [44,163–167]. Preliminary seismic collapse safety assessment has revealed that the collapse capacity of beam-column joints equipped with Fe-SMA is significantly improved [168]. More studies are needed towards establishing a systematic and standardized design and construction process for Fe-SMA-based components and structural systems, and the entire design philosophy has to be revisited in future. Similarly, for the application of Fe-SMA in strengthening, further work is required to develop an integral design approach including the selection of materials, installation process, activation process, and quality check standard.

Last but not least, when one utilizes both the excellent LCF resistance and SME of Fe-SMA, a completely new structural design philosophy, namely, fatigue free and in situ recoverable structural design, may be developed. In other words, even if small residual deformation exists after an earthquake, a further deformation recovery may be promoted by heating the material via either electrical resistance or infrared heating, where the entire process is safe, quick, and economical. In this regard, the recoverable strain/recovery stress of cyclically 'trained' Fe-SMA (rather than that of the material under monotonic loading) is worth further studies.

5. Concluding Remarks

Fe-SMA is an emerging high-performance metal with unique properties that make it well suited to many applications in the construction sector. Its shape recovery properties result from reversible martensitic transformation and have been utilized for prestressing/retrofitting structural components. It is foreseeable that the convenient prestressing process and sound re-activating properties would promote the material for wider use in the field of structural retrofitting. A series of anchoring systems for Fe-SMA strengthening solution have been developed for different structural types.

The use of Fe-SMA as seismic-resistant material began in the 2000s. Satisfactory ductility and energy dissipation capability are identified. The repeated tension and compression-induced martensite, generated through cyclic loading, suppresses the micro-crack initiation and/or propagation which directly enhances the LCF life of Fe-SMA. These features make fatigue-free and maintenance-free seismic design philosophy possible in the near future.

Challenges and opportunities do remain for a more confident application of Fe-SMA in construction. For example, the influences of forming process and heating effects on the mechanical properties of Fe-SMA have not been fully investigated. Technologies for welding of Fe-SMA are needed. Constitutive models applicable to Fe-SMA should be proposed and the fundamental fracture mechanism of the material under different stress states needs to be clarified. Existing design approaches may be revisited when designing Fe-SMA-based structural components since many unique mechanical behaviors exist in Fe-SMA. The highly nonlinear stress-strain response, especially the substantial strain hardening characteristic may be a key design challenge for Fe-SMA-based seismic dampers. It is advised that more experiments on the component level and even system level should be conducted, seeking for a more comprehensive understanding of this material and its behavior in a structural system.

Author Contributions: Conceptualization, Z.-X.Z. and J.Z.; methodology, Z.-X.Z.; software, Z.-X.Z. and J.Z.; validation, Z.-X.Z., J.Z. and H.W.; formal analysis, Z.-X.Z.; investigation, Z.-X.Z. and J.Z.; resources, Z.-X.Z. and J.Z.; data curation, Z.-X.Z., Y.J. and J.Z.; writing—original draft preparation, Z.-X.Z.; writing—review and editing, Z.-X.Z., J.Z., D.D.K. and H.W.; visualization, Z.-X.Z., D.D.K. and J.Z.; supervision, Z.-X.Z. and H.W.; project administration, H.W.; funding acquisition, H.W. All authors have read and agreed to the published version of the manuscript.

Funding: The financial support from the National Natural Science Foundation of China (NSFC) with Grant Nos. 52078359 and 51778456 are gratefully acknowledged. Support for this study was also provided by the Shanghai Rising-Star Program (20QA1409400).

Institutional Review Board Statement: Not applicable.

Informed Consent Statement: Not applicable.

Data Availability Statement: Data presented in this study are available in this article.

Conflicts of Interest: The authors declare no conflict of interest.

References

1. Sato, A.; Chishima, E.; Soma, K.; Mori, T. Shape memory effect in gamma-ebsel transformation in Fe-30Mn-1Si alloy single crystals. *Acta Metall.* **1982**, *30*, 1177–1183. [CrossRef]
2. Sato, A.; Mori, T. Development of a shape memory alloy Fe-Mn-Si. *Mater. Sci. Eng. A* **1991**, *146*, 197–204. [CrossRef]
3. Sawaguchi, T.; Kikuchi, T.; Ogawa, K.; Yin, F.X.; Kajiwara, S.; Kushibe, A.; Ogawa, T. Internal Friction of Fe-Mn-Si-Based Shape Memory Alloys Containing Nb and C and Their Application as a Seismic Damping Material. *Key Eng. Mater.* **2006**, *319*, 53–58. [CrossRef]
4. Lee, W.J.; Weber, B.; Leinenbach, C. Recovery stress formation in a restrained Fe–Mn–Si-based shape memory alloy used for prestressing or mechanical joining. *Constr. Build. Mater.* **2015**, *95*, 600–610. [CrossRef]
5. Cladera, A.; Weber, B.; Leinenbach, C.; Czaderski, C.; Shahverdi, M.; Motavalli, M. Iron-based shape memory alloys for civil engineering structures: An overview. *Constr. Build. Mater.* **2014**, *63*, 281–293. [CrossRef]
6. Fang, C.; Yam, M.C.H.; Ma, H.; Chung, K.F. Tests on superelastic Ni–Ti SMA bars under cyclic tension and direct-shear: Towards practical recentring connections. *Mater. Struct.* **2013**, *48*, 1013–1030. [CrossRef]
7. Fang, C.; Zhou, X.; Osofero, A.; Shu, Z.; Corradi, M. Superelastic SMA Belleville washers for seismic resisting applications: Experimental study and modelling strategy. *Smart Mater. Struct.* **2016**, *25*, 105013. [CrossRef]
8. Wang, W.; Fang, C.; Liu, J. Large size superelastic SMA bars: Heat treatment strategy, mechanical property and seismic application. *Smart Mater. Struct.* **2016**, *25*, 075001. [CrossRef]
9. Sawaguchi, T.; Maruyama, T.; Otsuka, H.; Kushibe, A.; Inoue, Y.; Tsuzaki, K. Design Concept and Applications of Fe–Mn–Si-Based Alloys—from Shape-Memory to Seismic Response Control. *Mater. Trans.* **2016**, *57*, 283–293. [CrossRef]
10. Janke, L. Applications of shape memory alloys in civil engineering structures—Overview, limits and new ideas. *Mater. Struct.* **2005**, *38*, 578–592.
11. Lu, M. *Handbook of Practical Mechanical Engineering Materials*; Liaoning Science and Technology Press: Shenyang, China, 2004.
12. General Administration of Quality Supervision Inspection and Quarantine of the People's Republic of China. *General Administration of Quality Supervision Inspection and Quarantine of the People's Republic of China, Stainless and Heat-Resisting Steels-Designation and Chemical Composition: GB/T 20878-2007*; Standards Press of China: Beijing, China, 2007.
13. Mohd Jani, J.; Leary, M.; Subic, A.; Gibson, M.A. A review of shape memory alloy research, applications and opportunities. *Mater. Des.* **2014**, *56*, 1078–1113. [CrossRef]
14. Dong, Z.; Klotz, U.E.; Leinenbach, C.; Bergamini, A.; Czaderski, C.; Motavalli, M. A Novel Fe-Mn-Si Shape Memory Alloy with Improved Shape Recovery Properties by VC Precipitation. *Adv. Eng. Mater.* **2009**, *11*, 40–44. [CrossRef]
15. Lee, W.J.; Weber, B.; Feltrin, G.; Czaderski, C.; Motavalli, M.; Leinenbach, C. Phase transformation behavior under uniaxial deformation of an Fe–Mn–Si–Cr–Ni–VC shape memory alloy. *Mater. Sci. Eng. A* **2013**, *581*, 1–7. [CrossRef]
16. Chen, J.; Wang, W.; Fang, C. Manufacturing, testing and simulation of novel SMA-based variable friction dampers with enhanced deformability. *J. Build. Eng.* **2021**, *45*, 103513. [CrossRef]

17. Maruyama, T.; Kubo, H. Ferrous (Fe-based) shape memory alloys (SMAs): Properties, processing and applications. In *Shape Memory and Superelastic Alloys*; Elsevier: Amsterdam, The Netherlands, 2011; pp. 141–159.
18. Ghafoori, E.; Hosseini, E.; Leinenbach, C.; Michels, J.; Motavalli, M. Fatigue behavior of a Fe-Mn-Si shape memory alloy used for prestressed strengthening. *Mater. Des.* **2017**, *133*, 349–362. [CrossRef]
19. Zirakian, T.; Zhang, J. Structural performance of unstiffened low yield point steel plate shear walls. *J. Constr. Steel Res.* **2015**, *112*, 40–53. [CrossRef]
20. Wang, Y.; Wang, Z. Experimental Investigation and FE Analysis on Constitutive Relationship of High Strength Aluminum Alloy under Cyclic Loading. *Adv. Mater. Sci. Eng.* **2016**, *2016*, 2941874. [CrossRef]
21. Shahverdi, M.; Michels, J.; Czaderski, C.; Motavalli, M. Iron-based shape memory alloy strips for strengthening RC members: Material behavior and characterization. *Constr. Build. Mater.* **2018**, *173*, 586–599. [CrossRef]
22. Yin, F.; Yang, L.; Wang, M.; Zong, L.; Chang, X. Study on ultra-low cycle fatigue behavior of austenitic stainless steel. *Thin-Walled Struct.* **2019**, *143*, 106205. [CrossRef]
23. Fang, C.; Wang, W.; Ji, Y.; Yam, M.C. Superior low-cycle fatigue performance of iron-based SMA for seismic damping application. *J. Constr. Steel Res.* **2021**, *184*, 106817. [CrossRef]
24. Gu, T.; Jia, L.-J.; Chen, B.; Xia, M.; Guo, H.; He, M.-C. Unified full-range plasticity till fracture of meta steel and structural steels. *Eng. Fract. Mech.* **2021**, *253*, 107869. [CrossRef]
25. Shi, G.; Gao, Y.; Wang, X.; Cui, Y. Energy-based low cycle fatigue analysis of low yield point steels. *J. Constr. Steel Res.* **2018**, *150*, 346–353. [CrossRef]
26. Rosa, D.I.H.; Hartloper, A.; Sousa, A.D.C.E.; Lignos, D.G.; Motavalli, M.; Ghafoori, E. Experimental behavior of iron-based shape memory alloys under cyclic loading histories. *Constr. Build. Mater.* **2021**, *272*, 121712. [CrossRef]
27. Koster, M.; Lee, W.J.; Schwarzenberger, M.; Leinenbach, C. Cyclic deformation and structural fatigue behavior of an FE–Mn–Si shape memory alloy. *Mater. Sci. Eng. A* **2015**, *637*, 29–39. [CrossRef]
28. Inoue, Y.; Kushibe, A.; Umemura, K.; Mizushima, Y.; Sawaguchi, T.; Nakamura, T.; Otsuka, H.; Chiba, Y. Fatigue-resistant Fe-Mn-Si-based alloy seismic dampers to counteract long-period ground motion. *Jpn. Archit. Rev.* **2020**, *4*, 76–87. [CrossRef]
29. Koyama, M.; Sawaguchi, T.; Ogawa, K.; Kikuchi, T.; Murakami, M. Continuous Transition of Deformation Modes in Fe-30Mn-5Si-1Al Alloy. *Mater. Trans.* **2010**, *51*, 1194–1199. [CrossRef]
30. Bouaziz, O.; Allain, S.; Scott, C.P.; Cugy, P.; Barbier, D. High manganese austenitic twinning induced plasticity steels: A review of the microstructure properties relationships. *Curr. Opin. Solid State Mater. Sci.* **2011**, *15*, 141–168. [CrossRef]
31. Lu, F.; Yang, P.; Meng, L.; Cui, F.; Ding, H. Influences of Thermal Martensites and Grain Orientations on Strain-induced Martensites in High Manganese TRIP/TWIP Steels. *J. Mater. Sci. Technol.* **2011**, *27*, 257–265. [CrossRef]
32. Khalil, W.; Saint-Sulpice, L.; Chirani, S.A.; Bouby, C.; Mikolajczak, A.; Ben Zineb, T. Experimental analysis of Fe-based shape memory alloy behavior under thermomechanical cyclic loading. *Mech. Mater.* **2013**, *63*, 1–11. [CrossRef]
33. Chen, J.; Zhang, W.-N.; Liu, Z.-Y.; Wang, G.-D. Microstructural evolution and deformation mechanism of a Fe-15Mn alloy investigated by electron back-scattered diffraction and transmission electron microscopy. *Mater. Sci. Eng. A* **2017**, *698*, 198–205. [CrossRef]
34. Tasaki, W.; Sawaguchi, T.; Tsuchiya, K. EBSD analysis of dual γ/ε phase microstructures in tensile-deformed Fe-Mn-Si shape memory alloy. *J. Alloys Compd.* **2019**, *797*, 529–536. [CrossRef]
35. Yang, Y.; Breveglieri, M.; Shahverdi, M. Effect of Phase Changes on the Axial Modulus of an FeMnSi-Shape Memory Alloy. *Materials* **2021**, *14*, 4815. [CrossRef] [PubMed]
36. Matsumura, O.; Sumi, T.; Tamura, N.; Sakao, K.; Furukawa, T.; Otsuka, H. Pseudoelasticity in an Fe–28Mn–6Si–5Cr shape memory alloy. *Mater. Sci. Eng. A* **2000**, *279*, 201–206. [CrossRef]
37. Sawaguchi, T.; Kikuchi, T.; Kajiwara, S. The pseudoelastic behavior of Fe–Mn–Si-based shape memory alloys containing Nb and C. *Smart Mater. Struct.* **2005**, *14*, S317–S322. [CrossRef]
38. Baruj, A.; Bertolino, G.; Troiani, H.E. Temperature dependence of critical stress and pseudoelasticity in a Fe–Mn–Si–Cr pre-rolled alloy. *J. Alloys Compd.* **2010**, *502*, 54–58. [CrossRef]
39. Lin, H.C.; Wu, S.K.; Peng, Y.T.; Cheng, T.C.; Lin, K.M. Pseudoelasticity of thermo-mechanically treated Fe–Mn–Si–Cr–Ta alloys. *J. Alloys Compd.* **2013**, *577*, S338–S343. [CrossRef]
40. Vollmer, M.; Segel, C.; Krooß, P.; Günther, J.; Tseng, L.W.; Karaman, I.; Weidner, A.; Biermann, H.; Niendorf, T. On the effect of gamma phase formation on the pseudoelastic performance of polycrystalline Fe–Mn–Al–Ni shape memory alloys. *Scr. Mater.* **2015**, *108*, 23–26. [CrossRef]
41. Arabi-Hashemi, A.; Lee, W.; Leinenbach, C. Recovery stress formation in FeMnSi based shape memory alloys: Impact of precipitates, texture and grain size. *Mater. Des.* **2018**, *139*, 258–268. [CrossRef]
42. Megdiche, M.; Sallami, A.; Thiébaud, F.; Bouraoui, T.; Ben Zineb, T.; Chirani, S.A. Experimental analysis of the pseudoelastic damping capacity of the Fe-30Mn-6Si-5Cr Shape Memory Alloy. *Smart Mater. Struct.* **2020**, *29*, 084002. [CrossRef]
43. Fang, C.; Zheng, Y.; Chen, J.; Yam, M.C.; Wang, W. Superelastic NiTi SMA cables: Thermal-mechanical behavior, hysteretic modelling and seismic application. *Eng. Struct.* **2019**, *183*, 533–549. [CrossRef]
44. Liang, D.; Zheng, Y.; Fang, C.; Yam, M.C.H.; Zhang, C. Shape memory alloy (SMA)-cable-controlled sliding bearings: Development, testing, and system behavior. *Smart Mater. Struct.* **2020**, *29*, 085006. [CrossRef]

45. Sato, A.; Chishima, E.; Yamaji, Y.; Mori, T. Orientation and composition dependencies of shape memory effect IN Fe-Mn-Si alloys. *Acta Metall.* **1984**, *32*, 539–547. [CrossRef]
46. Yang, J.H.; Chen, H.; Wayman, C.M. Development of Fe-Based Shape Memory Alloys Associated with Face-Centered Cubic-Hexagonal Close-Packed Martensitic Transformations Part I. Shape Memory Behavior. *Metall. Trans. A* **1992**, *23*, 1431–1437. [CrossRef]
47. Ohtsuka, H.; Kajiwara, S.; Kikuchi, T.; Ishihara, T.; Nagai, K. Growth Process and Microstructure of ε Martensite in an Fe-Mn-Si-Cr-Ni Shape Memory Alloy. *J. Phys. Colloq.* **1995**, *5*, C8-451–C8-455. [CrossRef]
48. Kikuchi, T.; Kajiwara, S.; Yomota, Y. Microscopic Studies on Stress-induced Martensite Transformation and Its Reversion in an Fe–Mn–Si–Cr–Ni Shape Memory Alloy. *Mater. Trans.* **1995**, *36*, 719–728. [CrossRef]
49. Baruj, A.; Kikuchi, T.; Kajiwara, S.; Shinya, N. Effect of Pre-Deformation of Austenite on Shape Memory Properties in Fe-Mn-Si-based Alloys Containing Nb and C. *Mater. Trans.* **2002**, *43*, 585–588. [CrossRef]
50. Baruj, A.; Kikuchi, T.; Kajiwara, S.; Shinya, N. Improvement of shape memory properties of NbC containing Fe–Mn–Si based shape memory alloys by simple thermomechanical treatments. *Mater. Sci. Eng. A* **2004**, *378*, 333–336. [CrossRef]
51. Wen, Y.H.; Li, N.; Xiong, L.R. Composition design principles for Fe–Mn–Si–Cr–Ni based alloys with better shape memory effect and higher recovery stress. *Mater. Sci. Eng. A* **2005**, *407*, 31–35. [CrossRef]
52. Koyama, M.; Murakami, M.; Ogawa, K.; Kikuchi, T.; Sawaguchi, T. Influence of Al on Shape Memory Effect and Twinning Induced Plasticity of Fe-Mn-Si-Al System Alloy. *Mater. Trans.* **2007**, *48*, 2729–2734. [CrossRef]
53. Koyama, M.; Sawaguchi, T.; Tsuzaki, K. Si content dependence on shape memory and tensile properties in Fe–Mn–Si–C alloys. *Mater. Sci. Eng. A* **2011**, *528*, 2882–2888. [CrossRef]
54. Koyama, M.; Sawaguchi, T.; Tsuzaki, K. Microstructure characteristic and its effect on mechanical and shape memory properties in a Fe–17Mn–8Si–0.3C alloy. *J. Alloy Compd.* **2013**, *573*, 15–19. [CrossRef]
55. Omori, T.; Okano, M.; Kainuma, R. Effect of grain size on superelasticity in Fe-Mn-Al-Ni shape memory alloy wire. *APL Mater.* **2013**, *1*, 032103. [CrossRef]
56. Pricop, B.; Mihalache, E.; Nicoleta-Monica, L.; Istrate, B.; Mocanu, M.; Özkal, B.; Bujoreanu, L.-G. Powder metallurgy and mechanical alloying effects on the formation of thermally induced martensite in an FeMnSiCrNi SMA. *MATEC Web Conf.* **2015**, *33*, 04004. [CrossRef]
57. Xiong, R.; Peng, H.; Wang, S.; Si, H.; Wen, Y. Effect of stacking fault energy on work hardening behaviors in Fe–Mn–Si–C high manganese steels by varying silicon and carbon contents. *Mater. Des.* **2015**, *85*, 707–714. [CrossRef]
58. Xu, Z.; Hodgson, M.A.; Cao, P. A comparative study of powder metallurgical (PM) and wrought Fe–Mn–Si alloys. *Mater. Sci. Eng. A* **2015**, *630*, 116–124. [CrossRef]
59. Mihalache, E.; Pricop, B.; Suru, M.-G.; Lohan, N.M.; Comăneci, R.I.; Istrate, B.; Ozkal, B.; Bujoreanu, L. Factors influencing martensite transitions in Fe-based shape memory alloys. *MATEC Web Conf.* **2015**, *33*, 04002. [CrossRef]
60. La Roca, P.; Baruj, A.; Sade, M. Shape-Memory Effect and Pseudoelasticity in Fe–Mn-Based Alloys. *Shape Mem. Superelast.* **2017**, *3*, 37–48. [CrossRef]
61. Tasaki, W.; Tsuchiya, K.; Sawaguchi, T.; Takamori, S. Comparison of Reverse Transformation Behaviors of Thermally- and Deformation-Induced ε-Martensite in Fe-28Mn-6Si-5Cr Shape Memory Alloy. *Mater. Trans.* **2016**, *57*, 1707–1713. [CrossRef]
62. Xu, Z.; Hodgson, M.A.; Cao, P. Effects of Mechanical Milling and Sintering Temperature on the Densification, Microstructure and Tensile Properties of the Fe–Mn–Si Powder Compacts. *J. Mater. Sci. Technol.* **2016**, *32*, 1161–1170. [CrossRef]
63. Watanabe, Y.; Yamamura, A.; Sato, H. Fabrication of Fe-Mn-Si-Cr Shape Memory Alloy Fiber/Aluminum Matrix Smart Composite by Casting. *Mater. Trans.* **2016**, *57*, 590–599. [CrossRef]
64. Suru, M.-G.; Lohan, N.-M.; Mihalache, E.; Pricop, B.; Mocanu, M.; Bujoreanu, L. AFM Evaluation of Pre-Straining Degree Effects on the Dimensions of Stress Induced Martensite Plates in Fe-Mn-Si Based SMAs. *J. Test. Eval.* **2016**, *45*, 419–427. [CrossRef]
65. Xu, Z.; Hodgson, M.A.; Chang, K.; Chen, G.; Yuan, X.; Cao, P. Effect of Sintering Time on the Densification, Microstructure, Weight Loss and Tensile Properties of a Powder Metallurgical Fe-Mn-Si Alloy. *Metals* **2017**, *7*, 81. [CrossRef]
66. Leinenbach, C.; Arabi-Hashemi, A.; Lee, W.J.; Lis, A.; Sadegh-Ahmadi, M.; Van Petegem, S.; Panzner, T.; Van Swygenhoven, H. Characterization of the deformation and phase transformation behavior of VC-free and VC-containing FeMnSi-based shape memory alloys by in situ neutron diffraction. *Mater. Sci. Eng. A* **2017**, *703*, 314–323. [CrossRef]
67. Lai, M.; Li, Y.; Lillpopp, L.; Ponge, D.; Will, S.; Raabe, D. On the origin of the improvement of shape memory effect by precipitating VC in Fe–Mn–Si-based shape memory alloys. *Acta Mater.* **2018**, *155*, 222–235. [CrossRef]
68. Pricop, B.; Söyler, A.U.; Özkal, B.; Bujoreanu, L. Powder Metallurgy: An Alternative for FeMnSiCrNi Shape Memory Alloys Processing. *Front. Mater.* **2020**, *7*, 247. [CrossRef]
69. Popa, M.; Mihalache, E.; Cojocaru, V.D.; Gurău, C.; Gurău, G.; Cimpoeşu, N.; Pricop, B.; Comăneci, R.-I.; Vollmer, M.; Krooß, P.; et al. Effects of Thermomechanical Processing on the Microstructure and Mechanical Properties of Fe-Based Alloys. *J. Mater. Eng. Perform.* **2020**, *29*, 2274–2282. [CrossRef]
70. Nikulin, I.; Sawaguchi, T.; Yoshinaka, F.; Takamori, S. Influence of cold rolling deformation mechanisms on the grain refinement of Fe–15Mn–10Cr–8Ni–4Si austenitic alloy. *Mater. Charact.* **2020**, *162*, 110191. [CrossRef]
71. Yoshinaka, F.; Sawaguchi, T. Characterization of crystallographic fracture surfaces in Fe–33Mn–6Si alloy. *Int. J. Fatigue* **2020**, *130*, 105271. [CrossRef]
72. Lee, W.J.; Weber, B.; Feltrin, G.; Czaderski, C.; Motavalli, M.; Leinenbach, C. Stress recovery behaviour of an Fe–Mn–Si–Cr–Ni–VC shape memory alloy used for prestressing. *Smart Mater. Struct.* **2013**, *22*, 125037. [CrossRef]

73. Gu, X.-L.; Chen, Z.-Y.; Yu, Q.-Q.; Ghafoori, E. Stress recovery behavior of an Fe-Mn-Si shape memory alloy. *Eng. Struct.* **2021**, *243*, 112710. [CrossRef]
74. Wang, W.-Y.; Liu, B.; Kodur, V. Effect of Temperature on Strength and Elastic Modulus of High-Strength Steel. *J. Mater. Civ. Eng.* **2013**, *25*, 174–182. [CrossRef]
75. Wang, B.; Zhu, S. Cyclic behavior of iron-based shape memory alloy bars for high-performance seismic devices. *Eng. Struct.* **2021**, *252*, 113588. [CrossRef]
76. Leinenbach, C.; Kramer, H.; Bernhard, C.; Eifler, D. Thermo-Mechanical Properties of an Fe-Mn-Si-Cr-Ni-VC Shape Memory Alloy with Low Transformation Temperature. *Adv. Eng. Mater.* **2012**, *14*, 62–67. [CrossRef]
77. Sawaguchi, T.; Sahu, P.; Kikuchi, T.; Ogawa, K.; Kajiwara, S.; Kushibe, A.; Higashino, M.; Ogawa, T. Vibration mitigation by the reversible fcc/hcp martensitic transformation during cyclic tension–compression loading of an Fe–Mn–Si-based shape memory alloy. *Scr. Mater.* **2006**, *54*, 1885–1890. [CrossRef]
78. Sawaguchi, T.; Bujoreanu, L.G.; Kikuchi, T.; Ogawa, K.; Koyama, M.; Murakami, M. Mechanism of reversible transformation-induced plasticity of Fe–Mn–Si shape memory alloys. *Scr. Mater.* **2008**, *59*, 826–829. [CrossRef]
79. Watanabe, Y.; Sato, H.; Nishino, Y.; Kim, I.S. Training Effects on Damping Capacity in Fe-Mn and Fe-Mn-Cr Alloys. *Mater. Sci. Forum* **2010**, *638–642*, 2201–2206. [CrossRef]
80. Nikulin, I.; Sawaguchi, T.; Tsuzaki, K. Effect of alloying composition on low-cycle fatigue properties and microstructure of Fe–30Mn–(6−x)Si–xAl TRIP/TWIP alloys. *Mater. Sci. Eng. A* **2013**, *587*, 192–200. [CrossRef]
81. Nikulin, I.; Sawaguchi, T.; Ogawa, K.; Tsuzaki, K. Microstructure Evolution Associated with a Superior Low-Cycle Fatigue Resistance of the Fe-30Mn-4Si-2Al Alloy. *Met. Mater. Trans. A* **2015**, *46*, 5103–5113. [CrossRef]
82. Sawaguchi, T.; Nikulin, I.; Ogawa, K.; Sekido, K.; Takamori, S.; Maruyama, T.; Chiba, Y.; Kushibe, A.; Inoue, Y.; Tsuzaki, K. Designing Fe–Mn–Si alloys with improved low-cycle fatigue lives. *Scr. Mater.* **2015**, *99*, 49–52. [CrossRef]
83. Nagashima, N.; Sawaguchi, T.; Ogawa, K. Low Cycle Fatigue Properties of Fe-28Mn-6Si-5Cr-0.5NbC Alloy. *Tetsu-to-Hagane* **2016**, *102*, 517–524. [CrossRef]
84. Nikulin, I.; Sawaguchi, T.; Kushibe, A.; Inoue, Y.; Otsuka, H.; Tsuzaki, K. Effect of strain amplitude on the low-cycle fatigue behavior of a new Fe–15Mn–10Cr–8Ni–4Si seismic damping alloy. *Int. J. Fatigue* **2016**, *88*, 132–141. [CrossRef]
85. Nikulin, I.; Sawaguchi, T.; Ogawa, K.; Tsuzaki, K. Effect of γ to ε martensitic transformation on low-cycle fatigue behaviour and fatigue microstructure of Fe–15Mn–10Cr–8Ni–xSi austenitic alloys. *Acta Mater.* **2016**, *105*, 207–218. [CrossRef]
86. Shao, C.; Zhang, P.; Liu, R.; Zhang, Z.; Pang, J. Low-cycle and extremely-low-cycle fatigue behaviors of high-Mn austenitic TRIP/TWIP alloys: Property evaluation, damage mechanisms and life prediction. *Acta Mater.* **2016**, *103*, 781–795. [CrossRef]
87. Tasaki, W.; Sawaguchi, T.; Nikulin, I.; Sekido, K.; Tsuchiya, K. Effect of Deformation Temperature on Low-Cycle Fatigue Properties of Fe-28Mn-6Si-5Cr Shape Memory Alloy. *Mater. Trans.* **2016**, *57*, 639–646. [CrossRef]
88. Ju, Y.-B.; Koyama, M.; Sawaguchi, T.; Tsuzaki, K.; Noguchi, H. Effects of ε-martensitic transformation on crack tip deformation, plastic damage accumulation, and slip plane cracking associated with low-cycle fatigue crack growth. *Int. J. Fatigue* **2017**, *103*, 533–545. [CrossRef]
89. Nagashima, N.; Sawaguchi, T. Influence of Transformation Pseudoelasticity and Accumulated Plastic Strain on Low Cycle Fatigue Characteristics of Fe-30Mn-4Si-2Al Alloy. *Tetsu-to-Hagane* **2018**, *104*, 393–399. [CrossRef]
90. Nikulin, I.; Nagashima, N.; Yoshinaka, F.; Sawaguchi, T. Superior fatigue life of Fe-15Mn-10Cr-8Ni-4Si seismic damping alloy subjected to extremely high strain amplitudes. *Mater. Lett.* **2018**, *230*, 257–260. [CrossRef]
91. Nikulin, I.; Yoshinaka, F.; Sawaguchi, T. Superior fatigue life of Fe-15Mn-10Cr-8Ni-4Si seismic damping alloy under asymmetric cyclic loading with tensile mean strain. *Mater. Sci. Eng. A* **2019**, *748*, 371–378. [CrossRef]
92. Nikulin, I.; Yoshinaka, F.; Takamori, S.; Sawaguchi, T. Effect of carbon on the low-cycle fatigue resistance and microstructure of the Fe–15Mn–10Cr–8Ni–4Si seismic damping alloy. *Mater. Sci. Eng. A* **2019**, *766*, 138321. [CrossRef]
93. Yoshinaka, F.; Sawaguchi, T.; Ilya, N.; Takamori, S.; Nagashima, N. Improved fatigue life of the newly developed Fe-15Mn-10Cr-8Ni-4Si seismic damping alloy. *Procedia Struct. Integr.* **2019**, *19*, 214–223. [CrossRef]
94. Yoshinaka, F.; Sawaguchi, T.; Nikulin, I.; Takamori, S. Fatigue properties and plastically deformed microstructure of Fe-15Mn-10Cr-8Ni-4Si alloy in high-cycle-fatigue regime. *Int. J. Fatigue* **2019**, *129*, 105224. [CrossRef]
95. Fang, C.; Ping, Y.; Chen, Y. Loading protocols for experimental seismic qualification of members in conventional and emerging steel frames. *Earthq. Eng. Struct. Dyn.* **2019**, *49*, 155–174. [CrossRef]
96. Guo, Y.; Fang, C.; Zheng, Y. Post-fire hysteretic and low-cycle fatigue behaviors of Q345 carbon steel. *J. Constr. Steel Res.* **2021**, *187*, 106991. [CrossRef]
97. Fang, C.; Liang, D.; Zheng, Y.; Lu, S. Seismic performance of bridges with novel SMA cable-restrained high damping rubber bearings against near-fault ground motions. *Earthq. Eng. Struct. Dyn.* **2021**, *51*, 44–65. [CrossRef]
98. Fang, C.; Wang, W.; Feng, W. Experimental and numerical studies on self-centring beam-to-column connections free from frame expansion. *Eng. Struct.* **2019**, *198*, 198. [CrossRef]
99. Wang, W.; Fang, C.; Zhao, Y.; Sause, R.; Hu, S.; Ricles, J. Self-centering friction spring dampers for seismic resilience. *Earthq. Eng. Struct. Dyn.* **2019**, *48*, 1045–1065. [CrossRef]
100. Fang, C.; Wang, W.; Zhang, A.; Sause, R.; Ricles, J.; Chen, Y. Behavior and Design of Self-Centering Energy Dissipative Devices Equipped with Superelastic SMA Ring Springs. *J. Struct. Eng.* **2019**, *145*, 04019109. [CrossRef]

101. Fang, C.; Zhong, Q.; Wang, W.; Hu, S.; Qiu, C. Peak and residual responses of steel moment-resisting and braced frames under pulse-like near-fault earthquakes. *Eng. Struct.* **2018**, *177*, 579–597. [CrossRef]
102. Fang, C.; Ping, Y.; Chen, Y.; Yam, M.C.H.; Chen, J.; Wang, W. Seismic Performance of Self-centering Steel Frames with SMA-viscoelastic Hybrid Braces. *J. Earthq. Eng.* **2020**, 1–28. [CrossRef]
103. Lamarche, C.-P.; Tremblay, R. Seismically induced cyclic buckling of steel columns including residual-stress and strain-rate effects. *J. Constr. Steel Res.* **2011**, *67*, 1401–1410. [CrossRef]
104. Fang, C.; Yam, M.C.H.; Lam, C.C.; Xie, L. Cyclic performance of extended end-plate connections equipped with shape memory alloy bolts. *J. Constr. Steel Res.* **2014**, *94*, 122–136. [CrossRef]
105. Wang, W.; Fang, C.; Liu, J. Self-Centering Beam-to-Column Connections with Combined Superelastic SMA Bolts and Steel Angles. *J. Struct. Eng.* **2017**, *143*, 04016175. [CrossRef]
106. Fang, C.; Yam, M.C.; Chan, T.M.; Wang, W.; Yang, X.; Lin, X. A study of hybrid self-centring connections equipped with shape memory alloy washers and bolts. *Eng. Struct.* **2018**, *164*, 155–168. [CrossRef]
107. Qiu, C.; Fang, C.; Liang, D.; Du, X.; Yam, M.C. Behavior and application of self-centering dampers equipped with buckling-restrained SMA bars. *Smart Mater. Struct.* **2020**, *29*, 035009. [CrossRef]
108. Dusicka, P.; Itani, A.M.; Buckle, I.G. Cyclic response of plate steels under large inelastic strains. *J. Constr. Steel Res.* **2007**, *63*, 156–164. [CrossRef]
109. Nip, K.; Gardner, L.; Davies, C.; Elghazouli, A. Extremely low cycle fatigue tests on structural carbon steel and stainless steel. *J. Constr. Steel Res.* **2010**, *66*, 96–110. [CrossRef]
110. Lu, B.; Silva, P.F. Estimating equivalent viscous damping ratio for RC members under seismic and blast loadings. *Mech. Res. Commun.* **2006**, *33*, 787–795. [CrossRef]
111. Maruyama, T.; Kurita, T.; Kozaki, S.; Andou, K.; Farjami, S.; Kubo, H. Innovation in producing crane rail fishplate using Fe–Mn–Si–Cr based shape memory alloy. *Mater. Sci. Technol.* **2013**, *24*, 908–912. [CrossRef]
112. Hosseini, A.; Ghafoori, E.; Motavalli, M.; Nussbaumer, A.C.; Zhao, X.-L. Mode I fatigue crack arrest in tensile steel members using prestressed CFRP plates. *Compos. Struct.* **2017**, *178*, 119–134. [CrossRef]
113. Hong, K.; Lee, S.; Han, S.; Yeon, Y. Evaluation of Fe-Based Shape Memory Alloy (Fe-SMA) as Strengthening Material for Reinforced Concrete Structures. *Appl. Sci.* **2018**, *8*, 730. [CrossRef]
114. Hong, K.; Lee, S.; Yeon, Y.; Jung, K. Flexural Response of Reinforced Concrete Beams Strengthened with Near-Surface-Mounted Fe-Based Shape-Memory Alloy Strips. *Int. J. Concr. Struct. Mater.* **2018**, *12*, 45. [CrossRef]
115. Michels, J.; Shahverdi, M.; Czaderski, C. Flexural strengthening of structural concrete with iron-based shape memory alloy strips. *Struct. Concr.* **2018**, *19*, 876–891. [CrossRef]
116. Rojob, H.; El-Hacha, R. Fatigue performance of RC beams strengthened with self-prestressed iron-based shape memory alloys. *Eng. Struct.* **2018**, *168*, 35–43. [CrossRef]
117. Rojob, H.; El-Hacha, R. Performance of RC beams strengthened with self-prestressed Fe-SMA bars exposed to freeze-thaw cycles and sustained load. *Eng. Struct.* **2018**, *169*, 107–118. [CrossRef]
118. Strieder, E.; Aigner, C.; Petautschnig, G.; Horn, S.; Marcon, M.; Schwenn, M.; Zeman, O.; Castillo, P.; Wan-Wendner, R.; Bergmeister, K. Strengthening of Reinforced Concrete Beams with Externally Mounted Sequentially Activated Iron-Based Shape Memory Alloys. *Materials* **2019**, *12*, 345. [CrossRef]
119. Dolatabadi, N.; Shahverdi, M.; Ghassemieh, M.; Motavalli, M. RC Structures Strengthened by an Iron-Based Shape Memory Alloy Embedded in a Shotcrete Layer—Nonlinear Finite Element Modeling. *Materials* **2020**, *13*, 5504. [CrossRef]
120. Soroushian, P.; Ostowari, K.; Nossoni, A.; Chowdhury, H. Repair and Strengthening of Concrete Structures Through Application of Corrective Posttensioning Forces with Shape Memory Alloys. *Transp. Res. Rec. J. Transp. Res. Board* **2001**, *1770*, 20–26. [CrossRef]
121. El-Hacha, R.; Soudki, K. Prestressed near-surface mounted fibre reinforced polymer reinforcement for concrete structures—A review. *Can. J. Civ. Eng.* **2013**, *40*, 1127–1139. [CrossRef]
122. Czaderski, C.; Shahverdi, M.; Brönnimann, R.; Leinenbach, C.; Motavalli, M. Feasibility of iron-based shape memory alloy strips for prestressed strengthening of concrete structures. *Constr. Build. Mater.* **2014**, *56*, 94–105. [CrossRef]
123. Abouali, S.; Shahverdi, M.; Ghassemieh, M.; Motavalli, M. Nonlinear simulation of reinforced concrete beams retrofitted by near-surface mounted iron-based shape memory alloys. *Eng. Struct.* **2019**, *187*, 133–148. [CrossRef]
124. Schranz, B.; Czaderski, C.; Vogel, T.; Shahverdi, M. Bond behaviour of ribbed near-surface-mounted iron-based shape memory alloy bars with short bond lengths. *Mater. Des.* **2020**, *191*, 108647. [CrossRef]
125. Schranz, B.; Czaderski, C.; Vogel, T.; Shahverdi, M. Bond investigations of prestressed, near-surface-mounted, ribbed memory steel bars with full bond length. *Mater. Des.* **2020**, *196*, 109145. [CrossRef]
126. Shahverdi, M.; Czaderski, C.; Annen, P.; Motavalli, M. Strengthening of RC beams by iron-based shape memory alloy bars embedded in a shotcrete layer. *Eng. Struct.* **2016**, *117*, 263–273. [CrossRef]
127. Czaderski, C.; Shahverdi, M.; Michels, J. Iron based shape memory alloys as shear reinforcement for bridge girders. *Constr. Build. Mater.* **2021**, *274*, 121793. [CrossRef]
128. Shahverdi, M.; Czaderski, C.; Motavalli, M. Iron-based shape memory alloys for prestressed near-surface mounted strengthening of reinforced concrete beams. *Constr. Build. Mater.* **2016**, *112*, 28–38. [CrossRef]

129. Rojob, H.; El-Hacha, R. Ductility behavior of RC beams strengthened in flexure with NSM Iron-based Shape Memory Alloy bars. In Proceedings of the Third Conference on Smart Monitoring, Assessment and Rehabilitation of Civil Structures, Antalya, Turkey, 7–9 September 2015.
130. Rojob, H.; El-Hacha, R. Self-Prestressing Using Fe-SMA for Flexural Strengthening of Reinforced Concrete Beams. *ACI Struct. J.* **2017**, *114*, 523–532. [CrossRef]
131. Cladera, A.; Montoya-Coronado, L.A.; Ruiz-Pinilla, J.G.; Ribas, C. Shear strengthening of slender reinforced concrete T-shaped beams using iron-based shape memory alloy strips. *Eng. Struct.* **2020**, *221*, 111018. [CrossRef]
132. Rojob, H.; El-Hacha, R. Strengthening of concrete structures with iron-based shape memory alloy elements: Case studies. In Proceedings of the SMAR 2019—Fifth Conference on Smart Monitoring, Assessment and Rehabilitation of Civil Structures, Potsdam, Germany, 27–29 August 2019.
133. Izadi, M.; Ghafoori, E.; Shahverdi, M.; Motavalli, M.; Maalek, S. Development of an iron-based shape memory alloy (Fe-SMA) strengthening system for steel plates. *Eng. Struct.* **2018**, *174*, 433–446. [CrossRef]
134. Izadi, M.; Ghafoori, E.; Motavalli, M.; Maalek, S. Iron-based shape memory alloy for the fatigue strengthening of cracked steel plates: Effects of re-activations and loading frequencies. *Eng. Struct.* **2018**, *176*, 953–967. [CrossRef]
135. Izadi, M.; Hosseini, A.; Michels, J.; Motavalli, M.; Ghafoori, E. Thermally activated iron-based shape memory alloy for strengthening metallic girders. *Thin-Walled Struct.* **2019**, *141*, 389–401. [CrossRef]
136. Hosseini, A.; Ghafoori, E.; Motavalli, M.; Nussbaumer, A.; Zhao, X.-L.; Al-Mahaidi, R. Flat prestressed unbonded retrofit system for strengthening of existing metallic I-Girders. *Compos. Part B Eng.* **2018**, *155*, 156–172. [CrossRef]
137. Izadi, M.; Motavalli, M.; Ghafoori, E. Iron-based shape memory alloy (Fe-SMA) for fatigue strengthening of cracked steel bridge connections. *Constr. Build. Mater.* **2019**, *227*, 116800. [CrossRef]
138. Wang, W.; Hosseini, A.; Ghafoori, E. Experimental study on Fe-SMA-to-steel adhesively bonded interfaces using DIC. *Eng. Fract. Mech.* **2021**, *244*, 107553. [CrossRef]
139. Wang, W.; Li, L.; Hosseini, A.; Ghafoori, E. Novel fatigue strengthening solution for metallic structures using adhesively bonded Fe-SMA strips: A proof of concept study. *Int. J. Fatigue* **2021**, *148*, 106237. [CrossRef]
140. Hou, H.; Qu, B. Duraion effect of spectrally matched fround motions on seismic demands of elasric perfectly plastic SDOFS. *Eng. Struct.* **2015**, *90*, 48–60. [CrossRef]
141. Hosseini, A.; Michels, J.; Izadi, M.; Ghafoori, E. A comparative study between Fe-SMA and CFRP reinforcements for prestressed strengthening of metallic structures. *Constr. Build. Mater.* **2019**, *226*, 976–992. [CrossRef]
142. Ghafoori, E.; Neuenschwander, M.; Shahverdi, M.; Czaderski, C.; Fontana, M. Elevated temperature behavior of an iron-based shape memory alloy used for prestressed strengthening of civil structures. *Constr. Build. Mater.* **2019**, *211*, 437–452. [CrossRef]
143. Bu, D.; Peng, H.; Wen, Y.; Li, N. Influence of ageing on wear resistance of an Fe–Mn–Si–Cr–Ni–Ti–C shape memory alloy. *Mater. Des.* **2011**, *32*, 2969–2973. [CrossRef]
144. Della Rovere, C.; Alano, J.; Otubo, J.; Kuri, S. Corrosion behavior of shape memory stainless steel in acid media. *J. Alloys Compd.* **2011**, *509*, 5376–5380. [CrossRef]
145. Della Rovere, C.; Alano, J.; Silva, R.; Nascente, P.; Otubo, J.; Kuri, S. Influence of alloying elements on the corrosion properties of shape memory stainless steels. *Mater. Chem. Phys.* **2012**, *133*, 668–673. [CrossRef]
146. Ma, R.; Peng, H.; Wen, Y.; Zhang, L.; Zhao, K. Oxidation behavior of an austenitic stainless FeMnSiCrNi shape memory alloy. *Corros. Sci.* **2013**, *66*, 269–277. [CrossRef]
147. Leinenbach, C.; Lee, W.J.; Lis, A.; Arabi-Hashemi, A.; Cayron, C.; Weber, B. Creep and stress relaxation of a FeMnSi-based shape memory alloy at low temperatures. *Mater. Sci. Eng. A* **2016**, *677*, 106–115. [CrossRef]
148. Lee, W.J.; Partovi-Nia, R.; Suter, T.; Leinenbach, C. Electrochemical characterization and corrosion behavior of an Fe-Mn-Si shape memory alloy in simulated concrete pore solutions. *Mater. Corros.* **2016**, *67*, 839–846. [CrossRef]
149. Silva, R.; Arana, C.; Malafaia, A.D.S.; Filho, A.M.; Pascal, C.; Otubo, J.; Sordi, V.; Rovere, C. Microstructure and surface oxidation behavior of an austenitic Fe-Mn-Si-Cr-Ni-Co shape memory stainless steel at 800 °C in air. *Corros. Sci.* **2019**, *158*, 158. [CrossRef]
150. Silva, R.; Vacchi, G.; Santos, Í.; Malafaia, A.; Kugelmeier, C.; Filho, A.M.; Pascal, C.; Sordi, V.; Rovere, C. Insights into high-temperature oxidation of Fe-Mn-Si-Cr-Ni shape memory stainless steels and its relationship to alloy chemical composition. *Corros. Sci.* **2020**, *163*, 163. [CrossRef]
151. Hosseini, E.; Ghafoori, E.; Leinenbach, C.; Motavalli, M.; Holdsworth, S.R. Stress recovery and cyclic behaviour of an Fe–Mn–Si shape memory alloy after multiple thermal activation. *Smart Mater. Struct.* **2018**, *27*, 025009. [CrossRef]
152. Baddoo, N.R. Stainless steel in construction: A review of research, applications, challenges and opportunities. *J. Constr. Steel Res.* **2008**, *64*, 1199–1206. [CrossRef]
153. Fang, C.; Ping, Y.; Zheng, Y.; Chen, Y. Probabilistic economic seismic loss estimation of steel braced frames incorporating emerging self-centering technologies. *Eng. Struct.* **2021**, *241*, 112486. [CrossRef]
154. Zheng, Y.; Fang, C.; Liang, D.; Sun, R. An innovative seismic-resilient bridge with shape memory alloy-washer-based footing rocking RC piers. *J. Intell. Mater. Syst. Struct.* **2021**, *32*, 549–567. [CrossRef]
155. Chen, J.; Fang, C.; Wang, W.; Liu, Y. Variable-friction self-centering energy-dissipation braces (VF-SCEDBs) with NiTi SMA cables for seismic resilience. *J. Constr. Steel Res.* **2020**, *175*, 106318. [CrossRef]
156. Wang, W.; Fang, C.; Zhang, A.; Liu, X. Manufacturing and performance of a novel self-centring damper with shape memory alloy ring springs for seismic resilience. *Struct. Control Health Monit.* **2019**, *26*, e2337. [CrossRef]

157. Fang, C.; Yam, M.C.H.; Lam, A.C.C.; Zhang, Y. Feasibility study of shape memory alloy ring spring systems for self-centring seismic resisting devices. *Smart Mater. Struct.* **2015**, *24*, 75024. [CrossRef]
158. Lin, H.C.; Lin, K.M.; Chuang, Y.C.; Chou, T.S. The welding characteristics of Fe–30Mn–6Si and Fe–30Mn–6Si–5Cr shape memory alloys. *J. Alloys Compd.* **2000**, *306*, 189–192. [CrossRef]
159. Dong, Z.Z.; Sawaguchi, T.; Kajiwara, S.; Kikuchi, T.; Kim, S.H.; Lee, G.C. Microstructure change and shape memory characteristics in welded Fe–28Mn–6Si–5Cr–0.53Nb–0.06C alloy. *Mater. Sci. Eng. A* **2006**, *438–440*, 800–803. [CrossRef]
160. Qiao, Z.; Li, L.; Wang, D.; Li, Z. Microstructure and shape recovery characteristics in a TIG-welded Fe-Mn-Si-Cr-Ni shape memory alloy. In Proceedings of the Third International Conference on Smart Materials and Nanotechnology in Engineering, Harbin, China, 1–4 July 2007.
161. Zhou, C.; Lin, C.; Liu, L. Study on CO_2 laser weldability of Fe-Mn-Si shape memory alloy. In Proceedings of the Third International Conference on Smart Materials and Nanotechnology in Engineering, Shenzhen, China, 11–13 November 2011.
162. Druker, A.V.; Perotti, A.; Esquivel, I.; Malarría, J. A manufacturing process for shaft and pipe couplings of Fe–Mn–Si–Ni–Cr shape memory alloys. *Mater. Des. (1980–2015)* **2014**, *56*, 878–888. [CrossRef]
163. Yam, M.C.H.; Fang, C.; Lam, C.C.; Zhang, Y. Numerical study and practical design of beam-to-column connections with shape memory alloys. *J. Constr. Steel Res.* **2015**, *104*, 177–192. [CrossRef]
164. Fang, C.; Wang, W.; He, C.; Chen, Y. Self-centring behaviour of steel and steel-concrete composite connections equipped with NiTi SMA bolts. *Eng. Struct.* **2017**, *150*, 390–408. [CrossRef]
165. Liu, Y.; Wang, H.; Qiu, C.; Zhao, X. Seismic Behavior of Superelastic Shape Memory Alloy Spring in Base Isolation System of Multi-Story Steel Frame. *Materials* **2019**, *12*, 997. [CrossRef]
166. Fang, C.; Liang, D.; Zheng, Y.; Yam, M.C.; Sun, R. Rocking bridge piers equipped with shape memory alloy (SMA) washer springs. *Eng. Struct.* **2020**, *214*, 110651. [CrossRef]
167. Qiu, C.; Zhao, X.; Zhu, S. Seismic upgrading of multistory steel moment-resisting frames by installing shape memory alloy braces: Design method and performance evaluation. *Struct. Control. Health Monit.* **2020**, *27*, e2596. [CrossRef]
168. Nahar, M.; Islam, K.; Billah, A.M. Seismic collapse safety assessment of concrete beam-column joints reinforced with different types of shape memory alloy rebars. *J. Build. Eng.* **2020**, *29*, 101106. [CrossRef]

MDPI
St. Alban-Anlage 66
4052 Basel
Switzerland
Tel. +41 61 683 77 34
Fax +41 61 302 89 18
www.mdpi.com

Materials Editorial Office
E-mail: materials@mdpi.com
www.mdpi.com/journal/materials

www.ingramcontent.com/pod-product-compliance
Lightning Source LLC
Chambersburg PA
CBHW040817250926
51850CB00007BA/1190
9783036566078